WONDER SHOWS

WONDER SHOWS

PERFORMING SCIENCE, MAGIC, AND RELIGION IN AMERICA

FRED NADIS

RUTGERS UNIVERSITY PRESS
NEW BRUNSWICK, NEW JERSEY AND LONDON

Library of Congress Cataloging-in-Publication Data
Nadis, Fred, 1957–
Wonder shows : performing science, magic, and religion in America / Fred Nadis.
p. cm.
Includes bibliographical references and index.
ISBN 0-8135-3515-8 (hardcover : alk. paper)
1. Science news—United States—History—19th century. 2. Science news—United States—History—20th century. 3. Magic shows—United States—History—19th century. 4. Magic shows—United States—History—20th century. 5. Revivals—United States—History—19th century 6. Revivals—United States—History—20th century. I. Title.
Q225.N335 2005
791.1—dc22
2004008245

A British Cataloging-in-Publication record for this book is available from the British Library.

Manufactured in the United States of America

To Kate

CONTENTS

ACKNOWLEDGMENTS

Like one of the inventions that emerged from Thomas Edison's crowded research department later emblazoned with the name Edison alone, a work of scholarship involves much behind-the-scenes help. My thanks to Jeffrey L. Meikle, my dissertation advisor in the American studies department at the University of Texas at Austin, who made extensive notes on the manuscript as it developed and continued to aid me well past its completion as a dissertation, as I shaped it into a book. Thanks also to Bruce J. Hunt, who helped guide me through the vast literature on the history and philosophy of science and technology; and to Melanie Halkias, my editor at Rutgers University Press.

My thanks to the Graduate School of American Studies at Doshisha University in Kyoto, Japan, who found "wonder" intriguing enough to invite me to their campus in 2003 to teach and complete this book. The University of Texas American studies department's Robert Crunden Awards as well as the American Popular Culture Association and American Culture Association's Marshall Fishwick Grant aided earlier research.

Throughout the writing process, I have been helped with comments from numerous mentors and colleagues, among them Alison Kibler, William Davies King, Hillel Schwartz, Cathy Gutierrez, Janet Davis, Gunther Peck, Robert Abzug, Linda Henderson, Bill Bush, Kelly Mendiola, Cary Cordova, Taylor Dark, and members of the American studies "Bumon Kenkyu" at Doshisha University. Peer reviews from the *Journal of Medical Humanities* improved my discussion of the history of stage hypnosis, while Rutgers University Press reviewers helped me sharpen the argument throughout. At Doshisha, Kazumi Ura helped format the final manuscript.

My parents, Martin and Lorraine Nadis, have given both emotional and practical support to this project. My brother Steve Nadis, a science writer, served as a sounding board for ideas early in the project, and my sister Susan Nadis always came up with medical advice when the stress mounted. The rest of my thanks, drawn from a deep well, go to Kate, who with pioneer fortitude has accompanied me from New York City to Texas and on to Kyoto, and also to Rose and Saul, who are, naturally, the best children in the world.

PREFACE

One of my more prominent childhood memories involves a visit to Chicago's Museum of Science and Industry. Besides being impressed by the neoclassical building next to the lake and all the stone ladies holding up the roof on their heads, I recall a science demonstration that a young, smocked woman offered in a small anteroom, possibly the gift shop. Among other tricks she placed a rose in a vial of liquid nitrogen, removed it, and struck it with a hammer—its petals then shattered as if made of stone or plaster. This unexpected and beautiful event evoked a sense of awe in me. I was fascinated, both by the completely unexpected shattering of the rose, and that a young woman could control such strange cosmic forces.

Her explanation of this and her many other demonstrations remain vague; but I now realize that this was the first performance I had witnessed of what in this work I call a "wonder show." It did not make a scientist of me, as may have been the intention, but it did instill an appreciation for the possibilities that technology and showmanship could provide. For me, that museum demonstration combined science and poetics. Such a demonstration fits the description of a wonder show that this work relies on. In such shows, science and technology create surprise then pleasure in the spectator whose day-to-day perceptions are shattered and opened to new realms of possibility.

Another childhood memory also relates to this performance genre. At a junior high school gym assembly in the 1960s, on a tilted screen, a visiting lecturer projected time-lapse films of plants growing and sprouting flowers. The footage was black-and-white, grainy, shot from not particularly dramatic angles, yet enthralling. The repetition of the event and absence of style only increased our interest. The speaker told us that a fellow townsperson, an engineer, had pioneered this technique decades earlier. I still recall my near-deification of the mysterious inventor, once in our midst, who had thought up such a great technique. The ungainly film, with its wavering, blooming specimens, offered a glimpse of a non-human perspective that made the simple fact of growth seem a miracle. The film's tracing of organic growth served, perhaps, as an antidote to the rose I had seen years earlier shattered at the Museum of Science and Industry. These linked memories also suggest that, as a child, I was not particularly provoked to wonder by gazing at an actual rose, but only one that came filtered through technology.

My intention in reciting these childhood memories is to establish the purpose of the wonder show—to cause the spectator to see the world through new eyes, much like those of a child. A spectator's sense of wonder may be provoked via unexpected transformations, shifts in scale from the microcosmic to the macrocosmic, or through glimpses of the previously unseen or seemingly impossible. The spectator's ensuing experience, depending on its emotional depth, can be a mere cheap thrill or have a religious dimension. The evocation of wonder can also be a powerful sales tool. Over the years, for example, similar choreographies have been used to promote the varied fortunes of General Motors and of fundamentalist Christianity. The well-scripted wonder show, then, is a strong concoction.

Most recent defenders of science reject wonder as an ally of ignorance and superstition. Yet wonder should be ignored only with caution in America. Many thinkers have argued that the capacity for wonder, although universal, is particularly well developed in Americans. Alexis de Tocqueville, the chronicler of the American scene during the antebellum period, commented, "The American lives in a land of wonders . . . everything around him is in constant movement, and every movement seems an advance."[1] Over a century later, another foreign observer, British literary critic Tony Tanner, also highlighted the American taste for innovation when he entitled his look at American fiction *The Reign of Wonder* (1965).

It was not my childhood memories of roses shattering and time-lapse views of flowers opening that initiated this project. It began after I heard a hypnotist interviewed on National Public Radio in 1999. He spoke of the psychotherapy he practiced with hypnosis, and he promoted the value of this technique, yet he seemed vaguely defensive, attempting to distance his practices from the public performances of hypnosis that encourage subjects to make fools of themselves. This radio interview suggested the uneasy terrain that hypnosis still inhabits. As a public, we view hypnosis as a blend of science, art, and the otherworldly, our conception filtered through the pulp imagination that once led, for example, to the beautifully drawn panels of the cartoon *Mandrake the Magician,* in which the dapper magician defeated thugs with his paralyzing hypnotic illusions.

A once-dignified hypnotic subject imitating a canary on stage and a frozen rose shattered by a hammer blow mark the two ends of the wonder show continuum. One terminus involves the wonders of the human mind, the other the wonders of engineering and invention. The wonder show unites the two in its definition of progress. If displays of marvelous technology can threaten our sense of human uniqueness and destiny, then displays of amazing mental powers can reassure us of our very uniqueness. Often there is a hint of real magic, a whiff of the vapors from a wizard's lair, in such presentations. This book will look at ways the idea of "magic" has been combined with notions of science.

The title of this work, *Wonder Shows: Performing Science, Magic, and Religion in America,* requires slight explanation. I use the words "science" and "technology" interchangeably, as I am most interested in public attitudes toward that amalgam of forces—compounded of the work and thought of natural philosophers, scientists, inventors, engineers, businesspeople, consumers, intellectuals, policy makers, and laborers—that have instigated "modernization," and so changed lives, social relations, landscapes, and modes of thought. Throughout, I use the term "science" as viewed from the outside; this book does not analyze the procedures and practices of working scientists but instead the rhetoric of promoters of science at many social levels. I treat "technology" according to its simple connotation of the mechanical apparatus, but also as a compound that emerges from scientific practices and utopian goals.

Likewise, I use the words "magic" and "religion" as virtually interchangeable.[2] I approach the overlapping realms of magic and religion as social enterprises largely antithetical to the mechanistic or materialistic worldview and its assumptions. The sort of "magic" that public performances of Spiritualism, telepathy, mind reading, and hypnosis offered often had a spiritual component. Such performances provided new marvels to an age stripped of the older order of miracles, while bolstering arguments for the realities of a spiritual realm and an expanded vision of human powers. This restoration of humanity to its "Adamic powers" has long been a millennial religious goal.

While this work traces the roots of "wonder" earlier, the history proper begins in the 1830s, when the wonder show became firmly established in America. This decade saw the emergence in New England of mesmerism, sophisticated stage magic, widespread science coverage in newspapers, and numerous lyceum lectures on science. The book examines historical embodiments of the wonder show up to the turn of the twenty-first century, when I attended the performances of several promoters of utopian devices and so ended my eavesdropping on this ongoing dialogue about technology, progress, and the human need for wonder.

It could be argued that on at least one level, every historical study is a veiled autobiography. My anecdotes of the museum demonstration and of the school assembly implicate me as one who is susceptible to wonder. The manufacture of wonder has also been a family fascination. My grandfather, long deceased, came of age in Chicago in the Roaring Twenties; he was a lawyer who had a passion for show business that led him to become a stage magician and hypnotist and serve terms as president of the Society of American Magicians and the International Brotherhood of Magicians. If not a great talent—our family particularly liked to joke about his attempts at ventriloquism—he was dedicated to the guild and its values. Such personal associations transformed into what seemed family biography the process of writing the second section of this book, which details the acts of "mystic vaudevillians."

This book isolates the "wonder show" as a performance genre that has offered a populist vision of science and technology: in this reading, science is an enterprise that fulfills public desires and conjures new fears. On stage, the scientific showman has emphasized the glories of the human intellect and its products and often suggested that a spiritual worldview—more particularly, a belief in the human soul or in elevated mental powers—can be supported by the latest technical research. These shows, then, served as a haven where contradictions are canceled, opposites collapsed, and, as in a comedy, everyone can live happily ever after.

PART I

ELECTRIC WONDERS

"But I don't understand," said Dorothy, in bewilderment. "How was it that you appeared to me as a great Head?"
"That was one of my tricks," answered Oz. "Step this way, please, and I will tell you all about it."

—L. Frank Baum, *The Wonderful Wizard of Oz,* 1900

The Most Wonderful Inventor of the Century, Melbourne "The Ohio Rain Wizard" Will Be At Goodland Fair Week And has contracted to produce a Heavy Rain, the last day of the Fair, September 26th.

—1891 poster for Goodland, Kansas, town fair

INTRODUCTION

Beyond the Z-Ray

L. Frank Baum published *The Wonderful Wizard of Oz* in 1900 at a time when the American public was confident that science and technology were evoking a modern world of wonders. Inventors and scientific wizards abounded, such as the imaginary Oz, the ventriloquist, balloonist, and circus promoter, as well as his real-life peers, the rainmakers of Kansas who offered to save drought-ridden towns with their mysterious equipment. This book is a study of scientists and inventors who enter the realm of showmanship and of showpeople who mimic scientists to promote wonders. Approached from both directions, the public of the past two centuries has been treated to a version of "magic science" that encourages utopian hopes and religious dreams. In these shows, tensions between science and its unruly twin, magic, have played out in theaters, trade shows, dime museums, world's fairs, and county fairgrounds.

Oz, the humbugging scientific wizard who manipulated technology and his public had many precursors, dating back centuries. To understand how the modern version of this archetype emerged, an ideal figure to begin with is eighteenth-century Prussian showman Gustave Katterfelto. Touting his show as the "wonder of wonders," in 1780, at Spring Gardens in London, Katterfelto offered a performance that blended magic and science—mixing what one writer called "hocus pocus" with discourses on natural philosophy.[1] Wearing a frock coat and powdered wig, Katterfelto parodied a pompous lecturer of natural philosophy. He relied on his thick German accent, and phrases of pseudoscientific significance such as "stynacraphy" and "caprimancy," as he tackled suitably serious topics—one evening's lecture, for example, offered to examine "the powers of the FOUR ELEMENTS, and explain the CAUSE of THUNDER, LIGHTNING, EARTHQUAKES, and different WINDS."[2]

Katterfelto exhibited a "solar microscope" that projected enlarged images of the "insects" he insisted caused the influenza then devastating Londoners. He also sold bogus medicine derived from the works of the Renaissance physician "Dr. Batta" to protect buyers from this plague. Unexpected staged events interrupted his lectures. For example, the natural philosopher would be startled at the sudden appearance from his apparatus of one of his many black cats that he called the "doctor's devils." At different moments these cats would lose and gain tails and emit electrical sparks. To add to the drama, Katterfelto also would plant hecklers in the audience with whom he would argue and duel, wielding an enormous sword. After the philosophical lectures, Katterfelto educated his audiences when he would "discover various arts by which many persons lose their fortunes by dice, cards, billiards, and E.O. table."[3] These lectures would include card tricks, while his shows included other feats of legerdemain, such as his apparent ability to catch fired bullets in his mouth.

Katterfelto was a tireless self-promoter who took out many advertisements in newspapers that boasted of his aristocratic audiences, remarkable cures, edifying experiments, black cats, and "Wonders surpassing all description." Thanks to his advertisements, by 1783, Katterfelto was one of the most well-known performers in London and the subject of numerous satirical illustrations; the poet William Cowper, in a poem detailing the promises and cures offered in newspaper advertisements such as "teeth for the toothless, ringlets for the bald," immortalized the doctor—in the midst of a shocking electrical experiment—as follows, "And Katterfelto, with his hair on end / At his own wonders, wondering for his bread."[4]

His fame was short-lived. Two years after his London shows, Katterfelto descended into poverty while touring the English countryside. He was jailed at least once in Yorkshire after one of his hot air balloons descended and set fire to a haystack. He died destitute in 1799 and was largely forgotten. How such a performer becomes fixed in public memory, if at all, however, is of historical interest. In 1831, a letter appeared in a London newspaper asking if anyone had ever heard of the Katterfelto mentioned in Cowper's poem. One respondent, who noted that the only information he could find on the performer came from a book on quackery, dismissed the performer as one who had "astonished the ignorant, and confounded the vulgar."[5] Another correspondent, who had seen Katterfelto perform, praised the quality of his shows, stating "his apparatus was in excellent order, and very well managed, he conducted every experiment with great certainty, never failing; and though much knowledge might be gained from his lecture, people seemed more in-

This "Sketch of the Most Wonderful Prussian Philosopher" depicts Katterfelto at the height of his popularity in London in 1783. One of his "doctor's devils" or black cats attends while the natural philosopher's solar microscope reveals the agents of disease. *European Magazine*, June 1783.

clined to laugh than to learn." This writer also insisted that Katterfelto's abilities at prestidigitation were superior to Breslaw, a contemporary stage magician.[6]

These dueling assessments demonstrate that the act of labeling a particular performer a "quack" or "mountebank" usually will end further assessment. The second letter writer, who had a personal memory of

Katterfelto's performances, chose not to assess him as a simple quack. Whether or not he deserved the label, removing such a rhetorical gloss can offer a new perspective on such showpeople and the culture that fostered them. Katterfelto, as a "quack," may indeed have caused mischief, but as the second letter writer insisted, he also offered value as an entertainer and educator. Likewise, from a historical perspective, Katterfelto's act was a fascinating cultural hybrid of science and magic, catering to public desires.

This book examines the work of Katterfelto's progeny: performers and exhibitors whose shows blended technology and magic to make their sales pitches. The commodity being sold might be a bottle of "Dr. Batta's Elixer," an electrical healing device, an exhibition of mind reading, a night of stage hypnosis, the latest concoctions of a corporate research and development department, the remarkable qualities of science itself, or a version of evangelical Christianity. The surrounding sales context requires that each of these products appear unique, singular, and if possible, miraculous. The evocation of wonders, and the miniature reordering of the viewers' universe that the experience of wonder entails, can then encourage a buying frenzy. Wonder showmen, in their pitches, transport their audiences into a land of desire in which science takes on magical, healing properties.

The Natural History of Wonder

Katterfelto serves as a convenient starting point for understanding performers who mixed natural philosophy—or science—with stage magic. He was one of the first generation of performers to blend the rational Enlightenment demonstration of natural philosophy with the delights of an evening of stage magic. While prestidigitators and "jugglers" had existed for centuries, often as part of small medicine shows, selling fake cures, it was only in early eighteenth-century England that showmen of this kind left the fairground for the more respectable theater.[7] In the mid–eighteenth century, in increasing numbers the jugglers left behind their exotic robes and bottles of elixirs and took to the stage dressed as gentlemen. Like Katterfelto, these magicians often cast themselves as natural philosophers offering an evening of "experiments" rather than entertainments, and they also educated the public by "unmasking" the frauds of confidence men, cardsharps, mesmerists, and Spiritualists. Throughout their shows, they relied on the apparatus of natural philosophy to create optical illusions and "wonders"—for example, of inexhaustible liquids gushing from small canisters.

Prominent stage magicians from Isaac Fawkes to Jean Eugène Robert-Houdin insisted on the ultimate rationality of their stage illusions to identify themselves as friends of science and to court respectability. Katterfelto, however, chose another course and indicated that he had access to otherworldly secrets and magic abilities. His use of black cats, his insistence on calling them "devils," and his reference to Dr. Batta of the Renaissance cast his act in a gothic, romantic mode. His buffoonery and bluster allowed his audiences to vent their hostility toward the natural philosopher as authority figure. And in order to sell tickets and bottles of medicine, he created wonders that blurred the line between the rational and the mythic, the modern and premodern. Although his presentation was somewhat tongue in cheek, this strategy added dramatic value by daring audiences to "unmask" his impostures. Avoiding the label of scientific respectability, Katterfelto ran greater risks, but also cloaked himself in older values, seeking kinship with the natural magicians of the Renaissance—such as the mysterious Dr. Batta—who believed that wonders were to be investigated, savored, and protected. These earlier investigators relied on vitalist notions of nature along with elements of the scientific method in their philosophies and investigations, and gained prestige through their access to secrets.[8]

Posing as a natural magician in an era when scientists were promoting a clockwork universe filled with clockwork organisms, Katterfelto's act offered his audiences the temporary appeal of a universe imbued with magical possibilities and meaning. On his stage, black cats, scientific apparatus, and mysterious effects blended. Katterfelto's work can be denounced as quackery, or regarded from another angle: his mixture of science and the fantastic embodied an impossible ideal, a "populist science" created in the image of public desire. Such a "populist science" moved beyond the sanctioned boundaries of "popular science" and on into the territory of myth.

The shows of Katterfelto and of Baum's fictitious Wizard of Oz capitalized on the long historical alliance of wonder with the scientific impulse with its many near-divorces and reengagements. For the elite collectors and thinkers of the Middle Age and Renaissance, wonder was one of the ruling passions. The experience of wonder was invaluable to the connoisseur for the speculations and vistas it opened.[9] The wonder-prompting phenomenon was variable: it might involve the appearance of a comet, the birth of a "monster" such as Siamese twins, a peculiar specimen of crystal, artifacts from an exotic species such as the unicorn, a trompe l'oeil art object that demanded that the viewer determine when nature stopped and art began, or a mechanical device such as a mirror with peculiar optical properties.[10] Renaissance thinkers believed wonder was processed via a faculty halfway between emotion and cognition, and, accordingly, excited both the passions and the intellect. When experienced, the wonder phenomenon would evoke awe, destroy previous conceptions, blur the opposition of nature and art, and provide tantalizing hints regarding the true order of nature.

Once appreciated, a wonder—whether "natural," "technological," or a mixture of the two—would prompt intellectual investigation. Many early natural philosophers heralded the importance of wonders to science. Francis Bacon, for example, based much of his program for a reformed natural philosophy, or his "Great Instauration," on the belief that investigators should begin by studying wonders or "strange facts." Such a pursuit would not spur further abstract philosophizing about nature's laws and regularities but instead lead to direct experimentation and hard-won knowledge of nature's creative powers. Following Bacon's program, founding members of the Royal Society of London and the Paris Académie Royale des Sciences sought data about such "strange facts" or irregularities.[11]

Throughout the scientific revolution, wonder continued to be seen as a powerful motivation for research, but by 1800, when magicians were taking to theater stages, wonder became vulgar, fit only for the masses, as a mechanistic, dispassionate vision of science emerged. New regularities had been imposed on nature, and investigation of the monstrous or the irregular became less valuable as a research program.[12] At that same time, philosophers began to apply an idea of the "sublime" to the aesthetic appreciation of nature and art but no longer to "technology"—which then became categorized as products of the mechanical arts.

At this point, wonder ceased to be an elite preoccupation, and the wealthy no longer amassed wonder cabinets—collections of wonder-provoking artifacts—but wonder did not disappear. First, it appeared in public museums open to all and in the fine arts and literature in their romantic and gothic strains;[13] second, wonder took refuge in the vulgar, where it emerged in theater, stage magic, commercial culture, and the art of the sale, refined by impresarios such as P. T. Barnum, Cecil B. DeMille and their progeny; third, wonder reemerged in what recent American historians have defined as "the technological sublime," that is, the modern tendency to view the technological object with the awe once reserved for dazzling displays of nature's majesty.[14] If American contemporaries of Katterfelto turned to mountain peaks and waterfalls as sources of nationalistic pride, by the late nineteenth century railroads and vast iron bridges became more common inspirations. Promoters and capitalists, elites cleansed of the passion for wonder, mounted large displays of the technological sublime at world's fairs, in heroic engineering projects, in gigantic piled displays of potted ham, and in rocket launches to enunciate and legitimize power and to provide unifying experiences to a mass democracy.[15]

The "technological sublime," or "wonder" in its vulgar phase, became a weapon available to all. Numerous performers and subculture representatives, including mesmerists, electrical healers, hypnotists, mind readers, techno-evangelists, and New Age salespeople, have relied on the technological sublime to promote their agendas. Like Katterfelto, they have presented a vision of a world that fuses the marvels of modern science with ancient, invisible forces. In this light, Katterfelto and his progeny take on greater significance than commentators generally allow; they were not just simple charlatans, dangers to the public health and morality, but in their own monstrous way guardians of older values. If the medicine they sold in bottles did not cure anything, their shows offered an antidote to the alienation and dehumanization that accompanied the mechanistic philosophy.

Wonder in America

Like a spark hesitating to jump between damp electrodes, the wonder show à la Katterfelto did not arrive in America until the early nineteenth century. Performance modes of all kinds made a stalled entrance to the colonies. The early Puritan distaste for theater, compounded by the republican philosophy of the revolutionary period that equated theater with English corruption, limited the quantity and quality of theater. Soon after the revolution, when the states lifted restrictions on drama, along with standard theater, variety performers began to test the new country's taste in entertainment. Magic lantern phantasmagoria shows that presented "apparitions of the dead and absent" appeared in the very early 1800s along with displays of mechanical androids or automatons; 1819 brought Mr. Charles, a "ventriloquist and professor of mechanical sciences to his Majesty the King of Prussia," to Boston with his "mechanical games and philosophical recreations."[16] Johann Nepomul Maelzel brought the long-renowned chess-playing automaton "The Turk" to America in 1826. In Philadelphia, lectures that included displays of electrical wonders were added to Charles Peale's Museum in Philadelphia, which included a portrait gallery, natural history displays, and displays related to the industrial arts.

It was not until the 1830s that the wonder show truly flourished in America. In that decade, skilled magicians such as Signor Antonio Blitz first arrived to offer sleight-of-hand and large illusions. Although their discipline had been founded in late eighteenth-century France, mesmerists also first arrived in America in the 1830s to demonstrate their skills in evoking trance states in their subjects. Electrical healers with lectures similar to Katterfelto became common in America in the 1830s and 1840s.[17]

The American wonder show emerged in the 1830s and 1840s along with a robust market economy, an explosion in printed materials explaining science, and strong public interest in scientific lectures and demonstrations. To complicate matters further, in antebellum America, interest in science and pseudoscience was diffused into a culture of fervent religiosity.

The religious fervor of the antebellum period can be found in the evangelical revivals of the Second Great Awakening but also in a wider culture of self-improvement and utopianism.[18] In these decades popular science intermixed with popular religion. The geology taught at most American universities, for example, pointed to the correlation of the book of Genesis with the geological record.[19] The new revelations of the microscope and telescope only added to the glory of God's creation, while technological and economic advances were matched with widespread optimism that a more-perfect society was dawning. New sciences such as mesmerism and phrenology, which encouraged the belief that each individual's mind or soul was capable of amazing powers, flourished. Self-improvement could be a profound undertaking that would remake society. The "harmonialist" philosophy of mesmeric subject Andrew Jackson Davis, as well as the transcendentalist philosophy of Ralph Waldo Emerson, expressed limitless possibilities for the individual's fusion with nature and the divine. Tidy distinctions between a "correct" scientific worldview and an "incorrect" worldview of magic-science largely were absent; wonder showmen and credentialed scientists vied for audiences, and both fed the public's millennial hopes.

The Civil War and its carnage, as well as the increased industrialization, immigration, and urbanization that followed, however, led to the erosion of the perfectionist vision of antebellum America. A new cynicism about the old truths arrived, along with marketing strategies and a mania for documentation, data collection, photography, statistical analysis, and other scientific methods for improving society. In tandem with this realism came a public taste for illusions and the awareness that marketing, like the promotions of P. T. Barnum, created an environment in which the buyer must beware. Show business, where illusions were the currency of the realm, thrived on an atmosphere of disclosure, of challenge, and of revelation. Performers presented their rivals as shams, as mere imitators, and responded to mock challenges of their skill and authenticity.[20]

This strategy of public challenges, however, rarely extended to the engines of change themselves. The prevailing public mood followed the preaching of Auguste Comte, who valorized technological innovations and their creators. By the late nineteenth century, engineers and inventors were culture heroes, featured in newspapers and novels as the embodiment of progress.[21] Progressive reformers seized on technology and scientific methods to mend social ills, while promoters and capitalists encouraged public worship of technology with awesome spectacles and pleasing displays. Both praise songs and critiques of the coming technological society also emerged.

Utopian novels based on the coming technological paradise had a vogue and encouraged progressive beliefs. Edward Bellamy's *Looking Backward,* written in 1888, foresaw a future paradise in the year 2000, when all corporations had merged to form a benign governing body. In Bellamy's year 2000, competitive capitalism and its wastefulness had vanished, and all of society's resources were directed toward the general good, with all workers engaged in a large industrial army. A technical elite kept the system efficient. Streets were clean, crime, poverty, and advertising disappeared, credit cards were used instead of money, pneumatic tubes efficiently distributed goods from warehouses to homes, and amusements were serene—the citizens of 2000 had a special taste for listening to nightly concerts or Sunday sermons via telephone broadcast. Bellamy's book was wildly popular. Dozens of utopian novels patterned after *Looking Backward* were published. Other utopians drew up plans for colonies and cities and a few led pioneers off to remake the world.

If the progressives spoke of technological and material progress as if it tallied with divine will, some critics were less certain of the validity of such progress. Walt Whitman took up the concerns of the romantic movement of the early nineteenth century when he sounded an alarm in *Democratic Vistas* (1871) and fretted over the soulless empire that the Enlightenment, along with technology and business, had ushered onto the American scene. Escape from the crude materialism of the Gilded Age into an imaginary pastoral past became a middle-class preoccupation. Readers of Lew Wallace's *Ben Hur* (1880), patrons of designer William Morris, admirers of the Gothic revival in architecture, and others took pleasure in imagining medieval times, exotic oriental locales, or golden pasts.[22]

The Rescue of "Progress"

While many intellectuals sought such escapes from modernity in the late nineteenth century, a much stronger public dissatisfaction with technology and modern life appeared in the wake of World War I with its seemingly meaningless sacrifice of soldiers' lives, horrific weaponry, and public scandals over the armaments industry's profiteering. Progress was no longer seen as inevitable—nor could it safely be linked simply to technological innovation.

Both progressive intellectuals and reactionaries began to question the materialist creed. Liberal notions that had underpinned the scientific revolution became dubious propositions. Observation and experiment, the free dissemination of information, and the refusal to accept theory on authority had wrought great changes—the rise of the merchant class, the Industrial Revolution, mass production, mass communication, mass democracy, and the business revolution of the late nineteenth century. But along with these came regimentation, class warfare in industry, faster-paced lives, media-saturated landscapes, and what turn-of-the-century neurologist George Beard had classified in 1881 as increased "nervousness." The post–World War I modernist movement in arts and letters, which included writers as varied as André Breton and William Carlos Williams, artists such as Pablo Picasso and Marcel Duchamp, and which drew from philosophers such as Henri Bergson and William James, promoted fascination with the irrational, spontaneous, and the primitive, as means to shake free of the nightmarish aspect of the machine age.

At the same time, a lowbrow modernism had long been under way—with wonder shows included—directing its own rebellion against the modernization rooted in the Enlightenment. Expressions of this lowbrow modernism included the stage acts of hypnotists, magicians, and mind readers, as well as depictions of mad scientists in early science fiction stories and films, Rube Goldberg cartoons that revealed the common "boob" enmeshed in surreal mechanisms, and the work of the critic of science Charles Fort, who spent decades collecting clippings from newspapers and scientific journals of oddities such as "black rains," "red rains," falls of slime from the heavens, and showers of toads and small fish to prove that scientists had no monopoly on a true understanding of the universe.[23] Fort questioned the "uniformity" of nature and the regularity of scientific laws and reveled in documenting the errors of scientists past, as when the French Academy of Sciences claimed that accounts of meteors or rocks falling from the heavens must be delusions of the provincial mind.

By the 1920s, when Fort was writing, the rise of the machine became an obsession among populists and intellectuals alike. Dramas such as Eugene O'Neill's *The Dynamo* and Elmer Rice's *The Adding Machine* depicted average man terrorized by the machine.[24] In this same era, along the same lines as Whitman, one of adventure novelist Arthur Conan Doyle's characters preached, "It is this scientific world which is at the bottom of much of our materialism. It has helped us in comfort—if comfort is any use to us. Otherwise it has usually been a curse to us, for it has called itself progress and given us a false impression that we are making progress, whereas we are really drifting very steadily backwards."[25]

As the previous quotation suggests, in the early twentieth century, many people lamented that scientific progress was accompanied by a decrease in spiritual values. Prior to Darwinism, theologians of the early nineteenth century, and intellectuals such as the transcendentalists, had used each scientific discovery as proof of the glory of nature, or alternately, of God's creation. Darwin's theory of natural selection, which suggested that creation was the product of random change, ultimately put an end to such "arguments from design."[26] Notions of "theistic evolution," which implied that a greater intelligence guided the evolutionary process, had their vogue and temporarily salvaged the relationship between science and religion.[27] In retrospect, however, notions of guided evolution resemble the crutches that appear throughout surrealist artist Salvador Dali's oeuvre propping up melted clocks, cabinet drawers, and human grotesques.

In a universe ruled by chance, humanity's central place was lost. Sociologist Peter Berger has pointed out that one of the prime functions of a religion is to erect a "sacred canopy" to provide a network of meaning to the cosmos.[28] When that canopy became tattered, as it did when the full impact of Darwinism became clear in the 1920s, following the discovery of genetic material, new sects and schisms arose. These responses often did not involve a complete rejection of science, but as in the antebellum period, involved efforts to bridge the realms of science and religion. Against the backdrop of the John Scopes "monkey" trial of 1925, which examined the legality of teaching Darwinism and created a split between the scientific and fundamentalist Christian worldviews, scientists such as American physicist Robert Millikan argued publicly that science need not detract from the values of religion and urged that these enterprises maintain separate realms and functions.[29]

Such a tidy separation could easily be prescribed but was not easily followed. Promoters of the scientific project had long hitched progress to the millennialist dream of redemption.[30] Likewise, fervor for science and technology had emerged on a mass scale because their appeal was as much emotional as rational. Uninitiated into the technical languages, the public sensed that the new technological forces formulated a quasi-mystical realm. Consumer culture's cultivation of a public taste for "positive thinking" and the technological sublime provided a quasi-religious, quasi-scientific cast to American popular culture that could admit many hybrids.[31] So religious sects with scientific trappings such as Theosophy and Christian Science emerged, psychical researchers attempted to investigate scientifically the reality of the soul and the afterlife, science fiction writers offered visions of futures that mixed technology with the mystical and paranormal, and performers offered dazzling "wonder shows" that combined demonstrations of "science" intermixed with religious or magic content.

Beyond the Z-Ray

Wonder showmen worked at the boundary of science and magic, relying on wonder to help their audiences suspend disbelief. To prop up their version of the fantastic, they continually cast scientific and technological breakthroughs in magical terms and remnants of a magical worldview in scientific terms. Especially useful from a showman's point of view were electricity and other new invisible energy forms scientists were uncovering that might be regarded as part of the same spectrum as the invisible powers of the ancient world. To give one example, public reception of the discovery of the Roentgen ray or X-ray in the 1890s points to the linkage of one such new electromagnetic "ray" with preternatural forces.[32]

Displays of X-ray apparatus drew crowds in a variety of venues. Shortly after its discovery, for example, a Roentgen-ray apparatus became a big draw at the National Electric Exhibit in New York City in 1896. Of this display, a reporter indicated, "A never-ending line of men and women patiently await their turn to have a 'glimpse of their bones' through a fluoroscope. . . . The Edison exhibit of Roentgen ray apparatus some evenings is presided over by the great inventor [Edison] himself, who affably explains its mysteries to visitors."[33]

What sort of thrill or insight did the spectator gain when finally placing his or her hand behind the screen? Many articles of the era dedicated to this new discovery included delicate X-ray photographs of the human hand. These plates showed white finger bones and wrist bones surrounded with darkness, and, perhaps, a small circle of metal around the metacarpus of the wedding ring finger. The spectator of that era would be amazed to see inside herself, and perhaps a little disconcerted to view

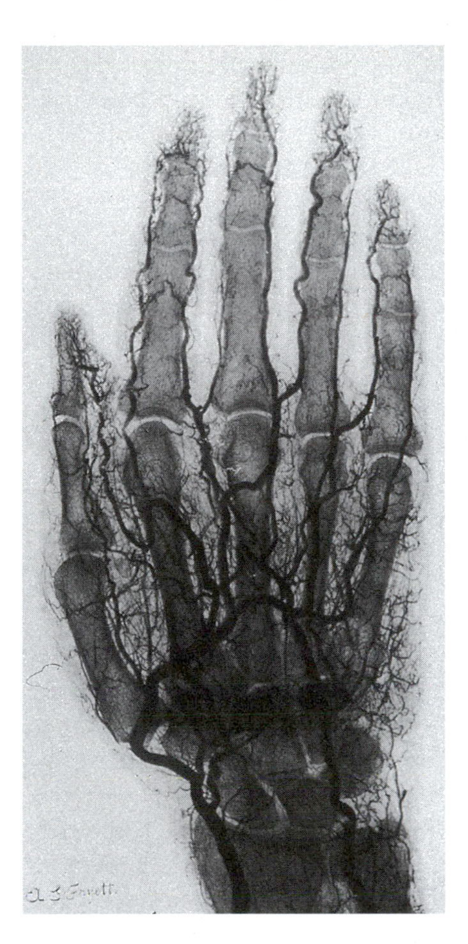

In the 1890s the X-ray or Roentgen ray became a metaphor not only for science triumphant but also for the possible reality of an unseen spiritual dimension. Courtesy of the National Museum of American History, Smithsonian Institution.

a portion of her own skeleton, a symbol of mortality. Conceivably, he or she might even reflect on how life and death were intertwined.

Such trade show demonstrations can be compared to a more theatrical use of X-ray effects that linked the X-ray to the spirit realm. In the 1890s, the Cabaret du Néant (or Tavern of the Dead) first opened its production in Paris and later in New York City. After entering the Cabaret, spectators followed a "monk" down a blackened hall to a café with candles on coffin-shaped tables where they could order refreshments from waiters in funeral garb. A lecturer called their attention to paintings of figures that dissolved into paintings of skeletons. While bells tolled and a funeral march played, the monk led the audience to a second chamber; here, a volunteer was asked to step up on a stage and enter a standing casket. After the volunteer was wrapped in a white shroud the spectators gasped at an apparent "X-ray" effect—actually a simpler optical effect[34]—as the man dissolved into a skeleton and then once again returned to plain sight as the skeleton disappeared. In the last chamber, using a similar optical effect, a live spirit appeared to walk around an audience volunteer who mounted the stage to sit at a table.

The Néant spectacle, which may have pleased the jaded tastes of fin de siècle audiences, contrasts sharply with the homely trade show atmosphere of Edison's demonstration of the fluoroscope amid potted palms and plants. Juxtaposed, these two presentations suggest how scientific discoveries became associated with mystical processes and with entertainment. If Edison's display prompted fascination and possible thoughts about mortality, the Cabaret du Néant show explicitly progressed from an apparent exhibition of X-rays to a spiritualistic exhibition. The X-ray, which revealed bones, and the deeper gaze that revealed spirit could appear as a simple progression along a continuum.[35] In both settings, the trade show and the cabaret, the X-ray reminded people of how narrow their sensory range was and how technology could extend and enhance the senses. The new ray also suggested the power of invisible forces that scientists and inventors were just beginning to tap.

Scientists were then investigating other mysterious rays. Several years after Roentgen's discovery of the X-ray, Henri Becquerel discovered that certain phosphorescent minerals emitted their own characteristic "rays." Soon after, Marie and Pierre Curie isolated several radioactive substances, including the new elements polonium and radium. The Curies spoke of radiation with awe. Of the mineral samples he and Marie Curie kept in a shed, Pierre Curie remarked, "I well remember the sense of excitement we felt when we used to enter our little world at night and saw on all sides the luminous products of our work glowing faintly in the darkness."[36]

Wonder showmen, pulp writers, and the technically minded alike had a fixation with new rays. The imaginary Z-ray, for example, was frequently invented and reinvented. An electrical journal from 1896, the year the English-speaking world learned of the Roentgen ray, reported that "an enterprising Bowery dime museum proprietor advertises an exhibition of Z-rays, announcing that they are much more wonderful in their properties than the X-rays, which will be relegated to the rear by the discovery, now for the first time exhibited, etc. The museum proprietor, however, is not alone."[37] The writer went on to describe a scientist who had recently announced variants of the X-ray that he chose to call the "X_1 and X_2" rays. The exasperated writer concluded, "When the Wurzburg professor [Roentgen], in an excess of modesty and with too great confidence in scientific etiquette, made use . . . of the letter which is held so mysterious by the non-mathematical, he could not have known how

Madame Curie, an early experimenter with radioactive substances, is depicted here as an alchemist. Stuffed crocodiles were once draped above the entrances of medieval churches to remind visitors of the wonders of the universe. Courtesy of the National Museum of American History, Smithsonian Institution.

attractive a bait it would prove to the mongers of sensationalism." This writer's point was proved again in 1903, when French physicist Prosper Blondlot announced to the world his discovery of "N-rays." Articles describing the properties of the mysterious new N-ray generated excitement until an American researcher determined that Blondlot, however sincere, was deluded.[38]

As new rays were posited, writers and researchers theorized that a sixth sense or enhanced senses might take advantage of the widening spectrum of such vibrations of the ether. Such speculations, which fit the wonder show's formula of enhanced human powers, were common in the late nineteenth century. In the 1890s, both William Crookes and another leading British physicist, Oliver Lodge, speculated that "thought transference" or telepathy could be explained via "brain waves" moving through the "ether." One mind, tuned to a similar resonance, might then receive thoughts, much in the manner of the radio prototype that Oliver Lodge had sketched out for others to pursue.

Magical rays—especially "Z-rays"—continued to energize pulp literature well into the twentieth century. For example, in 1920, the radio entrepreneur, publisher, and early patron saint of science fiction, Hugo Gernsback, ran the story "The Ultimate Ray," in which a mad scientist asks his captive, "Are we justified in concluding that X-rays are the *ultimate* rays; that is, the rays of highest frequency which it is possible to produce?" Before the scientist's captive could respond, the mad genius announced that this was not so and added, "I, Pax Marriote . . . have finally discovered the *ultimate ray,* the 'Z' ray, the long sought ray that would decompose matter utterly into energy alone, the disintegrating ray!"[39]

Gernsback writers continued to exploit the notion of an extended sensory apparatus. In 1923 Clement Fezandié produced "Dr. Hackensaw's Secrets: No. 15: The Secret of the Sixth Sense." In this tale Doc Hackensaw created various machines to enhance the senses, and after allowing a flapper-like heroine to put on a mechanical glove that made her melt with ecstasy, he told her, "These are radioactive waves that you perceive . . . you perceive them neither as flavor, scents, sounds or visions, but in an entirely new and hitherto undreamt of manner. What you have been listening to is a radio-active composition of my own composition."[40]

Wonder show purveyors, who also relied on the extended-consciousness model that scientists like Crookes and that Gernsback's writers explored, offered both miracles and scientific contexts for their acts. Performers claimed that hypnotism, telepathy, and Spiritualism involved channeling or transmitting subtle energies, fluids, or "rays." When attacked for offering degrading entertainments, these showmen defended their displays in terms of older values that modernity threatened. Hypnotists and mesmerists insisted they were revealing the subject's soul to the audience and helping to restore religious faith to the public; husband and wife mind readers claimed their abilities developed from love and domestic harmony; other showmen, returning to the argument from design, insisted that the wonders they revealed on stage merely reflected the glories of divine creation.

On Vanishing Acts and Historical Analysis

This book concentrates on the history of the wonder show after its delayed entry to America and demonstrates how such performances helped forge a populist science shaped by public desire. In order to establish the importance of the wonder show as a historical touchstone, chapters follow the genre from the early nineteenth century in America to the present, looking at both obscure performers and better-known historical actors such as P. T. Barnum, Thomas Edison, Nikola Tesla, and Harry Houdini.

The work is divided into three parts: the first looks at the history of electricity and examines how performers, inventors, and electrical technicians perceived and promoted electricity as a quasi-magical force. The second part focuses on hypnotists and mind readers who performed "mystic vaudeville" and differentiates these performers from mainstream stage magicians and their anti-Spiritualist crusades. The final part examines the wonder show as a species of modern salesmanship, and looks at corporate-sponsored science shows, evangelical science shows, UFO occultists and their wares, and New Age promotions of the twentieth and twenty-first centuries. It concludes with a discussion of the relationship between wonder and superstition and places this issue in the context of the "science wars" debates of the late twentieth century. This debate pitted "skeptics"—that is, scientists and professional debunkers of the New Age mentality—against intellectuals who valorized movements that contested scientific authority.

A scholar who relies on vanished spectacles like wonder shows as texts for analysis has to assume that these performances offer a glimpse into popular interests and beliefs. Although it is always difficult to prove how vanished audiences "received" such performances, the fact that simple entertainment formulas like those found in countless *How to Hypnotize* pamphlets emerged from the harsh evolutionary environment of show business suggests what audiences then "were buying." Likewise, although untrustworthy testaments to a showman's genuine beliefs and values, performers' pamphlets and memoirs do reveal the rhetorical strategies they employed to cater to audiences.

While this book explores the history of performance, religion, science, and technology, its primary contribution is in the realm of the history of popular science and technology. It shows how versions of the technological sublime have been employed to support differing visions of progress. It also complicates the common historical theory that the nineteenth-century public's democratic interest in science degenerated with the rise of consumerism in the twentieth century.[41] Instead, it emphasizes that a parallel vein of gee-whiz science promotions continually has surfaced to address the paradox of scientific progress as both wonder-erasing and wonderful.

The wonder show is not just a historical curiosity, however—not merely a quaint matter upon which antiquarians might speculate. From the outside looking in, the realms of the technical elites can take on many unusual colorings and hues. Science, as most people perceive it, is a land where new monsters emerge even as the old are vanquished. Playing the fool to science as its superior, the "absurd" contours of populist science remind us that science too rests on reductive concepts and acts of imagination. If Francis Bacon, four centuries ago, wished to use wonder in his "Great Instauration" to jar natural philosophy out of its then-imaginary triumphs, wonder can still function, in its small way, to suggest that science triumphant still depicts a "real world" made in its own image.

As the concluding chapters suggest, the themes this book inspects are not simple curiosities, but are at the heart of many of the current debates about technology and progress. Utopian and dystopian threads continue to inform public responses to breakthroughs in computer and communications technology, genetic engineering, medicine, and global commerce. Environmentalists' calls to place limits on technology, industry, and population clash with the no-limits notion that is built into the scientific project and global commerce. Obsessed scientists, greedy industrialists, strange mutants, new viruses, corrupt officials, and heroic hackers are common fixtures in popular films. Likewise, the battle of rationalist skeptics against what Carl Sagan once called the "dragons of Eden" continues. Yet the current fascination with alternative medicine, Eastern mysticism and the otherworldly, and the never-ending dream of the technological "fix" all clearly existed one hundred years ago when Baum concocted the *Wizard of Oz* and the Curies, watching the gentle blue light glowing from their mineral samples, saw it to be good.

CHAPTER ONE

The Electric Wonder Show

In July 1853 the Crystal Palace opened on Forty-second Street in New York City. Like its slightly older namesake in London, it was an enormous building of glass, its panes held in a framework of bronze-tinted iron, topped with towers, flags, and a soaring dome. This glass structure housed exhibits of industrial and fine arts that ranged from sculptures of Cupid and a "stag in zinc" to soda-water apparatus, gas meters, mechanical chairs, steam-operated machinery, false limbs, and "a specimen of the banana plant."[1] For the opening, much of New York was draped with bunting and flags to greet President Franklin Pierce, who, accompanied by other Mexican War heroes, rode on horseback up Broadway and Fifth Avenue; amid marching troops, he braved a rain shower, then arrived to make his speech at the Crystal Palace. Pierce's talk followed an opening prayer from Reverend Bishop Wainwright, who insisted that God's hand could be seen in "all this fertility of invention."[2] Pierce emphasized that the exhibition should be seen as a symbol of national pride and unity despite the nation's regional divisions over slavery.

Wainwright and Pierce spoke before an audience that included governors of several states, senators, and members of New York's elite. Outside the palace, in a neighborhood that still had its shanties, soldiers dragged citizens into "grogeries" to drink alcohol, while others consumed ice cream or sodas. A reporter noted that in the turmoil, military drums and tubas soon were draped on the backs of stagecoaches. Some members of the ticketless crowd were lured into a museum that featured a bearded French girl, others into an establishment with a placard indicating that the "President and his suite had been invited, and would probably attend the performance." The newspaper also reported that among the throng were hawkers of patent medicines and a lecturer

The New York Crystal Palace Exhibition of 1853 was a modern wonder cabinet—the building itself was an engineering gem, filled with marvels of art, crafts, science, and manufacture. The building survived as a venue for concerts and events until fire destroyed it in 1859. Courtesy of New-York Historical Society, New York.

who exhibited a "novel machine, made 'to test the strength and capacity of the lungs.'"[3]

While business at the Crystal Palace seemingly abounded, with throngs buying season passes or single tickets to walk through its turnstiles, after Pierce's departure from the Astor House another visitor to the city came with no fanfare—an electrical exhibitor named Charles Came who hoped to make his fortune by presenting to the public "The Great Physical Sleeping Phenomenon of the Nineteenth Century." With the help of a lithograph of his Sleeping Man, posed in sleeping, sitting and standing postures, a book about the Sleeping Man's life, and a phrenological reading of the sleeper from the firm of Wells and Fowler, Came hoped "to make a strike" and eventually sell his attraction to showman P. T. Barnum. Although the newspapers reported his attraction, Came was eventually to leave New York embittered, having barely met expenses, and loaded down with one thousand unsold copies of his life of the Sleeping Man.

Charles Came was just one of many scientific exhibitors of the nineteenth century who inhabited the borderland of respectable society and educated the public with homespun versions of science. Before promoting his Sleeping Man, Came had traveled upstate New York, putting on electrical demonstrations while lecturing about astronomy, phrenology, and other topics. Although Came may have mixed his science with fancy, he clearly had more genuine dedication to science than the man outside the Crystal Palace with a lung machine that could predict one's life span when one blew into its tube. Came was a self-taught physician who offered electrical cures for various illnesses and sold remedies such as "Dr. C. Came's Vegetable Strengthening Syrup" and "Dr. C. Came's Magnetic Stimulating Drops," yet he also was a sincere student of science and medicine who believed in the efficacy of his treatments. Indeed, in the many small towns he repeatedly visited, he gained as much or more esteem from his doctoring as he did from his showmanship.

The mid–nineteenth century featured an explosion of public information about science, presented in print and in lectures. Just as the Crystal Palace exhibited the latest products of science and technology in a manner meant to prompt thought and reflection, outside such grand expositions, learned men from academies and itinerants like Came offered scientific demonstrations and lectures to genteel and working-class audiences. A commentator of the era examining this trend noted, "These are the days of popular lectures and familiar treatises on scientific subjects."[4] These lectures could range from very capable demonstrations by "natural philosophers" who hesitated to lower themselves to public speaking, to the performances of men like Charles Came who thrived on exciting his audiences about astronomy and electricity before selling them medicine, to the spiels of the hawkers of patent medicines and lung machines on the streets of cities.

The rural and working-class audiences who attended the lectures of performers like Charles Came found not only an educational experience but also a healing spectacle. Such performances were an amalgam not easily pigeonholed. Came and other electrical exhibitors were offering popular science as part of a variant of the ever-popular medicine show with its ancient roots. Despite criticism from scientific and social elites, the electrical wonder shows of Came and others thrived because of their mixed purposes: to educate, to sell, and to heal. Came's blending of healing with electrical displays helped establish his authority and enhanced the possibility for cures—electrical or otherwise.

The Advent of the Forty-foot Flea

Among the few books in Charles Came's library that survived him were several treatises on phrenology, a medical handbook, and several published lectures on electricity, including two by well-known scientific popularizers of the day: Benjamin Silliman, a Yale science professor who gave frequent public lectures in the 1830s and 1840s, and Dionysius Lardner, an English scientist who edited a multivolume encyclopedia of science and industry and who made an American lecture tour in the 1840s. That Came, who traveled a small-town circuit of his own devising, often with no advance publicity, relied on the publications of Silliman and Lardner, both of whom lectured to large audiences in big cities, illustrates how scientific knowledge was diffused in this age from elite to plebian settings.

That diffusion, often enough, began in England, which then had a better-established scientific culture. In England both the practice and popularization of science divided along social class lines. In *Frankenstein's Children* (1998), historian Iwan Morus has argued that the flowering of electricity in nineteenth-century England relied on two cultures: that of the middle-class or elite natural philosophers—theorists following in the tradition of Newton and other gentlemen scientists—and that of the mechanics—hand workers who fashioned the scientific (or "philosophical") apparatus for experimenters and often made new discoveries of their own. Often artisans successfully forged new identities as natural philosophers. Two such "self-made" natural philosophers, electrical experimenters Humphrey Davy and Michael Faraday, saw their goal not primarily in technological development, but in the development of abstract principles. As gentlemen, they denied their talents in devising equipment and adopted an ideology that encouraged the pursuit of knowledge for its own sake, not for the sake of manufacturers and the marketplace. Those experimenters who remained among the popular lecturers and mechanics, however, highlighted their instruments and skills, and proudly claimed title to inventions.

Yet the natural philosopher, no matter how aloof, had to cultivate an audience and "perform" for it in order to gain backing for research.[5] If anything, this dynamic was more obvious in the nineteenth century. The author of an 1867 article in the *Nation* lamented that the scholar of his era, in order to build an observatory or found a museum, had to "appeal . . . to masses of people; he must fascinate two or three hundred average American legislators . . . he must exhibit his disinterestedness, enthusiasm, and learning before large audiences; he must be constantly before the public in newspapers, periodicals and popular books. For the noble patron we have substituted the long subscription list."[6] Science did not "just happen" but was subject to market forces and dependent, in this case, on the judgment of an educated public and lawmakers.

The natural philosopher's performances were often highly theatrical affairs. Humphrey Davy, for example, first gained a reputation as a chemist when he began to experiment with "dephlogistic nitrous air," or laughing gas. He published discussions of his responses to the gas in 1801 and soon after staged demonstrations before large audiences at the Royal Institution in which natural philosophers inhaled the gas and reacted with varying responses of hilarity.[7] While Davy's demonstrations and writings secured his place in the British scientific establishment, itinerant lecturers with lesser credentials began giving laughing gas demonstrations to audiences in England and America.

Michael Faraday, who began as one of Davy's assistants, became one of the most important electrical experimenters of the early nineteenth century—the first, for instance, to use a rotating magnet to induce a steady current in a separate circuit. Yet to gain acceptance as a natural philosopher, Faraday had to overcome his own artisanal background and prove he was not just clever with his hands. When first hired, Faraday helped Davy build his apparatus, and he soon began to contribute to Davy's research. Eventually Faraday reported his findings independently and was admitted to the Royal Society. Faraday also helped cement his reputation by performing and giving popular lectures that added much-needed income to the Royal Institution.

In the 1830s Faraday began to give "Friday Evening Discourses" that became popular among members of the elite fortunate enough to secure a ticket. As the Royal Institution's founding premise was to bring science to the aid of agriculture, electricity was not intrinsic to every performance. For example, as part of an 1837 lecture on "Early Arts: The Bow and Arrow," Faraday demonstrated how to use a blowgun and delighted his audience by shooting darts at a target.[8] Undoubtedly, Faraday, who had studied elocution as part of an active self-improvement campaign, did so while maintaining his dignity.

Such showmanship was also available to the less privileged classes. Electrical experimenters who maintained their artisanal status often lectured in the popular galleries of practical science that had a vogue in England from the 1830s to the mid-1840s. One such popular lecturer was William Sturgeon, a leading electrical experimenter and deviser of the first electromagnet in 1825. He was also a master mechanic and builder of electrical apparatus, and editor of the journal *Annals of Electricity*. Sturgeon's position was, of necessity, populist. He did not lecture at the Royal Institution and never gained admission to the Royal Society. Instead he built equipment and lectured at the popular galleries.

Galleries such as the Adelaide Gallery and the Polytechnic Institution in London where Sturgeon and other popularizers lectured were similar to today's science and industry museums: the galleries offered spectacles and examples of industrial arts and working machinery. Attractions might include a large orrery that replicated the movements of the planets, an electric eel, an "oxyhydrogen" projecting microscope of three million power that could project an image of a flea forty feet wide, and a working diving bell that took volunteers underwater. At such galleries, lecturers such as Sturgeon demonstrated the decomposition of water via electric current and other electrical effects. Along with education and uplift, these galleries also blended in entertainments such as dioramas, comic impersonations, and the appearances of celebrities such as Tom Thumb.[9] These galleries were popular with middle-class and working-class audiences, and did not limit tickets to subscribers.

Early nineteenth-century America lacked the galleries of mechanical science of England. One long-running exception, however, was Peale's Museum in Philadelphia, a precursor of both contemporary science museums and natural history museums. Charles Willson Peale was a gifted portraitist with strong interests in natural history. In 1784 he added a portrait gallery to his studio, and mounted panoramas; yet visitors were even more interested in his natural artifacts. Over the following decades he created the prototype for the natural history museums eventually to come. His eclectic museum, the finest of American museums at the turn of the nineteenth century, featured the skeleton of the first mastodon unearthed in America; it also included hundreds of artfully preserved birds and insects, wildlife posed before painted dioramas, mineral displays, American Indian artifacts, waxwork figures representing the different races of humankind, examples of the industrial arts, and model machinery.[10]

To combat rising rent, Rubens Peale, who headed the museum from 1810 to 1821 after his father retired, added pure entertainment to the museum's former "scientific entertainment" approach. Under his direction, the museum offered concerts and popular lectures. On Tuesday and Thursday evenings, Rubens arranged for lectures that might include chemical experiments offering visual delights; ungainly electrical apparatus would also be rolled into the hall, and Rubens or another performer would work the static electricity generators to create sparks, cause dolls to dance, detonate soap bubbles, knock down a "thunder house," and fire a brass gas cannon.[11]

Catering to prevailing tastes, such entertainment had to provide "uplift." Republican ideology, long promoted by theorists of democracy, insisted that to maintain a flourishing democracy, all citizens, particularly those entrusted with the vote, needed to be virtuous, informed, and of sound moral judgment. Public lectures, accordingly, should have a sound moral core and be of educational value. After the religious fervor of the Second Great Awakening swept through America's middle and upper classes in the early nineteenth century, the demand for moral entertainment grew. If a strong educational—if not evangelical—bent had always been Charles Peale's goal, such currents also affected lowbrow entertainment forums.[12] For example, sensing the temperament of the era, P. T. Barnum determined to make his new American Museum in New York City a comfortable place for middle-class women to visit with their children. To do so, he surrounded his human oddities and wondrous frauds with natural history specimens and portraits purchased from the collections of Peale's defunct museum.[13]

Even more obviously than Barnum's American Museum, the Crystal Palace in New York continued Peale's premise of offering visitors a version of republican uplift and education. Its walls of glass invited inspection and promised to educate, not mystify. Highbrow critics were thrilled at the Crystal Palace's aesthetic and educational value. One such critic suggested that a diligent traveler could spend years or even lifetimes trying to see all the marvels and gain all the knowledge of the history of art and industry that the exposition easily offered to a visitor in a few visits.[14]

The lyceum lecture circuit of the antebellum era also appealed to the republican virtues of its audiences. The lyceum idea had spread to America from England in 1826, primarily through the efforts of Josiah Holbrook, who had studied with Benjamin Silliman at Yale. The purpose was to create local societies that would sponsor lectures, establish libraries, build collections, and improve schools.[15] The duty of the lyceum lecturer—Ralph Waldo Emerson later became one of the circuit's most prominent—was to educate and uplift. What were the science lectures like? At best, educational and "uplifting" but not dull. Benjamin Silliman, who was one of the eminent American science lecturers of the nineteenth century, did not hesitate to entertain his audiences and add a moral message. A prominent chemist and geologist, Silliman taught at Yale, founded and edited the *American Journal of Science,* superintended several collections at the Crystal Palace exhibition, and helped found the American Geological Society as well as the American Association of Geologists.

Silliman graduated from Yale in 1796, planning on a career in law. In 1802, he was converted during an evangelical revival; that same year Yale president Timothy Dwight appointed him to be a lecturer in chemistry and natural history. Silliman incorporated a Christian outlook into his understanding of nature and of science, which he saw as a moral undertaking. His lectures on geology and chemistry at Yale and on the lyceum circuit continually stressed the harmony between science and Christianity. Silliman's religious faith made him lean toward the then-popular geological theories of Abraham Gottlob Werner, who insisted that the geological record proved the accuracy of the biblical creation story of Genesis. Silliman's Christian-inflected view of science increased his popularity when he began to tour.

By most accounts, Silliman was a remarkable lecturer and showman. While teaching at Yale, and in the 1840s before audiences in Boston, New York, and numerous other towns, he gloried in demonstrations—then termed "experiments"—that involved the use of the oxyhydrogen blowpipe, an early version of the blowtorch that Silliman's friend, the chemist Robert Hare, had devised. With it Silliman would melt and burn metals and create pyrotechnic displays. He also had numerous mineral samples and large-scale drawings to illustrate his lectures.

One of his assistants recalled, "During the lecture hour there was no lull or intermission; all was rapid movement, a constant appeal to the delighted senses. Here were broad irradiations of emerald phosphorescence, there the vivid spangles of burning iron, or the blinding effulgence of the compound blow-pipe, or the galvanic deflagrator. Strange sounds saluted the ear, from the singing hydrogen tube, the crackling decrepitation up to the loud explosions of mingled gases and detonating fulminates."[16] Silliman's rapid lectures, with their nonstop pyrotechnics, displays of natural wonders, and reassuring religious message, had great appeal. In his lectures he liked to insist on the harmony between science and religion, with phrases such as, "Truth is always truth, and no one truth . . . can ever be in conflict with any others."[17]

A contemporary of Silliman, Dionysius Lardner was another tireless science popularizer with a more secular outlook. Prior to touring America, he had taught mathematics and astronomy at London University and edited the multivolume *Cabinet Cyclopedia*. To this encyclopedia, he contributed numerous volumes. In 1840, a scandal over an adulterous love affair brought him to the United States and his new role as a public lecturer. He ridiculed the stuffy performances of most lyceum lecturers and instead traveled with a show that approximated the offerings of one of England's galleries of practical science. Lardner's "philosophical apparatus" included electromagnets, an oxyhydrogen microscope, an oxyhydrogen blowpipe, and numerous magic lantern slides that canvassed astronomy, meteorology, and even episodes from the life of George Washington. Lardner performed, often with musical accompaniment, on a stage decorated with props appropriate to a natural philosopher—globes and philosophical instruments in the foreground, for example, before a backdrop of library shelves. Lardner gained popular acclaim but disdain from America's scientific elite—this despite his status as a member of the Royal Society.[18] Able to overlook his moral "offenses," average Americans eagerly learned from Lardner's shows and his published lectures.

"I Bid You Hope"

Charles Came began his career as a wandering lecturer and healer in the 1840s.[19] The opening of the Erie Canal, which made small towns in New York accessible to large-scale manufacture, decreased the profitability of his cabinetmaking business in Pittsford, New York, in the Rochester region. The canal also helped spur the growth of towns that made ideal venues for traveling showmen. Came decided to leave his businesses at home to hired hands and take to the road. In high spirits, he once wrote to his wife of the assistant who was maintaining the cabinet business, "Tell Mr. Hall to keep all things wright and keep up good courage till I come he may make what he is a mind to and I will make lightning and see which will do the best."[20] Details of Came's small-time show business career are known because he wrote numerous such letters—full of his idiosyncratic spelling—to his second wife while touring New York for a period of about eight years.

Self-taught in science and medicine, Came's affiliation clearly was with artisanal culture. In addition to his furniture making, he carved

Charles C. Came toured upstate New York with his horse and wagon, selling patent medicines and offering scientific lectures and electrical healing demonstrations in the 1840s and 1850s. Courtesy of the National Museum of American History, Smithsonian Institution.

headstones, often prepared his own promotional materials, and built much of his static-electric equipment. Some sources also credit him with building a working telegraph as early as 1830, but his letters only mention that he commissioned such a device in the mid-1840s. Came's plebian background made him confident when lecturing to rural audiences, but hesitant in big cities, where natural philosophers and medical men might share stages. In New York City, for example, he hired a trained physician to lecture for him.

Came traveled New York State in a wagon that held all his bulky equipment and promotional material. His lectures were accompanied with music from his self-playing mechanical organs, while he stood below a proscenium canvas that included painted images of an eagle at its top, two flanking columns, and beside the columns, urns issuing flames. These flames suggested ancient mysteries and contributed to the awe Came attempted to evoke. The other equipment he traveled with included electrostatic generators and Leyden jars that had been common to electrical demonstrators from Benjamin Franklin to Rubens Peale. These machines produced and stored energy that could then be released as sparks, shocks, or small bolts of "lightning" that would break down a small model "thunder house." He also could induce bells to ring, and small figures to dance, while his Colt submarine battery caused loud explosions.[21] Came also had an orrery that modeled the revolution of the planets around the sun, pneumatic devices that showed that in a vacuum a feather and coin fell at the same speed, and a model telegraph with which members of his audience could send messages to one another. He also traveled with two "dissolving" magic lanterns (which allowed one projection to fade into the next) and slides to give lectures on astronomy, phrenology, biblical subjects, "spirit landscapes," architecture, and the battles of the Mexican War. He also used painted magic lantern slides of microbes, fungi, and miniatures such as the "mouth of a beetle" to emulate the projecting "gas microscopes" that performers like Dionysius Lardner traveled with. One handbill announced Came would exhibit "one of the most Powerful Lucernal microscopes, the latest improvement of the age, for assisting the Human eye to discover the inhabitants of a Drop of Water . . . this ingenuous Instrument, which alone is worth the price of admission."[22]

With his horse Fanny as his sole companion, Came traveled on muddy roads, warmed his feet on a charcoal burner during cold weather, and wrote home often of his loneliness, the difficult weather and road conditions, and about the unpleasant bedding and meals. He bragged about cures he had managed, reported whether or not a town had pretty girls, and sent home money ranging from one dollar to four dollars from

This woodcut shows Came on stage with his thunder house soon to receive an electrical lightning bolt from a static generator. Unseen on the painted proscenium canvas are the urns belching flames. Courtesy of the National Museum of American History, Smithsonian Institution.

his latest shows. He generally charged twenty cents admission to men, less for children and women, and his modest profits suggest he at times gained audiences of about fifty people. The modest income also suggests he was not profiteering as a hawker of patent medicines. For some of his tours his advance man Longfellow put up posters, but usually Came traveled on off-days to post handbills in towns he then traveled to with his equipment. He was very fond of his horse Fanny, though he grumbled when it cost more to stable her than to house him. He liked to describe the landscapes, occasionally quite lyrically. A caring father, he constantly suggested remedies for his children's ailments, including "strong beer to fat up" one of his sickly girls. He also wrote of his longing to be home to work his small farm and urged his oldest son at home to weed, to provide the cow and calf with hay, and otherwise tend to the gardening and farming chores. Occasionally he sent home, via the canal or railroads, groceries and lumber he had accepted in trade for his doctoring. His letters reflect his mercurial changes in moods as he quickly altered from boasting to worrying. Often he would start a letter ordering a member of his family to take a particular action but later in the same letter would change his mind. The tone of his letters was not that of a hard-boiled confidence artist eager to humbug the public. Nevertheless, he had no qualms about stretching the truth at times, as when he called his magic lantern images the magnifications of a "Lucernal Microscope."

Came's electrical wonder show was similar to that of other itinerant performers, some of whom he corresponded with, and others of whom he borrowed poster copy from. Came sought the prestige and profits of performers such as Dr. Gardiner Quincy Colton, who in the 1840s toured Erie Canal towns and at twenty-five cents a person unveiled a giant temperance painting, *Court of Death*, to audiences. Colton also offered telegraph demonstrations, science demonstrations, and laughing gas entertainments.[23]

Other demonstrators highlighted electrical thrills and cures. A poster circa 1849 for the "Scientific Exhibition! of Messrs. Howig & Langdon," like those of Came, mixed copy about electrical pranks with a variety of demonstrations and offers of electrical healing. Howig and Langdon boasted of their new "apparatus for illustrating the LIGHTNING's power," as well as their "Electric Telegraph." Their satirical act called "The Magic Gold Piece" encouraged locals to try to reach into a bucket of electrified water to get a gold coin fresh from the goldfields of California; they also encouraged audience members to try to wear "Magnetic Slippers! Which produce involuntary dancing by the person wearing them." Finally, they mention their electrotherapy devices: "Electricity will be applied through the Medical Coil to persons who desire it, for the cure of Rheumatism, Deafness, Paralysis" and other ailments.[24]

In a similar vein, B. A. Bamber, who ran the "Great Dime Show," offered magic lantern slides and comic sketches, and promised "Electricity without Extra Charge" from "a very fine galvanic battery." He insisted this galvanic treatment was an "excellent remedy for rheumatism, neuralgia and headache" and urged customers to "Be sure to come before the show begins if you want to try it."[25] Another performer of the era, Mr. J. St. John, offered "Electrical and Magnetic Experiments." These included medical treatments: "shocks administered to those who desire them for medical use free of charge." Likewise, he promised "shocks given for amusement first to the ladies, then to the gentlemen."[26] Magnetic Slippers were also featured, as well as a "Miser's Cup" offered for free to anyone able to take the charged cup away from a performer standing on an insulated pad. J. St. John assured parents that his show had moral and educational value and concluded his poster with the slogan, "Deprive them of moral, and they will seek immoral entertainments."

Perhaps the most involved electro-medical poster, which Came relied on for one of his own medical posters, was printed by Professor C. F. Bolles, who made great promises for cures. His poster was addressed "TO THE DISEASED," and after mentioning many ailments, its copy insisted, "I BID YOU HOPE!" His poster then elaborated the connection between electricity and vitality, insisting that they were not identical but that electricity was one of the components of vitality and could help induce health. Bolles offered to cure numerous ailments electrically, including pulmonary consumption, liver disease, paralysis, palsy, and conditions that other physicians deemed incurable. He stressed, however, "I am as opposed to every species of quackery, and the deceptions which mere pretenders to medical science practice upon the people, as any living man."[27]

Came had a poster printed up almost identical to that of C. F. Bolles. It also was headed "TO THE DISEASED" and promised to help restore to health those "considered incurable." Paraphrasing Bolles, Came's poster added, "The highest and most active of all the elements constituting life is ELECTRICITY; it is the organizing, the vitalizing, and equalizing Agent of Nature's God." The poster explained that medical lectures would be offered and encouraged the public to visit both Charles Came, "The Great Electrician and Successful Operator," and his partner in this venture, M. L. Vosburg, a physician of the eclectic school that generally offered herbal cures. Their poster included a long list of diseases they could cure, such as pulmonary consumption, liver disease, paralysis, bronchitis, rheumatism, ulcers, spinal complaints, piles, St. Vitus's Dance, and what would be more telling for Came when he attempted to exhibit his Sleeping Man, "suspended animation."

Came's interest in electrical cures was not a fringe obsession in his age. As early as the mid–eighteenth century, electricians tried to cure paralytics or to revive drowned people through Leyden jar discharges. The connection between electricity and healing became even more pronounced in 1791 after anatomist Luigi Galvani reported that an amputated frog's leg would twitch when exposed to atmospheric electricity. He theorized that he had discovered a new fundamental force, that of "animal electricity." Such energy, he reasoned, had been stored up in the frog to be discharged under appropriate conditions. Galvani's theory, which connected electricity to the life force, helped bolster a vitalist model for medical science. More specifically, his theory, which indicated that humans, like other animals, utilized electric currents, led to the conclusion that human health might be enhanced electrically. Many experimenters applied electricity to patients to attempt to restore health. Galvani's nephew, Giovanni Aldini, even applied electricity to the corpses of executed convicts and made them briefly breathe, twitch, grimace, and kick.

Shortly after Galvani's findings, another researcher, Alexander Volta, reasoned that it was the contact between the metals that Galvani's frog had been attached to that created the charge—not a buildup of charge within the frog's nervous system. This led to Volta's development of the first battery, the Voltaic pile, in 1796. Voltaic piles and other chemical batteries, somewhat ironically, also aided the further development of electrical healing apparatus. On the institutional level, in the 1830s, Guy's Hospital in London included an "electrifying room" where both inpatients and outpatients received regular electrical therapy, primarily for nervous disorders.[28]

Itinerant electricians such as Came, however, used electrical devices for a myriad of illnesses. One such device was Samuel B. Smith's "torpedo electro-magnetic machine." Came owned one of Smith's pamphlets, published in 1849, which explained that "the human body acts on the principle of the galvanic battery. . . . So long therefore, as the integrity of this circuit is maintained health will be enjoyed."[29] Smith claimed that his torpedo could both test for and treat various forms of rheumatism. Came used a similar instrument for electrotherapy that included a battery, induction coil, and two electrodes that could be applied to the patient's body. His posters made wide claims for the sorts of illnesses this therapy could aid. Came's medical apparatus also included a healing crystal, numerous cures both "botanical and etheric," and such heroic medicines common to the era as calomel, a mercury compound.

Such promotions brought Came and his colleagues close to the camp of out-and-out mountebanks. Numerous "quacks" toured the country then, putting on makeshift performances and selling elixirs of questionable value. Such elixirs were extremely common, and their manufacture had become a profitable business. Patent medicine advertising helped fund the penny press just as department store advertising later would fund tabloids. Much of the page space in the *New York Sun* of the 1840s, for example, was full of long-winded advertisements for irregular doctors, as well as wonder medicines such as "Dr. Wheeler's Balsam," which could cure cramps, spasms, and dysentery, and "Parr's Life Pills," which could dispatch dyspepsia, bilious complaints, and cholera in the early stages. Many of these advertisements pointed out the shortcomings of "regular" doctors, as with Dr. Morrison's promise, "No Cure No Charge No Mercury," and an advertisement for "Richardson & Company's Celebrated American Panacea," which treated rheumatism, scrofula, fever sores, sore eyes, "and all diseases or pains arising from an injudicious use of mercury."[30]

Nostrums, frauds, irregular doctors, and medicine salesmen abounded in the mid–nineteenth century. Rogan Taylor has argued that the traveling mountebanks who sold nostrums—common in Europe for centuries and also plentiful in the Americas from their first settlement—

were modern analogues to tribal shamans.[31] If their compounds did not necessarily heal, their shows offered entertainment and the possibility of miracles. The mountebank presented wonder shows much like the rituals of "death and resurrection" that shamans offered. At the beginning of their curing ritual, like the commencement of a sales pitch, shamans traditionally dramatized the story of their initiation to explain how they had gained their awesome powers. In such initiation journeys, the shaman traveled into the world of spirits and went through an often-grisly death and dismemberment. This death was followed by a resurrection in which the shaman's body was reassembled with the aid of helping spirits. The death and resurrection motif of the initiation was recreated in subsequent healing rituals. Shamans would offer "miracles," whether performing great leaps, walking on fire, offering sleight-of-hand tricks, hacking up and resurrecting an animal or a human assistant, or providing other evidence of extraordinary powers.

As Taylor has pointed out, mountebanks and traveling showmen mirrored such narrative patterns in their stagings.[32] European mountebanks who sold medicines, often dressed as wizards or in Orientalist garb, offered tricks like those of shamans to evoke awe in audiences. A historian of the medicine show described one mountebank who handled poisonous snakes and others who "gashed their arms with knives and mysteriously healed them again." Likewise, a performer named John Brenon who traveled New York in 1787 with his wife offered balloon ascensions, slack wire performances, songs, and sleight-of-hand tricks, and also would urge an audience member to cut off the head of a fowl, which Brenon would then restore to life.[33] Such acts conferred on performers the supernatural powers that shamans claimed to gain through their symbolic deaths and resurrections—the healing powers granted by spirits.

Came and his peers offered an updated variant of the medicine show with their substitution of modern technological wonders for the older entertainment forms of juggling, sleight of hand, and singing. Instead of a voyage into the world of spirits, they offered a glimpse into the secrets scientists had teased from nature. Came offered genuine education, for example, with his orrery and other devices that showed the movements of the planets and explained the seasons. The vacuum pump that he used to demonstrate that a feather and coin dropped at the same speed in the absence of air established the truth of Galileo's insights into falling bodies. Yet, Came and his peers also presented electricity as the latest wonder force, with healing power similar to that once offered by denizens of the spirit world.

Though it is tempting to label Came as a simple confidence man, with his patent medicine sales and apparatus of wonder, he also took his lecturing and doctoring seriously. Unlike the traveling quacks who set up small consulting tents beneath their stages to offer quick fixes, Came's letters indicate that he carefully tended patients on his travels, often staying a week or more to treat isolated patients. During the era that Charles Came was doctoring, most American physicians, particularly in rural areas, were lay healers. Some had attended eclectic or homeopathic schools. They tended to rely on herbal remedies and folk cures, which were less harmful than bloodletting and the poisonous compounds of sulfur and mercury that academy-trained physicians administered to wealthier patients.

Came prided himself on his skills as a folk healer. In the summer of 1850, he traveled to rural Michigan to help his ill sister Lucretia, on whom the local doctors had given up. When his treatments seemed to help her, he lectured several times and gained esteem in the community. He wrote home, "There is a good chance for a good Doctor here there is 5 or 6 of them within two or three miles but they are not verry good they thought that Lucretia could not be helped and by my coming here and helping her it brought others here and I now have about as much as I can attend to."[34] In a letter the following week he repeated how he was gaining the confidence of the people with his treatments and proudly described curing one family of "fits." Often prone to new inspirations, he began to toy with the idea of moving to Michigan, where he had an appreciative audience. A few days later he reported that "the people will not hear about my going home."[35]

Came also relied on his own remedies for doctoring himself and his family. In the winter of 1852 he wrote to his wife that he had been feeling sick. "I finally went to my medicines and took a good dose of my Elixer—and went to bed and got up in an hour and did not feel much of any better I then went and took another dose and went to bed and sweat some I have got up this morning and feel a little better."[36] Though not the stuff of a grand testimonial, this letter does reveal he took his own medications seriously. In another letter he wrote just prior to his venture to New York City, he advised his ten-year-old daughter, Mary Eliza, to "take freely of the [Nervium?] and Elixer be Electrised everyday rub her with the brandy."[37] Later, he encouraged his wife to consult with a local physician and with some hesitation told his wife to allow the doctor to let some of their daughter's blood. All of this suggests that by the day's standards, Came was a competent doctor, and more likely a better doctor than showman.

"A Curious Case"

Considering that Spiritualism came to prominence in upstate New York precisely when he was working as a showman, it is surprising that Came never incorporated references to Spiritualism in his lectures. A movement that swept America and later Europe, Spiritualism was launched in upstate New York in 1848. That year the Fox sisters of Hydesville began to hear strange rapping sounds in their house that allowed them, once decoded, to communicate with spirits. Soon after, many Spiritualists began to offer séances in homes and theaters. Stage magicians of the era quickly incorporated references to Spiritualism in their acts, adding rapping hands and strange materializations and escapes in their performances. Likewise, Came's contemporary, the electrical lecturer and healer J. St. John, included a séance reference in his act. "Raps distinctly audible to the audience will be produced describing personal appearances, uses, and cetera, exciting great curiousity in the minds of people."[38]

Although Came's letters reflect no strong interest in Spiritualism, in 1862 he did receive a letter from a fellow showman, Mr. Crump, who wrote of Spiritualism in a tone that suggests they had many times before discussed such matters. After a report on the entertainers that had arrived in his small town in Alleghany County, New York, Crump commented with interest on news from Boston of spirit photography, and how it seemed improbable but undoubtedly lucrative: "This is a little ahead of my time for the spirit of the dead to be so natural as to reflect light enough to produce chemical changes in the Picture. I think that if I could only get so as to take the likeness of our dear Fathers or rather their spirits I should neither want for Money or Business." Crump added that if spirit photography were genuine, "I should have to knock under and spiritualism would pass from speculation and Theory to positive Science."[39]

Like his correspondent Crump, Came seemed to have a rational, nonreligious mind, untouched by revivalist fervor. Nor did he, like Silliman or Colton, the exhibitor of the dramatic temperance painting *The Court of Death*, rely primarily on the religious fervor of his audiences. Came seldom mentions God in any of his letters, except when he was under extreme duress, as for example after the death of his sister in 1860. He also notes a few Sunday church visits, but this pursuit seemed largely a way to pass the time and meet locals. Likewise, he records staying at a temperance hotel on one occasion but seems less interested in temperance ideology than in the fact that the room was warm and the bed comfortable. His letters home also had a mildly ribald streak that would not have been appropriate to an evangelical household. He remarked in one early letter to his second wife that "there is a great many girls out here I do not know what to do with them all there is some that is about as nice as anything I have seen yet The Lightning Man is all the go with them . . . the way I send the Lightning through them out this way is not ways slow I will assure you But I shall keep shady about some things."[40]

Came's religiosity was relegated primarily to his shows and most likely was a ploy for gaining the approval of the town leaders where he performed. Came's use of Christian rhetoric, for example, is evident in one poster for his astronomy lecture. The poster's centerpiece was his "Splendid Golden Apparatus." Came's engraving of this orrery was flanked with quotes in high-biblical mode, with one reading, "The Heavens declare the glory of God, and the Firmament sheweth His handiwork." Likewise, one of his magic lantern lectures included "fifty splendid paintings" of "Sacred and Ancient History" as well as "Bible Scenes" from the "Land of Palestine." He often offered these lectures at schools and added the notice on his poster, "Clergymen and trustees invited to attend, free."[41] Conceivably, Came presented his magic lantern shows of "Spirit Scenes" in a Spiritualist context. However, it is more likely that this was his way of promoting the capabilities of his "dissolving" magic lanterns and the effects they produced when he exhibited views of landscapes and ancient sites.

Despite his apparent lack of interest in religion and failure to take advantage of the Spiritualism fad as had J. St. John and the era's stage magicians, Came does indicate some interest in the mysteries of the mind as presented in phrenology and mesmerism. Came paired his showings of "spirit scenes" with lectures on astronomy and phrenology, and he also offered phrenological readings. Phrenologists believed that the careful examination and measurement of an individual's skull and physiognomy could indicate the individual's character type. The client would then have to take the initiative to improve his or her character in areas of deficiency. This theory offered a scientific basis to the self-improvement campaigns common to the era, and to the wonder show formula of heightened powers of the mind.

Phrenology also was directly related to mesmerism in the then-popular technique of "phreno-mesmerism." Phreno-mesmerists would stimulate different parts of the skull to encourage improvement in character traits found wanting. Mesmerism was also considered a medical technique in its own right. Mesmerists of the era would place maidens into trances, and these subjects would then offer evidence of their telepathic abilities and clairvoyance. These somnambulists often would diagnose illnesses and suggest cures to audience members or patients. In

This poster for Came's lecture on the "Beautiful and Sublime Science of Astronomy," which employed such biblical rhetoric as "the Heavens declare the glory of God, and the Firmament sheweth His handy-work," shows Came catering to the assumed intimacy between religion and science in the antebellum era. Courtesy of the National Museum of American History, Smithsonian Institution.

the 1850s a faddish interest in "electro-biology"—another name for mesmerism— was sweeping America, and this may have been yet another lure to Came's attention.

Though none of his posters mention hypnotism or mesmerism as part of his performances or healing, a scrap of a lecture note in his handwriting and one of his letters point to his interest and apparent proficiency in mesmerism.[42] In the lecture notes he promised to perform an "amusing experiment" that involved "biology or the electrical science of Life." He continued, a "person in a perfectly wakeful state . . . shall voluntarily come forward from among the audience will be experimented upon shutting their eyes they will be unable to open them." This indicates that he indeed did practice stage hypnotism, which also promoted dramatic yet "scientific" powers of mind.

A reference to mesmerism in a letter from 1850 was even more dramatic. In this letter, Came described how he was performing three nights successively in the town of Oak Orchard, New York. He learned of a woman who had drowned while attempting "to cross the Creek in the place which is large and deep." For over a week, as many as a hundred men had been searching for her body. Her husband had given up hope when Came volunteered to help. "So I immediately went to person who I never saw before three miles from the place [of drowning] mesmerized him and put him in a clairvoyant state and got all the particular minute description of the place where she lay. . . . They went amediately to the place and took her out of the water with hooks fastened to long poles." Came reported home that both the husband and "the whole of this community is astonished at this circumstance it will be published in all of the papers. . . . I will give a more full account of it when I get back."[43] He also asked his wife to report his triumph to his friend Mr. Crump. Unfortunately, Came never commented again about this incident or any other forays into mesmerism.

His interests in phrenology, and these sparse mentions of mesmerism, do suggest that Came was eager to place self-improvement and the mysteries of mesmerism within a quasi-scientific context. And, curiously, Came began to mimic the act of the traveling mesmerist when he began his promotion of his Sleeping Man. An itinerant mesmerist generally would travel with a subject known as a somnambulist who could be placed in trances. In the summer of 1853 Came took up with his own Sleeping Man; this sleeper, however, showed no enhanced powers of mind while in his trance. He remained lost to the world, a case of "suspended animation." Cornelius Vroman, a hired man on the farm of Moses Jennings, in Clarkson, New York, had been in a coma or cataleptic state for nearly five years when Came visited. All hopes for a cure had long since vanished, so Came made a deal with the farmer Jennings to exhibit Vroman, or the Sleeping Man, as a great medical attraction.

In his letters, Came made excited pronouncements to his family about the Sleeping Man's lucrative promise and energetically set out to make a "strike" with his new attraction. He began exhibiting the Sleeping Man to upstate crowds along the Erie Canal and gained audiences in cities such as Albany and Syracuse. He wrote to his wife, "I think I can make something worth laboring for if I can control the motor although I do not get but a third but I think I can make a Thousand Dollars with him."[44] This was a grand promise for a man who seldom gained more than three dollars profit from a performance.

Came was eager to seize this opportunity. Life on the road had become less enjoyable as he aged, and he had been greatly saddened when his horse Fanny had died in an accident the previous winter. He persuaded himself that the Sleeping Man was his ticket to renown and status. His letters took a grandiose turn. Of his difficulties negotiating with Vroman's "master," the farmer Jennings, he remarked, "He finds I am not to be trifled with."[45] To gain status and an edge in promoting the Sleeping Man as a moral entertainment, Came began to collect letters of introduction from leading citizens in Syracuse. In Albany he commissioned a "great artist" to illustrate a lithograph of the Sleeping Man in the several postures he could hold while in his coma: lying down, sitting up, and standing. In Saratoga Springs, Came told of gaining letters of introduction from "a number of the greatest men of our State, and they just begin to find out who they are dealing with and consider it quite a privilege to talk with me."[46] He urged his teenage son Raphael to keep out of bad company and reminded him, "I have never had anyone to lead me through the world since I was your age and you see what I now can do with those who think themselves the highest class in society. They are obliged to take notice of what I say."[47] He also improved his wardrobe to meet the cosmopolitan standards of the resort town of Saratoga Springs when exhibiting there. Noting that his feet were "all blistered," he bought new socks and a pair of light pumps in Saratoga Springs; he remarked on the pleasures of being waited on and dressed in luxury and suffered no apparent twinges of conscience at benefiting from the institution of slavery. "A slave bought me a nice stock and shirt collar they are so verry fashionable."[48]

But Came was ultimately more comfortable with rural life and rustic audiences. He was shocked at the high costs of room, board, and business in Saratoga Springs and at its opulence. Grand hotels that held two thousand boarders "all paying from two Dollars to three a day" seemed both a wonder and an outrage. Because of high costs, he barely broke even in Saratoga Springs, yet he remained optimistic and began short trips to New York City to prepare the way for a Sleeping Man show. As in Saratoga Springs, Came felt out of place in cosmopolitan New York. It was crowded and full of rowdies waiting to prey on strangers. Traffic continued all night long. During one dismal spell in his hotel he wrote his wife, "It is now two o clock at night and the street is full of carriages just as though it was in the day time they are going all night long and the stores full of people all night every thing going on that can be thought of and somethings that cant be hardly the city never gets still."[49] This lament is a far cry from his happy reports on the richness of soil, or the promising stands of timber and abundance of game he wrote of in his letters from upstate New York and Michigan. If Vroman, his exhibit, opposed nature by doing nothing but sleep, Came felt the city opposed nature with its never-ending commerce and its tireless citizenry.

Despite such discomforts and trials in the city that never sleeps, Came remained optimistic. He visited the Crystal Palace but did not go in—he either did not have the time or found the fifty cents admission price steep. He continued to gain letters of introduction, worked on a pamphlet, and planned to have lithographs "in every window of him all along the streets. I believe there is better days coming."[50] On August 13 he hired a physician to present the Sleeping Man to the "New York Medical Faculty," asserting that "when tomorrow the results will be known it will forever establish the subject of the Sleeping Man and I think so much that it must raise an excitement."[51] Edward R. Dixon, a prominent New York physician, wrote Came a letter extolling the virtues of his exhibit, and Came incorporated the letter in an advertisement, including a long quote ending with, "The physiologist and the philosopher will find in this case, now at Academy Hall, 663 Broadway, a subject of profound interest."[52]

Came continued publicity efforts by arranging a phrenological reading of the Sleeping Man with L. N. Fowler of the firm of Fowler and Wells. Yet the renowned phrenologist was unable to say anything conclusive. Fowler's reading noted of Vroman that "under favorable circumstances he would exhibit a fair share of practical common sense . . . but would not be brilliant, versatile, pliable, jealous, or enthusiastic in any sense of the term. . . . His phrenology throws no light upon the phenomenon of continued sleeping. The organ of sleep as recognized by Dr. Buchanan is developed in an average degree and there is no reason phrenologically speaking why he should remain in this state so long; or if there is such a reason we are not able to point it out at present."[53]

Though he could not entirely admit it, Came was in over his head. Suddenly stage-shy in the big city, he hired a lecturer. Hotels, printing, and hall rentals were all costly. He began to see fabled showman P. T. Barnum as his out. "I want to make a bargain with Barnum if I can but he is not in the city now but will be next week."[54] Came was not able to arrange to exhibit his Sleeping Man until the middle of September. The hall would cost him $150 a week, and he thought he would earn at least $50 after dividing with Jennings. On September 13 he offered a free exhibition of the Sleeping Man to clergymen, faculty, and the press.

Two press notices appeared just prior to this first exhibition and one week following. The first, from September 9, was titled "A Curious Case."[55] It was as favorable as could be imagined. The writer noted, "Medical men regard this case with the profoundest interest." Struggling to find a narrative in which to place Vroman, the reporter described in detail Vroman's lack of response to stimulus; he also mentioned that Vroman had blisters and scars from various attempts—medical or otherwise—to awaken him. Other details mentioned about Vroman included such facts as "once he was left standing for three days"; and, on another occasion, he went without food for five days. Since falling asleep his weight had dropped from 160 pounds to 90 pounds. Occasionally he sighed or stirred. Vroman was capable of waking up on occasion. This offered some dramatic possibilities. The author said, "The last time he awoke was while he was in Rochester, some ten weeks since, which gives us a hope that his waking hour now approaches, and that we may see him in his wakeful condition." The author, following the lead of Dr. Dixon, thought Vroman a fascinating medical display. Also important from the standpoint of "moral entertainment," the reporter attested that the exhibit lacked fraudulence: "There is not the slightest chance for any collusion or deception in the matter."

A week later, a reporter from the *New York Daily Tribune* penned his own less-than-thrilled reactions to the Sleeping Man in an article titled "Disgusting Exhibitions."[56] He pointedly described the wonderful displays of culture that the Crystal Palace offered to the public uptown, then questioned that same public's low tastes for weird exhibitions. "The normal and beautiful are inadequate: the unnatural and hideous must be called into view. Hence it is that Broadway is never without one or more damnable monsters on exhibition." The article went on to describe one such monster, a seven hundred pound "Fat Woman." Even worse was the Sleeping Man. "Here is a poor wretch, who, as Dr. Dixon says, has less vitality than an oyster, placed before the public gaze. . . . He is woefully emaciated . . . a fierce degradation of manhood—not living, not dying, not dead . . . a thing that should be kept out of sight and notice, and yet he is pushed into the van of publicities." This reporter refused to consider Vroman as a medical curiosity as did the *Times* reporter, and also insisted that the Sleeping Man could never be fit moral entertainment. In fact, Vroman was a living symbol of moral degradation that should be kept out of sight. He also stood for the low tastes of the public and the assault on culture that resulted from a sensationalist press. Both reporters took a highbrow approach to the Sleeping Man, but only the first accepted Came's strategies for exhibiting Vroman in a "moral" and "educational" manner.

Despite the press coverage, Came's hope of causing a public stir with his medical spectacle failed. He suffered from poor timing, upsets, and an exhibit that lacked the flash and drama likely to draw in those with "low tastes." Watching a sleeping man at most "sigh" when turned onto his side did not make for grand theater. Came believed the Sleeping Man was a promising attraction, a true wonder that could provoke awe and reflection in audiences. He wrote his wife the week after opening that when he would "get the books and lithographs ready," his luck would change. Came apparently spent one thousand dollars to print one thousand copies of a book detailing the life history of the Sleeping Man. He did not receive these books until the first week in October when he had already decided to close the exhibition. In the meantime he was stuck in a cold hotel with the Sleeping Man and an unlikely senior partner, the elderly farmer Mr. Jennings, who, Came reported, "is sick and what is worse than all he seems to be out of his right mind."[57]

True to his mercurial nature, in his darkest hour, when his hope for a big strike had soured, Came reported a happy dream, one that convinced him his luck would turn. "I dreamed last night that I was away from home and found Fanny with her leg all well as ever and she was fat and sleek and that I was going to carry my apparatus and three men and I dreamed of seeing some of the most beautiful silver money that I ever saw and that a part of it was going to be mine. I feel a little better in my mind."[58]

Came continued to put the best face on his enterprise even after closing the exhibit. Always the optimist, he awaited the expected offer from Barnum that never arrived, then insisted the thousand copies of the Sleeping Man's life story would be of value. "I think I make something on them if nothing on the Sleeping Man."[59] A week later he gave up all hope and remarked, "I have made up my mind not to trouble my mind any more with the Sleeping Man so as to prevent me from attending to my own affairs, I am verry sorry I ever had any thing to do with it."[60]

Came's chronicle of life on the road largely ended with his failed efforts with the Sleeping Man. He continued as a showman into the 1860s, as his handbills for magic lantern shows of Civil War scenes indicate, but it is probable he spent more time at home and as a country doctor. Census information would indicate that if he never saw all the "beautiful silver money" that he could dream of, he did modestly prosper. The value of his real estate and personal property increased in total value from $1,250 to $3,800 between 1860 and 1870. When he died in 1881, his obituary emphasized his skills as a physician and suggested that if "he had more push," his development of a telegraph system in 1830 could have gained him fame and fortune. The obituary also praised Came as an important disseminator of scientific ideas. The writer insisted "many testimonials by teachers and professors of science all over acknowledged a great debt to Dr. Came for their start in science."[61]

Selling to Barnum

If these testimonials to Came's capabilities as a science educator have any validity, then it is worthwhile to reconsider the usual unstated assumptions of historians about how scientific knowledge is transmitted in a democracy. According to the standard historical narrative, nineteenth-century lecturers such as Silliman offered earnest artisans and the middle class alike a solid grounding in science, which, presumably, made them better, more informed citizens. Plebian performances, like those that Came offered, with their strange mix of science education and "quackery," if taken note of at all, could only indicate the continuing reign of superstition on a popular level.[62]

The assumption that only "genuine" science shows should be inspected because of their greater contribution to the spreading of democratic values leads to a simplified scheme of the varieties of popular science available in the nineteenth century. The separation between highbrow and lowbrow culture, the divide between education and entertainment, and that between "good" science and "bad" science were not that clear-cut.[63] Both Silliman and Came were offering science "diluted" by their own biases, their methods for captivating audiences, and their need to earn a living. For Silliman the bias came from a devout Christianity emphasized in his institutional relationship to Yale; for Came the bias came from his dedication to a therapeutic vision and his loose affiliation with performers such as his friend Crump and his fellow lay healers. Silliman's Christian vision led him to promote a hackneyed brand of geology, while Came's therapeutic vision led him to promote dubious medical practices. Yet both of these lecturers sparked their public's interest in science.

Both these scientific showmen also attempted to navigate in the tricky waters of the antebellum marketplace. In his worst hour, when bogged down in his efforts to promote his Sleeping Man, Came clung to the dream of "making a strike" by selling his attraction to the great showman P. T. Barnum. In that era, the phrase "Barnumization" had become common in essays and editorials decrying the spread of commercial culture, yet the public awaited Barnum's newest promotions eager to be "deceived."[64] In a curious twist, the Crystal Palace, whose science exhibits were supervised by Silliman, making it everything that Came's "Disgusting Spectacle" of the Sleeping Man was not, also tried to sell to Barnum. For a brief moment, Barnum, the savvy marketplace manipulator, became the happy medium between Silliman and Came. Barnum reigned as president of the Crystal Palace only briefly before complaining of financial mismanagement and resigning. That leading citizens had urged Barnum to take over the Crystal Palace management underlines how ambiguous was the separation between highbrow and lowbrow culture, how permeable the line between education and entertainment in the nineteenth century. The difference between "high" and "low" had less to do with content than with social context and promotional methods.

Even if Came can be dismissed as a quack and promoter of pseudoscience, his performances, like those of Katterfelto before him, are worthy of inspection. His mixture of scientific lecture, showmanship, and folk healing cohered in a version of "populist science" that fulfilled the antebellum public's desires. Came relied on his knowledge of science, however imperfect, his scientific apparatus, his sparking machinery, and his impressive stage trimmings to create authority and instill a sense of wonder in his audiences. His proscenium painting with its urns belching flames provided imagery usually associated with a wizard or shaman and his heightened spiritual powers. His show, like the exotic label on a patent medicine bottle, prepared his audiences for the miracle of healing. Yet he also offered audiences a basic understanding of the scientific worldview.

Came and his fellow itinerants can best be viewed as early promoters of a populist science catering to audience beliefs about progress. Like promoters of other pseudosciences such as Wernerian geology, mesmerism, phrenology, Spiritualism, and mind-cure, Came's electrical healing and phrenological beliefs sprung from the antebellum audience's desire for moral uplift. Silliman tended toward Wernerian geology because of his Christianity and conservatism, whereas Came offered the alternative "uplift" of the therapeutic with its democratic premises that progress should buoy everyone.

Came's shows, however flawed, also introduced the rural public to tokens of modernity. The knowledge that these shows represented ultimately filtered down from centers of modernity such as London, New York, and Boston. Women attended his lectures on health and anatomy, gaining access to knowledge about taboo topics. Other audience members could manipulate a telegraph and learn its protocols, or consider ways to improve their moral character through the science of phrenology. The shows of Came and of other electrical healers established that sparking electrical apparatus and cosmopolitan innovations need not "shock" but could heal one's ills and improve one's life. In this sense, Came's shows were an elixir that calmed potential anxieties about the dawning age of electricity; while Came's phrenological readings and mesmeric demonstrations, true to the wonder show script, encouraged an audience's beliefs that human potential would unfold alongside technological progress.

CHAPTER TWO

The Techno-Wizard

On December 31, 1883, William J. Hammer, one of Thomas Edison's chief assistants, threw a New Year's Eve party at his all-electric house in Newark. Dinner guests at twenty-five-year-old Hammer's "Electrical Diablerie" were offered nonstop delights. As they walked up the first step to the house, the address blazed in tiny lights, the next step caused the doorbell to ring, the third opened the door and lit the gas jets in the hallway. An electrical device brushed snow and mud from the guest's shoes, then administered a shock to the shoes' wearer. All the furniture in the house also was electrically booby-trapped. Sit in one chair and the lights would go out. Sit in another and drums would thunder or strange "rapping" noises would come from the floorboards. To lie down on the bed would cause the gas jets to go out and a phosphorescent moon to cross the ceiling. Electric telephones, cigar lighters, bells, and other devices could be found in every room. A glowing pitcher of lemonade gave electric shocks to anyone foolish enough to pick it up. When a storyteller began to amuse friends with an anecdote, a giant dunce cap descended from the ceiling and covered him from head to toe.

During the New Year's feast that followed, an automaton named Jupiter sat at the head of the table shouting welcomes via an Edison cylinder. Hammer's menu included "Electric Toast," "Wizard Pie," "Magnetic Cake," "Incandescent Lemonade," "Ohm-made Electric Current Pie," and "Electric Coffee." At midnight cannons discharged on the front porch, bricks rained down the chimney, the "Sheol Pudding" (i.e., "hell pudding") blazed with flames, the silverware was charged with electricity, and the "thunderbolt pudding" discharged black bolts (on springs) about the room. Jupiter raised his glass to drink, the room darkened, a luminous skeleton paraded the room, and Jupiter then shouted, "Happy

New Year! Happy New Year!" The guests left the table to file past Hammer's younger sister, posed on a pedestal as "The Goddess of Electricity" with various electrical lights on her gown and dangling from her ears; they then went on the front porch to shoot fireworks by push button. A reporter commented that "the guests departed with a bewildered feeling that somehow they had been living half a century ahead of the new year."[1]

Hammer's electric ghost house suggests both the prankish atmosphere that prevailed at Edison's Menlo Park, New Jersey, laboratory, and Hammer's awareness of public fascination with electricity in the late nineteenth century. The inhuman shouting of the automaton Jupiter, the parading skeleton, diabolical player piano, and other "haunted house" effects played upon the public perception that electricity was a magical force. Despite its sophomoric tone, Hammer's pageant commented on a serious Progressive Era debate—and one with even greater relevance today in the postatomic age: in what way is technology fundamental to progress and how might its diabolical potential be contained? Though arguably a civilizing force, electricity had its destructive aspects.

This debate pervaded the culture of the electricians and the general culture at the turn of the century. Whether the technological future could be read as tragedy or comedy was still open for debate. Trade show exhibits, pageant floats, and articles in electrical trade journals all explored the dual vision of electricity and technology as an outgrowth of heavenly or demonic sources. The debate over the ethics of electrocution shifted these concerns to a pragmatic realm. To civilize electricity, it seemed necessary to associate the force with femininity. The lovely Goddess of Electricity was a regular visitor to the pages of electrical journals, just as she had appeared at Hammer's party.

The evolving electrical industry, not surprisingly, was an all-male enclave. Legendary male inventors often pushed the Goddess of Electricity from the spotlight. First and foremost was Thomas Edison. After earlier work introducing an improved stock ticker and other telegraphic equipment to the business realm, Edison unveiled his phonograph in 1878 and the press hailed him as the new "wizard" of the age. Men like Edison, to a lesser extent his assistants like Hammer, as well as other engineers and inventors to follow such as Nikola Tesla and Charles Steinmetz, became culture heroes. At the vanguard of technology, these men had seemingly wrestled with the earth's demonic forces to offer up new devices and possibilities. Edison was presented as a homespun Titan, Tesla as a dapper yet otherworldly seer, and Steinmetz as General Electric's crippled genius who tamed lightning. Just as wizards might be practitioners of "black magic" or "white magic," so could the modern

The Goddess of Electricity in a rendering clipped from an electrical journal of the turn of the century. Trim and sensuous, she stands astride the globe in a pose likely to appeal to such journals' largely male readership. Courtesy of the National Museum of American History, Smithsonian Institution.

technical elite be envisioned as wizards or destructive mad scientists. Edison, Tesla, and Steinmetz all toyed with these prescribed dramatic roles to further their ambitions, while industrialists and inventors offered dramatic pageantry to valorize the technological revolution then under way.

Historian John C. Burnham has explained the correlation of electricity with magic as the product of public relations experts in the early twentieth century.[2] Publicists were looking for an easy way to interest the public in the scientific capabilities of corporate laboratories. Although twentieth-century public relations experts undoubtedly took up this theme of "electricity as magic," it had been long nurtured at nineteenth-century electrical industry trade journals and industrial trade shows, whose audiences were members of the technical elite. To dismiss this interest as simple irony would ignore the psychoanalytic insight that jokes reveal genuine fears and concerns.

The connection of magic and science emerged from the public and the technical communities as much as it was foisted on the public by the corporate imagination. The inventor as wizard was a favored journalistic motif from the 1880s through the 1930s, and it implies the priestly esteem the public then granted the technical elite. In the public eye, the inventor could blend the traits of the scientist, the artist, and the mystic. Nikola Tesla's public persona was more that of a romantic artist than that of the rough-and-tumble man of affairs, and Steinmetz's and Edison's eagerness to comment on religious issues all illustrate the odd symbiosis between science and the supernatural at the century's turn. In their roles as salespeople, these turn-of-the-century scientists and engineers actively constructed a populist science that encouraged a belief in science magic.

Toward the Electrical Wedding

A reporter for the *New York Sun* first dubbed Edison the "Wizard of Menlo Park" in an interview in 1878, shortly after the world was stunned by Edison's wonderful "talking machine," the phonograph. The appellation stuck. An 1879 illustration of Edison titled "The Wizard's Search" shows the inventor in a full wizard's gown and cap, his costume inscribed with scientific emblems. Edison is shown climbing up a perilous path, seeming to bring along with him the illumination of the sun, perched behind his shoulder.

The concept of technology as not only a new source of progress, but also an expression of sorcery or magic appeared throughout the popular press at the turn of the century. Though often tongue in cheek, such references suggested both the utopian possibilities of science and a healthy misgiving about the inevitability of progress. Utopian visions of a dawning Age of Electricity dueled with a vision of the world as Hammer's "Electrical Diablerie" written large. Numerous writers debated whether electricity and technology were inherently "civilizing" forces. In its trade journals, the electrical industry struggled to offer an image of electricity triumphant over darker earthly powers, arguing that electricity, despite its immensely destructive potential, would advance civilization.

Late nineteenth-century technical journals often explored electricity's dual nature and emphasized its mysteries. William Clerk Maxwell's

THE DAILY GRAPHIC

AN ILLUSTRATED EVENING NEWSPAPER.

39 & 41 PARK PLACE

All the News. Four Editions Daily.

NEW YORK, WEDNESDAY, JULY 9, 1879.

$12 Per Year in Advance. Single Copies, Five Cents.

THE WIZARD'S SEARCH.

This newspaper illustration from 1879 depicts Edison as both a wizard and an explorer bringing enlightenment to a dark and mysterious landscape. Courtesy of the National Museum of American History, Smithsonian Institution.

mathematical explorations of electromagnetism published in the 1860s and 1870s were not understood or widely accepted until the turn of the century.[3] The working engineer of 1890 could still think of electricity as a "subtle fluid" as in Franklin's time. Electricity was invisible, weightless, flowed instantaneously, animated dissected frog's legs, perked up neurasthenics, and imparted force.

As with gravity, it was better to describe what electricity could do than attempt to explain its nature. Writers of articles titled "What Is Electricity?" would admit that no one, not even the most brilliant scientists, had any idea of its ultimate nature. As a result, even among the technical elite electricity could have a quasi-mystical status. A speaker for the Jovians, an electrical entrepreneurs' trade organization, claimed as late as 1913 that "electricity occupies the twilight zone between the world of spirit and the world of matter," and then added, "Electricians are all proud of their business. They should be. God is the Great Electrician."[4]

If the trade journals at times supported mystic associations with electricity, they also criticized the uninitiated who fell for sales scams in which the word "electricity" was linked to miracles. One writer commented, "There is a tendency on the part of certain people to contemplate it [electricity] as an African does his gree-gree, a wonderful and ineffable something that need only be invoked to produce almost any result that can be named."[5] Such articles would then go on to make fun of a gullible public willing to buy "electric nostrums, electric hair-curlers, electro soap, electric tooth-brushes, electric knife-sharpener and hundreds of others . . . articles not at all connected in any manner with anything electric."[6]

While pronouncements like those above tended to separate the gullible public from the technically astute readers and members of the electrical industry, they also sought to alleviate their own readers' fears about the seemingly inhuman forces that they as technicians were ushering into the social realm, the landscape, and the home.[7] However mysterious its fundamental nature, electricity was not simply another "gree-gree." It was compatible with progressive desires for a more efficient and scientific culture. And yet it was a mysterious power as great as any known to ancient times. To grapple with a magical force was to take on some of its nature. The trade magazines portrayed the taming of electricity by associating this struggle with imagery of femininity and gentle magic.

The Vienna Electrical Exhibition of 1883, for example, included an electrical ballet that explored the tension between the "earthly" and "heavenly" aspects of the electrical age. During this ballet "fantastic goblins"—presumably ballerinas—dragged electromagnets, dynamos, telegraph apparatus, and telephones onto the stage. A journalist on the scene noted that these machines were "handled by the graceful danseuses with as much ease as if they were especially trained in the mysteries of electricity. One of the prettiest scenes is the telegraph polka, which is danced by two ladies in the costume of telegraph boys."[8] The goblins of this ballet suggest dark powers wrested from the earth, while the graceful female telegraphers suggest that same power civilized. As they are one and the same, this staging implied that electricity had—at the very least—two faces.

A float in a Columbus Day parade in New York City in 1892 also encoded such tensions. Called "Electra," this Edison float contained both pagan and Christian iconography. The float, ostensibly a celebration of Columbus's discovery of the New World, instead stressed the triumphant emergence of a Brave New Electric World. Horses pulled the float, and batteries powered its three thousand incandescent lightbulbs; at the front were fierce dragons, and in the middle nubile maidens in two-tiered rings, as if part of a wedding cake, looking outward at the crowd. Above them, on the cake's top tier, caryatid statues held up a glowing globe labeled "Electra." On the float, not only women but female angels were shown harnessing the power of the ocean and electricity. Towering above the dragons in the front, a female angel held a lit wand, while below her a woman held a large palm frond and a large medallion emblazoned with Edison's portrait. Trumpeting female angels also faced from the back of the float above a grouping of Triton horns, while Roman soldiers marched alongside.

The float was a deliciously mixed metaphor, proclaiming that just as Columbus had braved the ocean and dragons to discover America, electricians had tamed the demonic power of electricity. The dragons were harnessed and the emperor Edison and his angels were now creating a noble electric world. Similar narratives would be employed fifty years later to persuade the public that atomic energy, though having destructive potential, was ultimately friendly and peaceable.

A panorama at an electrical exhibition in New York in 1896 followed the same theme of electricity's dual nature. Called "Mischief Brewing," it featured "a witch rather singularly portrayed, young and handsome," bending over a cauldron. Her cat had fiery electric eyes, the red flame of electric lights heated the cauldron, and "at intervals fiery sprites appear and disappear in the background." The writer mentioned that this exhibit attracted "much attention."[9]

In a less-muddled fashion than the float "Electra" or the witch at her cauldron, the image of the Goddess of Electricity suggested how femininity could tame electricity. William J. Hammer lightened the "diablerie" of his dinner party by directing his younger sister May to personify

The Edison float "Electra" in the 1892 Columbus Day parade in New York City offered a heroic vision of the electrician and his heavenly troops triumphant over powerful earthly forces. R. F. Outcault, the artist, worked as an illustrator for Edison and for *Electrical World* prior to cartooning for Pulitzer and Hearst. *Electrical World*, October 22, 1892.

this goddess. Dressed in a classical gown, perched on a pedestal, her hair and ears decorated with small electrical lightbulbs, she held an electric wand topped with a glowing star. Hammer's sister helped depict electricity as innocent and virginal—a harsh contrast to the mad bellowing of the automaton Jupiter—the God of thunder and lightning—at the dinner table.

The electrical journals also insisted that electricity had been made safe for the domestic sphere and reported on women's interest in things electrical. These publications often described, with approval, refined women who were gaining technological savvy. In so doing the journals yoked electricity to the nineteenth-century cult of domesticity, which insisted that women—even when constricted to the home—had great power as civilizers and enforcers of morality and religiosity. At the same time, the journals suggested that electricity had broadened the domestic sphere. In 1883, a columnist described a mother who held her baby to the telephone so a doctor could listen to its cough and declare it was not the "dreaded croup."[10] The woman's world had been enlarged. Entering that larger world was the duty of the age's bright, youthful New Woman.

Electrical journals also delighted in reporting on society women who decked themselves in incandescent garb for social events and fundraisers. Electricians could congratulate themselves for making electrical embellishments status symbols on par with expensive jewelry. The "help" could also be electrified. In 1884, for example, the Electric Girl Lighting Company rented out "Electric Girls" as hostesses or servants for parties.[11]

Not only electrical parties but also electrical weddings became fashionable. Reports on such events implied that electricity was a progressive force that could only add to human happiness. The coverage of such weddings also suggested that like love, the electrical age belonged to the youthful, to those prepared for the new. A journal outlined with approval one such courtship. It began when a contributor to the *Electrical Review,* Miss Gretchen Van Tassel, of an old Knickerbocker family, met Edouard Constant, a member of "one of the most ancient Huguenot families in New Rochelle" and a technically steeped member of New York's Board of Electrical Control. Miss Van Tassell was very much a New Woman, enthusiastic, bright, and knowledgeable about electricity and all things modern.

The two met when Miss Van Tassell called Mr. Constant to ask a technical question. After expressing mutual admiration at each other's publications, he insisted upon visiting her to discuss the technical issue. The description of their meeting is full of puns on electrical activity. For example, Miss Van Tassel's enthusiasm "was electric," her eyes "sparkled like a Leyden jar," her golden hair like the "pale golden light of an Edison lamp." When they telegraphed and telephoned one another, "Cupid, like a rope-dancer, traveled by wire." The suitor became a "frequent visitor at the happy electric fireside" of Miss Van Tassell. Ultimately their nuptials were arranged. The wedding scene conjured the dialectical opposite of Hammer's "Electrical Diablerie." The bride and bridegroom entered the drawing room "welcomed by a chime of electric bells. Then an automatic electric piano played the wedding march from 'Lohengrin.'" The bride wore a comb arched with tiny electric lights in her hair. Bouquets of flowers glowed with electric lights. They stood before the minister. "The words were spoken, there was an electric kiss, and the continuous current of their happiness had begun."[12] The electrical wedding was a happy event that God smiled down upon, quite unlike Hammer's offering of electrical diabolism.

The dualism of electricity as powerful and primitive and as civilized and safe also informed a species of reporting that reminded technicians of the potential dangers of electricity. Electricians wished to prove that their medium was safer than gas lighting systems. Horrors ensued when electricity escaped or existed outside its civilized containment—in accidents involving live wires or lightning. Such articles often offered grisly details to remind readers of the dangers of electricity—particularly in the untamed form of lightning. These pieces suggest that there was no protection from lightning's wrath. One article from 1883 described the death of a man hit by lightning while reading his Bible and noted that "his clothing was stripped from his back and his flesh lacerated."[13]

Human-built power sources could also lead to grisly accidents. A journal interviewed a lineman who survived a five-hundred-volt shock. He described the experience in near-mystical terms. He reported that after being knocked off his feet he felt that he was flying, "soaring away, just as one feels when put under the influence of ether or chloroform. Then all was blank." Upon waking he reported a "strong taste of brimstone in my mouth."[14] Electricity, like the ecstasy of a shaman's ceremony, could transport one both to heaven and to hell.

Yet electricity, a protean and plastic force, ultimately would tame nature. Artificial, electrically run waterfalls figured in many electrical exhibitions—to point to the value of hydropower but also to provide spectacles of nature tamed. These waterfalls, first replicated carefully, and surrounded with boulders, shrubs and trees, slowly became theatricalized. In 1888 William Hammer designed a miniature Niagara with colored lights and an electric rainbow that could fade in and out of its mists.

And at the Buffalo Pan-American Exposition of 1901, a Niagara replica gushed from the base of the neoclassical Electricity Tower, suggesting that nature was no longer a force apart but one that engineers had assimilated fully into human culture.[15]

Ideally electricity was a civilizing force for everyone, but many envisioned it as a weapon to be used against those who then were thought to be less civilized. In 1891, when news agencies were reporting the "Ghost Dance" of the Sioux Indians at the Pine Ridge Reservation, one journal reported one crank's solution for dealing with the Indians. This crank recommended that the government "surround hostile camp with wire. The 'juice' having been tuned into the wire, lo, the poor Indian, is to be driven down to it in herds and electrocuted."[16] Though the journalist did not entirely approve of the plan, his use of the word "herds" indicated entrenched racism. Such tendencies were typical of an era when "ethnic congresses" like those at the World's Columbian Exposition of 1893 underlined racial hierarchies culminating with the Anglo-Saxon at the pinnacle.

The belief that electricity could protect civilization's borders also could take on a direct Darwinian twist. In 1892, a scientist with interests in evolutionary theory intended to use electrical devices in his studies of the language and "social life" of the "hut building ape" of the Congo River. R. L. Garner planned an expedition to Africa with a load of electrical devices. Reversing the formula of the zoological park, Garner intended to live in the jungle in a metal cage, electrified for his protection, while he took flash photographs and made recordings of various animals day and night. With his electrical aids, he intended to get views "never before seen by savage or civilized man."[17]

Mr. Brown's Crusade

Whether electrical power should assist in state executions was a fierce debate of late 1880s America that points to the progressives' difficulty in insisting that science and technology were inevitably civilizing. The progressive philosophy encouraged the use of science and technology to heighten efficiency, eliminate wastefulness, and improve social conditions. Could electricity provide a more painless, humane execution, or would it be a cruel and unusual punishment?

Electrical trade journals first began the debate in the 1880s. In 1883, a trade writer suggested electricians were leery of state electrocutions with his remark, "In all probability capital punishment will be abolished before electricity is summoned to the aid of the executioner."[18] By the late 1880s, however, New York's reform-minded legislators passed a bill calling for electric executions. Most electrical professionals doubted that this would be good for their trade. Yet some gave it their blessing. In 1888, one trade journalist expressed his delight: "This method of execution is a tremendous realization of the old notion that Jove struck the guilty with his sudden thunderbolts."[19] Here, as so often, the writer finds it necessary to make a reference to the archaic—Jove—to support the new—electricity. This habit of mind indicates writers found a certain pleasure in the irony, using it to remind readers of how different the "modern" world was from the "ancient" world, but this strategy also suggests an awareness of the haunting of the technological world by old ghosts.

New York's lawmakers' approval of electrical executions stemmed largely from the agitation of a previously obscure electrician, Harold P. Brown. His work in the arc-lighting industry convinced him of the dangers of high-voltage alternating current. Brown would have remained obscure if his hatred of alternating current had not prompted him to prove its dangers by executing dogs and other animals in demonstrations before fascinated witnesses. This theater of cruelty gained the interest and covert backing of Thomas Edison, an opponent of capital punishment who nevertheless reasoned that if state executions used his rival Westinghouse's alternating current, the populace would not want that same hateful "executioner's current" in its home.

Brown, who had invented several devices that alternating current would make obsolete, first gained public attention when he wrote a letter to the editor of the *New York Evening Post* in the spring of 1888. Brown described alternating current as "damnable" and urged New York, like Chicago before it, to forbid by law the use of high-voltage alternating current.[20] That summer, assisted by a member of New York's Medico-Legal Society, Brown demonstrated the dangers of alternating current before an audience at Columbia College's School of Mines. They wired a dog, trapped in a cage, and gave it direct-current shocks of 300, then 400, 500, 700, and 1,000 volts. The dog, neither silent nor pleased at such torments, survived. Brown then killed the dog with 330 volts of alternating current. When members of the Electrical Board of Control and representatives of Edison and Westinghouse argued over the meaning of the results, Brown offered to continue the experiments with other dogs. A humane society officer stepped in and forbade any further demonstrations. Brown concluded the night by telling the audience that alternating current was only fit for the "dog pound, the slaughter house, and the State prison."[21]

At the time of this performance, Edison and his corporate interests were locked in battle with George Westinghouse, who then was championing alternating current. Edison preferred direct current. Edison argued, with some validity, that the direct-current systems he had devised in New York City were safer, as transmission relied on lower voltages. Likewise, Edison noted that alternating current lacked metering devices and efficient motors; further, engineering problems in direct current required simpler calculations. However, direct current could only work well in heavily populated areas as it could not be transmitted more than several miles without losing much of its efficiency. Long-distance transmission would require the great expense of using extraordinarily thick copper wires or pipes. Alternating current, however, could be stepped up to high voltages and transmitted for what would prove to be hundreds of miles at a high efficiency, making it more appropriate for the electrification of rural areas, and also for the development of dynamos at remote hydropower stations. In his stage demonstrations and pamphlets Brown countered that alternating-current interests were cruel profiteers, concerned far more with profit than public safety.

Although most younger, university-trained electricians believed alternating current would win the day, Edison's opposition and influence were great obstacles. The debate over the currents became intensified in 1887 as a result of Nikola Tesla's work. In 1887 and 1888, Tesla, who had briefly apprenticed with Edison, took out patents for an electric motor, transformers, and other devices that could make alternating-current power a viable alternative to direct current. Tesla came to prominence when he presented a paper in 1888 to the American Institute of Electrical Engineers that explained the principle behind his alternating-current motor and generator.

Tesla's polyphase motor, considered by many to be the most important technological contribution of his career, created a rotating magnetic field by phasing in alternating currents in a circular pattern; the magnetic polarity rotated and could induce a metal bar to follow. Tesla's polyphase motor was an improvement on the earlier direct-current motors that changed alternating current into direct current and required that sparking brushes maintain a contact with a revolving drum. When news of his breakthrough circulated, the popular—and technical—press hailed Tesla as a new "wizard."

George Westinghouse recognized the importance of Tesla's concepts, purchased his patents, and offered him royalties. The "war of the currents," formerly something of an academic issue, increased in intensity. Both sides waged this war economically, through propaganda, with threatened or actual acts of industrial espionage, and also with theatrical presentations. Though Brown was never on Edison's payroll, Edison encouraged Brown's macabre theatrical efforts. In July 1888, two months after Tesla had presented his paper on alternating current, Brown electrocuted the dog at Columbia College. With the encouragement of Edison, Brown went on to execute horses and calves at Edison's new laboratory in West Orange, New Jersey. Impressed New York State legislators voted in favor of adopting electrocution as its new method of capital punishment, and the New York Medico-Legal Society appointed Brown to head arrangements for the first execution. Brown insisted that Auburn State Prison use one of Westinghouse's generators. He and Edison's publicity team also attempted to make "to Westinghouse" a new verb for electrocution. Edison was still at it five years later, when he arranged for the execution of the elephant "Topsy" at Coney Island, relying on alternating current.

Not delighted at the adverse publicity, Westinghouse provided a lawyer for the first proposed victim, William Kemmler, and they appealed the constitutionality of electrocution. When Edison appeared at the hearings, the *New York Times* titled its coverage "Testimony of the Wizard."[22] Edison testified that electrocution would be neither cruel nor unusual and that an alternating current of one thousand volts would instantly kill a human. (Most of these debates ignored the fact that electric power is calculated by multiplying voltage, or electric "pressure," with amperage, or "current" and did not provide values for amps used.) The court ultimately agreed. Kemmler was killed in 1890, and Edison mounted a campaign that asked the public, "Do you want the executioner's current in your home?"[23]

Press coverage of the electrocution of Kemmler and ensuing electrocutions was divided. Many electrical experts mourned the adoption of electricity for executions—especially after grisly press reports appeared of electrocutions that went far from smoothly. Surely the Goddess of Electricity was intended for finer affairs. The *Electrical Review* ran many articles denouncing the executions—first because they were run by amateurs, and second because no expert would stoop to taking on the role of executioner. One of the *Electrical Review*'s writers commented, "The matter is but an experiment, a horrible experiment, upon human flesh, with no advantages whatever over the old method of hanging. . . . The law was conceived by cranks, has been carried out under the supervision, very largely, of theorists, and these would-be reformers should now be set to the right-about to employ their little minds upon subjects less revolting to public decency and modern thought."[24]

Gala Nights at the White City

While Edison's doomed campaign to discredit alternating current continued, Westinghouse and Tesla seized on theatrical tactics to promote their power system's safe and miraculous nature. Tesla as showman proved himself a formidable proponent of alternating current. He first stepped onto the stage to defend alternating current when he lectured about light and high-frequency phenomena for the American Institute of Electrical Engineers in May 1891 at Columbia College. Here, Harold Brown had also debuted several years earlier.

At the lecture, Tesla unveiled a generator attached to an induction coil (or "Tesla coil") that together could create frequencies of 20,000 alternations per second and voltages as high as 250,000 to 1,000,000. This apparatus created an electrical field that charged the room. Tesla displayed the potential value of such arrangements. He ran devices that were not wired to outlets or power sources. He held up empty glass tubes and bulbs and they gave off a bright light. Geissler tubes filled with gases lit up brilliantly in different colors. A reporter from the *Electrical Review* insisted, "Here Mr. Tesla seemed to act the part of a veritable magician."[25]

During the lecture, Tesla also offered dramatic proof of the safety of alternating current—at least at high frequencies. He let currents as high as 250,000 volts pass into his own body so that sparks shot from his fingertips and his entire body glowed with violet electrical flames. Here was dramatic proof that alternating current need not be thought of as the executioner's current. The *Electrical Review*'s reporter insisted that while Tesla offered "a brilliant exhibition of fireworks," the performance also revealed that "a distinct advance had been made in scientific research."[26] During this and ensuing lectures, Tesla discussed the possibility of using such electrical fields for lighting systems, for remote control of devices, for medical therapy, and for the wireless transmission of universal time, information, and electrical power.

Tesla repeated his performances for scientific and technical societies. In February of the following year, 1892, Tesla went to London's Royal Society to lecture on "Experiments with Alternate Currents of High Potential and Frequency." The next night he duplicated the talk at the Royal Institution before an audience of eight hundred. The English scientist Lord Rayleigh thanked him, noting, "Mr. Tesla has taken us into some of the dark—metaphorically dark—places in nature. These fields have been little trodden . . . it does not require any great capacity to see that Mr. Tesla has the genius of a discoverer."[27]

Tesla's theatrics and apparatus made a great impression on the British scientists. An electrical journal reports how these scientists explored

Tesla's effects at a Royal Society conversazione later that year. At the salon, J. T. Bottomley exhibited discharges "à la Tesla" from vacuum tubes in one room. In another "Professor Crookes played with a vibratory current of 100,000 volts pressure and a million alternations per second, and offered his audience shocks from the apparatus free." A journalist commented that Crookes—one of the era's preeminent scientists—created "phosphorescent effects" of great beauty.[28]

These lectures of the 1890s made Tesla a celebrity. When he lectured in St. Louis at the National Electric Light Association Convention, four thousand copies of the biography of Tesla inserted in the program were sold on the streets. Several thousand people came to hear his lecture that night at the Grand Music Entertainment Hall in St. Louis, paying four to five dollars for the ticket.[29] Audience members were fascinated with Tesla's light effects, particularly when he touched a high-voltage electrode and his body burst into flames and, reportedly, continued to glow long after the demonstration. He seemed no ordinary man. The *New York Sun,* the first newspaper to dub Edison a "wizard," soon after ran a front-page illustration of Tesla, his body glowing with light, with the caption "Nikola Tesla, Showing the Inventor in the Effulgent Glory of Myriad Tongues of Electric Flame After He Has Saturated Himself with Electricity."[30] The "New Wizard of the West," as *Pearson's Magazine* called him, had arrived.[31]

Public approval of alternating current reached its apotheosis at the World's Columbian Exposition in Chicago in 1893. Outmaneuvering his rivals, Westinghouse gained the contract to install the fair's electrical system. The wonders that the fair's designers wrought with electricity all but decided the outcome of the battle of the currents—and helped clinch Westinghouse's bid to build and install electrical dynamos at Niagara Falls. Never before had technology created such an impressive spectacle. The Columbian Exposition with its floodlit white Beaux Arts buildings suggested an amazing advance for humankind. If the buildings were classical in design, embodying timeless virtues, electricity brilliantly lit the fairgrounds, electric searchlights plied the heavens, incandescent bulbs created a halo around the Ferris wheel, and, with the aid of engineers such as William Hammer, colored lights turned spouting water into dazzling displays then known as "electrical fountains."

At the Westinghouse exhibit in the Palace of Electricity, Tesla gave lectures, presented his "Egg of Columbus," which spun and then stood on end as it responded to a polyphase current like that of his induction motor, and the inventor doused himself with electricity in huge voltages to champion the safety of alternating current. Tesla also designed a small electric sign that outlined the name Westinghouse in glass and that pe-

This illustration of inventor Nikola Tesla, "in the Effulgent Glory of Myriad Tongues of Electric Flame," helped promote him as the latest "wizard," capable of seemingly supernatural feats with electricity. *New York Sun*, July 22, 1894.

riodically emitted high potential discharges, creating a miniature display of lightning and the deafening crash of thunder. Its noise could be heard throughout the Electricity Building.[32]

In the building, exhibitors displayed thousands of devices, such as an automatic electric door, a diver's suit equipped with a telephone, and an array of Edison phonographs that could be used for business dictation or language study, and others equipped with nickel slots to play music. Elsewhere, the businessman was offered a seven-day clock in which pins could be placed at any day, hour, or minute to ring alarms and remind him of an upcoming appointment. Visitors also eagerly toured the fair's powerhouse where forty steam engines ran over a hundred dynamos. Such displays made a convincing argument for the utopian possibilities of electricity.

On a grander scale, the entire fairgrounds served as a theater for a nightly illumination ceremony. On "gala nights" crowds gathered at twilight to watch fireworks over Lake Michigan, the electric fountains foaming with colored lights, and burning torches above the fairground's otherwise dark buildings. Then, "once the last rocket has been shot into the sky and the last string of flambeaux has collapsed into darkness,"[33] electrical illumination began transforming the buildings, canals, and concourses into the "White City" of its popular nickname.

In 1894, John P. Barrett, who published a technical account of the fair's electrical displays, insisted that the fair offered electricians a chance to catch up with advances in their field, but also had created a stunning promotion. "It dissolved much of the mystery that had pervaded its domain; it brought electricity to the people in the light of a servant not as an awful master; and finally it created an impression of stability and soundness among the thinking and progressive element of the people that will mean wider commercial development."[34]

The issue of electricity acting as "servant" or "awful master" also had inflected the rhetoric of President Grover Cleveland at the fair's inauguration. As Cleveland tapped a gold telegraph key to start the electrical dynamos and machinery, he proclaimed, "As by a touch the machinery that gives life to the vast Exposition is set in motion, so at the same instant let our hopes and aspirations awaken forces which in all time to come shall influence the welfare, dignity and freedom of mankind."[35] The remark reflected the progressive notion that the country must harness knowledge and technology for the goal of social betterment. It also connected electricity—the new form of power that flooded the exposition—to the life force and to larger realms whether political, geographical, or historical. Electricity was coming of age, the United States was a world power, and even a business-minded president could imagine the nation's—and nature's—new powers misused.

Electricity's dual nature, however, was barely expressed at the Columbian Exposition. Unlike the float "Electra" from the year before that had connected the advent of electricity to Columbus's discovery of America, this World's Fair celebration of Columbus had no obvious dragons to tame. The foils for "modernity" instead were the ancient models used in the Beaux Arts architecture and the "exotic" cultures on display in the different national villages of the midway. One stunned visitor, Henry Adams, eventually came to the conclusion that the dynamos themselves were the dragons. In his autobiography, *The Education of Henry Adams,* he remarked that the Columbian Exposition for him had been a rude awakening. Its technological wonders forced all thinking men to "sit down on the steps and brood as they had never brooded on the benches of Harvard College, either as student or professor."[36] Adams was again drawn to the dynamo exhibit at the Paris Exposition of 1900 and, after lengthy brooding, was prepared to answer that which "Chicago asked in 1893 for the first time . . . whether the American people knew where they were driving."[37] His conclusion was that technology was spinning culture into a meaningless future and that his countrymen were slavishly worshipping the machine.

In his famous chapter "The Dynamo and the Virgin," Adams contrasted the religious power that once emanated from female deities such as Venus and the Virgin Mary to the power that emanated from the dynamo. Unaware of the electrical industry's attempt to fuse the two in the image of the Goddess of Electricity—and even if he had chanced upon her in an electrical advertisement, Adams would likely have been unimpressed—Adams chose to posit fervor for the old religion and new as complete opposites. "All the steam in the world could not, like the Virgin, build Chartres."[38] For Adams, the progressive assumption that science and technology could lead to a rational, well-regulated society seemed unlikely. He feared that modernity was ushering in chaos—a society of splintered social values and vision. Even though the machine was inherently rational, the spirit underneath the new age was not that of rationality, but a new form of worship and awe. For Adams, the dynamo was the modern era's golden calf.

Thomas Edison Conquers Mars

The electrical inventors who had brought forth Adams's feared dynamo were objects of public fascination. Although Edison's image did not easily fit the paradigm of a wizard or mad scientist, Nikola Tesla was ready-made for such roles. Edison, sloppily dressed, amiable, folksy, ruthless, able to get along better with America's pragmatic businessmen than with uppity men of science, managed to remain a folk hero whose products and industriousness, not persona, represented the new age. In contrast, Tesla spoke with a European accent, was tall, dandified, erudite, celibate, and given to high-flown poetic speeches about the importance of his inventions. He embodied many of the contradictions that Adams read into world's fairs. Tesla's displays in which he soaked up voltage and emitted electric flames suggested that he was a hypermodern man. He reflected the Progressive creed in many of his endeavors, but the publicity surrounding him frequently linked him to more archaic, archetypal roles, as those of the ascetic saint, the wizard, the seer, and his tragic finale as a mad scientist.

Edison had rehearsed for the role of mad scientist with limited success. In 1890 he collaborated with George Lathrop to coauthor a futuristic book similar to Edward Bellamy's *Looking Backward*. Edison wrote thirty-three pages of notes that offered descriptions of antigravity devices and suspended animation machines for space travel, descriptions of a lost world at the Arctic pole, and numerous ways that people of the future might redesign the earth's weather and topography.[39] After his initial enthusiasm waned, Edison dropped the book project. He later permitted Garrett P. Serviss to write and serialize "Edison's Conquest of Mars" for Hearst's *New York Evening Journal* in 1898, in which the inventor saved the earth from Martian perils.

Though Edison's image was that of a pragmatic man of the people, he cultivated eccentric strains of thought. Edison signed on with the Theosophical Society shortly after it was launched in 1875. Though not an active member, his interest was highly prized. Particularly in his old age, Edison was willing to speculate on metaphysical matters in his homespun manner. But Edison was far more comfortable planning ways to extract metals from ores, improve telegraph recording devices, develop an electric automobile, and otherwise serve industry and increase his fortune. Edison presented himself as a Yankee tinkerer who had no otherworldly pretensions.

In contrast, mystic-minded writers have often tried to claim Tesla as one of their own. Tesla's own anecdotes about his early life gave cues to biographers seeking to describe him as a superman—as did his first biographer, John J. O'Neill—or as a saint or an odd amalgam of mystic and scientist. Tesla easily topped Charles Came and other electrical wonder showmen in promoting himself as a miracle worker.

Tesla's statements about his early life imply he periodically underwent tremendous ordeals that would neatly fit the life story of a tribal holy man. A man of science with an intense spiritual life would appeal to many at the turn of the twentieth century. Tesla's life story, as told by himself and his disciples, focused on his initiation into the cult of the electricians, his agony and ecstasies, and his eventual fall from grace as an electrical wonder worker. In later generations, fanatic worshippers would add tales of his posthumous ascension to a higher plane or orbiting spaceship.

Such a rendering suggests the public's desire that religion somehow mesh with the scientific currents of the age. William James wrote *Varieties of Religious Experience* in 1902, at a time when Henry Adams and many other Americans were full of doubts about the cold materialism then prevailing. James filled the book with historical narratives of conversion experiences. To give just one example, he described a simple farmer who came upon an evangelical camp meeting with its "terrible noise" of those seeking conversions. The man declared that "I fell on my face by a bench, and tried to pray, and every time I would call on God, something like a man's hand would strangle me by choking." The man continued to struggle with this invisible hand, then heard a voice that warned him, "Venture on the atonement, for you will die anyway if you don't." Ultimately he was revived and felt himself flooded with light and glory for several days. Everything was new. In short, he had experienced a death and rebirth.[40]

Anthropological descriptions of the initiation of tribal shamans—holy men or sorcerers—follow an outline similar to this version of an evangelical conversion experience, yet they tend to be more elaborate. First, the candidate often has a tenuous outsider status in the tribe. Social ostracism is common for the eventual shaman. Second, though repeatedly called by "terrible" voices, he or she may try to refuse these invitations of the spirits. Usually the candidate will fall deathly ill, particularly after refusing the call. This illness will bring the candidate to the brink of death. The shaman, like the camp meeting attendee, has to "venture on the atonement" for the shaman "will die anyway" if he or she does not.

Tesla and his followers cast his life story in a manner that made him fit the dual role of scientist and primitive holy man. As is typical of tribal candidates to be shamans, he was an outcast as a child—someone who seemed abnormal and disturbing to others. He also admitted to peculiar mental gifts. He claimed that he had the ability, for example, of sharply visualizing scenes, people, and things, making it difficult to separate these images from reality. He later credited many of his inventions to this visualization ability, insisting that he had no need of models, drawings, or experiments. "When I got an idea, I started right away to build it up in my mind. I changed the structure, I made improvements, I experimented, and I ran the device in my mind";[41] he insisted he would even work out the correct dimensions of parts in his mind, adding, "It is immaterial to me whether I run my machine in my mind or test it in my shop."[42]

Accounts of Tesla's life are loaded with initiation crises. Tesla, a Serb, was born in what is now Croatia; his father was an orthodox priest. When Tesla was five, his older brother, the family favorite, died in a riding accident. After the accident Tesla recalled constantly trying to please and impress his disinterested parents. Throughout his childhood and youth his father steered him toward the clergy, but Tesla continually longed to be an engineer. At age eighteen, Tesla fell deathly ill with cholera. Tesla claimed that, visiting his bedside, his father asked him if he would get well, and Tesla had responded, "I will get well if you will let me study engineering." His father agreed. Tesla spent a year in the mountains, regaining his strength and avoiding compulsory military service, and soon after began his education at technical schools elsewhere in Europe.[43]

As the preceding anecdote suggests, Tesla's autobiographies and biographies show him undergoing a crisis before arriving at his scientific vocation. New and disturbing abilities appeared after his father's death, while Tesla was beginning his career. In his student days, while searching desperately for the method to create a more efficient alternating-current motor, Tesla again fell ill. Tesla described his own peculiar syndrome as follows: his heart raced to as high as 250 beats a minute, he twitched and trembled, and his senses became enhanced so that the sound of a fly landing nearby or a clock ticking in another room caused him agony, the force of the sun's rays would stun him, and "in the dark I had the sense of a bat and could detect the presence of an object . . . by a peculiar creepy sensation on the forehead."[44] Physicians believed he was not long for the world.

Relief from these otherworldly symptoms came with a vision. According to Tesla, the vision appeared to him while he was walking in a park in Budapest and watching a sunset with a friend. He began to quote lines from Goethe's *Faust* about the setting sun, and perhaps incited by the image of the sun's daily revolution, Tesla solved the problem of the induction motor. He saw the principle of the rotating magnetic field, induced by staggered circuits arranged in a circular pattern. According to the legend, he drew a sketch of the device's workings in the dust for his friend—a sketch similar to his eventual patent drawings.

Following this great vision, Tesla continued on the road to becoming a great wizard. After his initiation, he apprenticed with another "wizard"—in this case Edison, for whose concerns he worked both in Europe and later in New York City. Although he had a falling out with Edison, he later described Edison as an "extraordinary" man. According to their biographers, both were tireless workers who could put in twenty-hour workdays for months at a time. Other anecdotes suggest that Tesla wished to model himself after Edison. One such anecdote has Tesla asking the more established inventor what his breakfasts consisted of and Edison replying "Welsh rarebit." Tesla dutifully ate Welsh rarebit every morning for some time before he recognized the jest.

The apprenticeship concluded with a dispute over fifty thousand dollars that Tesla claimed Edison had promised but never paid for improvements Tesla had introduced to Edison's power plants. Finally, after the break with Edison, and after becoming a master inventor himself, Tesla put on displays to reveal his otherworldly abilities. Tesla, on stage, subjecting himself to a million volts of alternating current and then bursting into flames, created a spectacle that undoubtedly outdid those of most tribal shamans. And Tesla certainly believed in the healing power of electricity, particularly high-frequency electricity, and doused himself daily as part of a regimen that he insisted would prolong his life.

Admiring biographers and popular press articles furthered the "Tesla as wizard" formula. In an 1899 article, a journalist described his visit to Tesla's Manhattan laboratory. The *Pearson's* writer reported, "A tall, thin young man walks up to you, and by merely snapping his fingers creates instantaneously a ball of leaping red flame, and holds it calmly in his hands. As you gaze you are surprised to see it does not burn his fingers. He lets it fall upon his clothing, on his hair, into your lap, and finally, puts the ball of flames into a wooden box."[45] Tesla would also play a variant on the "death and resurrection" motif of the shaman's voyage for laboratory visitors—possibly picking up a few tips from the performances of Harold Brown. Tesla would remove a small animal from a cage, kill it with one thousand volts of electricity, then let his audience view the meter as he allowed two million volts to pour through his own body.[46]

Though Tesla encouraged the public to think of him as a great wizard, he paradoxically also denied any interest in the occult and ultimately promoted an extremely materialistic worldview. Tesla attended at least one of the charismatic Swami Vivekenanda's lectures in Brooklyn in the 1890s, and became interested in Vedic teachings; he occasionally alluded to this philosophy in his letters but pointedly avoided the psychic research tendencies of many of his peers—particularly in the British scientific community. Tesla greatly admired the scientists Oliver Lodge and William Crookes but refused to share their enthusiasm for psychic investigations. He gently satirized those who believed in telepathy, premonitions, communication with spirits, or other psychic phenomena. An interview in the *New York World* from 1894 flatly declared that "he does not believe in telepathy, which is, according to its exponents, a sort of psychical electricity."[47]

In a bitter letter to the Westinghouse Company in 1899, Tesla poked fun at Crookes's then somewhat notorious advocacy of telepathy. The letter included a check from Tesla to pay for equipment he had borrowed for experiments. Tesla wrote of recalling a dream in which he was sent a check for fifty thousand dollars from the Westinghouse Company to thank him for his past efforts on their behalf. Having instead received a demand for payment, Tesla thanked them for confirming his belief that Crookes was mistaken and there were "no transmissions of mind efforts."[48]

Tesla combined his interests in utopian technology with an aggressively materialistic worldview. As such, he was a prototypical progressive. His clearest expression of these aims are in his article "The Problem of Increasing Human Energy," which ran in 1900 in *Century Magazine*. It reads at times like a lampoon of a technical thinker's approach to social problems—similar to later efforts of Buckminster Fuller—as when he states throughout that his goal is determining how to increase the "force" operating in the "human mass" in order to ensure progress. Tesla even gave a variant of the physics formula for momentum to indicate the total amount of human energy, indicating it was equal to MV^2, with M as the total "human mass" and V as a hypothetical velocity, equivalent to "*F—R,* or force minus resistance."[49] Social progress could be made by: (1) increasing the human mass (or M); (2) reducing "retarding" or "frictional" forces (or R) on velocity, primarily by ending ignorance, fanaticism, and warfare and realizing peace; and (3) increasing the "accelerating" force (V) by more efficiently harnessing the power of the sun for work.

His solution to these difficult problems reads as a prospectus for his various planned inventions. He believed the "human mass" could be increased with better food sources and improved health and hygiene, and proposed using an electrical method for fixing atmospheric nitrogen to fertilize soil. Warfare was also a wasteful practice that decreased human mass. The "frictional" force of warfare could be ended with his research into radio-guided ships and other robotic war machines that he thought would make human soldiers obsolete, or convince humanity of the futility of warfare. The acceleration of the human mass could be increased if engineers more efficiently harnessed the sun's power, developed renewable energy sources like wind, water, and sun, began transmission with wireless power, built perpetual motion devices that he called "self-acting engines," or used virtually free power sources that might usher in a utopian age.

Tesla's ultimate goal sounded identical to that of Edward Bellamy, the author of the influential utopian novel *Looking Backward* (1888), which depicted a future in which all corporations had merged into one, ending competition and wastefulness. In Tesla's version, the utopian society would appear "when all darkness shall be dissipated by the light of science, when all nations shall be merged into one, and patriotism shall be identical with religion, where there shall be one language, one country, one end, then the dream will have become reality."[50]

Emphasizing his technocratic dreams, Tesla's happy vision of man as a machine is a chief component of his thought. The supernatural and superstition had to be swept out of the path of progress. When describing his development of a radio-guided boat, or his "teleautomaton," which he demonstrated at Madison Square Garden in 1898, he insisted that the design had been based on his observations of his own behavior. In envisioning the future field of robotics, he outlined a philosophy similar to that of the behaviorists, insisting, "I remember only one or two cases in all my life in which I was unable to locate the first impression which prompted a movement or a thought, or even a dream."[51] According to this formulation, Tesla himself was an automaton, endowed with the power of movement and ability to react to external stimuli.

This robotic view of individual consciousness was quite opposed to the romantic sensibility Tesla also exuded. A solution to this paradox may be simple. For Tesla, if not for Crookes or Lodge, science and the supernatural were impossible bedmates. Tesla often eagerly remarked that it was only through severe discipline that he had learned to curtail his abilities to imagine distant landscapes and cities, or receive premonitions, as if he had corrected himself of outmoded manners of thought. With this in mind he also was able to explain to himself his own strange visualization abilities. His hopes for improving humankind were similar to those that behavioral psychologist B. F. Skinner was later to elaborate in *Walden Two*—people could be conditioned to coexist and create a productive, peaceful society.

While Tesla's article expressed interest in power sources such as wind and sun, his ambitions led him to the pursuit of more glamorous schemes. By the early 1890s, Tesla fixed on a plan to consolidate all communication and power distribution systems in a "world wireless" broadcast grid. In one pamphlet, he indicated that his first transmission center would broadcast electrical power; connect all telegraph exchanges, stock tickers, and phone exchanges; distribute news from the news industry; transmit intelligence on separately tuned frequencies; offer private telegraph services to the government; transmit facsimile copies of documents and drawings; and broadcast universal time.[52] Heralding his plans, a *New York World* article in 1896 reported "Electricity soon [will be] as free as air. . . . The end has come to telegraph, telephone companies . . . and other monopolies . . . with a crash."[53]

With financial backing from J. P. Morgan, Tesla began to work on his experimental broadcast station. Refusing offers of free electricity from Niagara power officials, Tesla chose the site of Wardenclyffe, Long Island, so that he could commute daily along with a chef from the Waldorf-Astoria Hotel in Manhattan and continue to enjoy the high life.[54] He hired socialite Stanford White as his architect and spent three years building an enormous coal-fueled powerhouse and a tower topped with a fifty-five-ton sphere (or condenser) that could store electrical charge; below the tower a winding staircase spiraled down a shaft that plunged

120 feet into the earth, with grounding pipes dug in another 300 feet. Tesla explained that "it is necessary for the machine to get a grip of the earth, otherwise it cannot shake the earth."[55] J. P. Morgan lost interest in Tesla's costly scheme, however, after Guglielmo Marconi succeeded in his more modest experiment of broadcasting telegraph signals across the Atlantic. Never tested, Tesla's power broadcasting station on Long Island was abandoned. Despite the inventor's frantic efforts through the decades to raise money to power the station, it was eventually torn down in 1917. A year prior, in 1916, the *Brooklyn Eagle* wrote, "The place has often been viewed in the same light as the people of a few centuries ago viewed the dens of the alchemists or the still more ancient wells of the sorcerers."[56]

When Marconi captured the public's imagination as the inventor of radio, Tesla's reputation was eclipsed. His standing in the scientific community also plummeted when he began to make wild announcements in the press of his plans to tap cosmic rays, communicate with other planets, create perpetual motion machines, change weather patterns, devise an early prototype of a "Star Wars" defense system against rocket and airplane attack, and develop other devices of destruction—for example, a mechanical oscillator that would set up standing vibrations that could destroy structures and ultimately even split open the earth. His grandiosity and propensity for imagining destructive devices helped a new image emerge of Tesla as mad scientist.

Tesla's devotion to his world system and the messianic hopes he had placed in it began to alienate his peers, many of whom assumed he was mentally deranged. Tesla, whose sparking, discharging "coils" helped create the visuals for the screen version of *Frankenstein,* provided a new prototype of the mad scientist—diverging from the images that had haunted the nineteenth century. In contrast to medical tinkerers like Victor Frankenstein, or H. G. Wells's mad vivisectionist Dr. Moreau, or the insane, vampire-like mesmerists of scare tales, Tesla's post–world system interests made him an early model for the mad scientist as physicist, a controller of "death beams," "disintegrating rays," and other manifestations of energy; a mad science popularized in the pages of Hugo Gernsback's magazines of the early twentieth century—magazines that also featured profiles of Tesla and his accomplishments. Predictably, a 1941 Superman comic book, *The Mad Scientist,* featured Tesla in the title role.[57]

In the meantime, Tesla's rivals were stealing the eclipsed wizard's thunder. Marconi, for example, in 1912, gave a Tesla-esque interview that resulted in an article titled "Marconi's Plan for the World." Referred to as a "wizard," Marconi talked up Tesla's favorite ideas as his own: the use of wireless power to light cities, the possibility of contacting other planets with radio, and the use of electricity to fertilize soil. Marconi concluded that his new innovations would have vast political ramifications. "It will be necessary to sweep out all the present privileged corporations of power. . . . In the future the government will be the owner of all energy. Individuals will use it to a certain amount free of any charge."[58]

For his many eccentricities, haughty ways, and his decision to work outside the corporate structure of technological research, Tesla's peers began to write him out of the story of electrical development. As late as 1956, an engineer who had known Tesla in the early days remarked that after Tesla's laboratory in Manhattan burned down in 1895, "he was getting odd. His ideas had become very visiinary at that time and I regarded him as rather unbalanced mentally. . . . I do not think his mind was ever perfcetly balanced and that is why I ceased seeing much of him. I was afraid of his going crazy at any time. Fortintely this did not take place and he only becmae very queer and impossible. . . . In other words he was a crank genius" [spelling in original].[59]

In Pursuit of "Entity *X*"

Even in the years of Edison's waning creativity and Tesla's long fall from grace, the American public had not tired of the metaphor of "scientist as wizard." The next electrical worker in line for canonization was Charles Proteus Steinmetz. In 1888, a youthful Steinmetz had left his technical studies and fled Germany to Zurich to avoid possible imprisonment for his involvement with a utopian socialist group. He emigrated to the United States in 1889, the year after Tesla had presented his paper on the polyphase motor to the engineering community. General Electric eventually employed Steinmetz in 1892.

Steinmetz made his name as a theoretician able to bridge the academic world of science and the grittier craft industry of electrical engineering. One of his first assignments at GE was to find ways to work around Westinghouse's Tesla patents on alternating current.[60] Such a start in his career ensured his later tendency to deny Tesla's importance to the history of electricity. Steinmetz was the first to codify the mathematical theory of Tesla's polyphase engine. He worked on the mathematics of the magnetic interference of inductance in transmission and hysteresis in transformers and generators. He also thoroughly developed the theory of alternating current for engineering use.[61] By 1902 Steinmetz had been elected president of the American Institute of Electrical Engineers.

Steinmetz's prominence in the engineering field was genuine, but GE's publicity department helped engineer his eventual fame as a "wizard." Steinmetz was a genial man whose humanity and clear thinking shone out in the popular writings he indulged in during the 1910s and 1920s. His status, despite his physical disability as a hunchback, made him a fascinating human interest story. His personal lifestyle was full of charming eccentricities, his socialist yet pro-corporate views made him a curiosity, and he was articulate and spontaneous in interviews, making him, along with Edison, a favorite of the press. Further, GE pressed him forward, as his biographer Ronald Kline has argued, to create a genial and individualistic face for a vast corporation facing ongoing antitrust litigation.

Most important, Steinmetz could not be accused of being a crank. If anything, he was too conservative in his engineering views to please his cohorts at GE. He was the anti-Tesla. He first rose to prominence about the same time that Tesla's world system crumbled after Marconi's successes with radio. Steinmetz was profiled for *Success* magazine in 1903. A 1904 article in *World's Work* outlined a rags-to-riches scenario in its subtitle, "His Rapid Advance from Poor German Student to an American Industrial Leader . . . [and] One of the Greatest Inventors in the World."[62] Many other such stories followed. In 1911, a *New York Times* reporter listed Steinmetz's attributes as follows: "He is not one of the limelight prophets; he makes no dazzling predictions; he announces no startling inventions to be brought forth on a morrow which never comes. He is of that other variety of scientist, the kind that do things."[63] In the ensuing interview, Steinmetz showed little sympathy for Tesla's predictions of wireless power transmission or electrical weaponry, labeling them "trash." Throughout the late 1910s and early 1920s, Steinmetz was profiled frequently, and he also contributed articles to popular magazines defending a mixture of socialism with American corporate culture. His articles described a utopian future of virtually free hydroelectric power, four-hour workdays, smokeless cities, and electric automobiles in every basement. He also insisted that science and religion could be compatible, though best kept in separate spheres. Compared with the wilder effusions of Edison and Tesla, Steinmetz's pronouncements were quite tepid.

Yet, like Tesla, Steinmetz could be promoted as a lovable eccentric. He bicycled from his Schenectady home to GE headquarters daily. As a young man he and his roommate Ernst Berg kept pet cranes, owls, crows, alligators, and gila monsters, maintained a greenhouse for orchids, and called their Saturday night poker game the "Society for the Adjustment of Differences in Salaries," with Steinmetz elected permanent president. Berg married and moved to another city. A decent but lonely man, in 1906 Steinmetz adopted one of his young protégés, the engineer Joseph L. R. Hayden. Steinmetz convinced Hayden and his bride to move into his large house and then helped raise their three children as their official grandfather. A lovable eccentric, Steinmetz cared little for money. Steinmetz's salary was trumpeted at $100,000 a year, when, in actuality, he had no set salary but simply asked for checks when he needed them for his research labs, personal needs, or home laboratory. He was also a bit of a prankster. Despite GE's nonsmoking policies, for example, Steinmetz was a frequent smoker of long, thin cigars. Newspaper reporters were also delighted to interview Steinmetz at his idyllic country retreat on the Mohawk River. A frequently distributed GE publicity photograph depicted Steinmetz, bearded, spectacled, hard at work on mathematical calculations, while leaning over the center thwart of a canoe floating near the shore of Camp Mohawk.

Even his socialism was nonthreatening. In 1919, at the height of the Red Scare in the United States, when the government was destroying the radical Industrial Workers of the World (IWW) union and deporting radicals, Steinmetz penned an article, "The Bolshevists Won't Get You—But You've Got to Watch Out!" In it, he argued that American capitalism was healthy and rarely exploitative. However, there were "industrial plague spots" that needed to be cured. Bolshevism would die out naturally if industries provided a decent standard of living for workers. He proposed a vague system of "industrial cooperation" that would end cutthroat competition and bring down costs for industries. "Industry," he said, "must be permitted to organize for economy, efficiency, and social responsibility."[64] To do so would require either socialized industries, or what he admitted might not appeal to American democratic impulses—a benign fascist state, controlling all industries, similar to that which Bellamy had earlier envisioned.[65] Steinmetz's politics had a more practical side; though childless, he was a longtime school board member in Schenectady and occasionally took over mayoral duties when the socialist mayor George Lunn was out of town.[66]

Steinmetz reached the peak of his fame in the early 1920s. Newspapers adopted him as their expert on the role of science in modern society. For eager reporters he explained the theory of relativity, discussed science and religion, and evaluated the future of science and technology.[67] GE was in the midst of antitrust litigation, and public relations expert Bruce Barton, who took over GE's advertising in 1922, saw great value in further promoting Steinmetz. Rather than focus on his theoretical work, GE realized that Steinmetz's work on lightning arresters, devices to protect power lines from lightning strikes, would make for a simpler narrative. Finally the mathematician and theorist could be cast as an inventor. Barton arranged a photo-op visit between Steinmetz and Edison, and the older inventor came to look at the laboratory in which Steinmetz created "artificial lightning." The *New York Times* ran an article titled "Modern Jove Hurls Lightning at Will—Million Horse Power Forked Tongues Crackle and Flash in Laboratory."[68] Others described the "hunchback taming nature" or "Little Wizard with Big Brain."[69]

Steinmetz's rise to fame in the 1920s coincided with public fascination with the apparent conflict between science and religion. Scientists of the 1920s frequently wrote articles on religion. The public, thrilled yet bewildered by Einstein's theories and their destabilizing consequences to basic categories of thought, and intrigued at the fundamentalist Christian challenge to Darwinism, was turning to new authorities for guidance. In 1923, the prominent physicist Robert Millikan arranged a "Joint Statement upon the Relations of Science and Religion," signed by scientists, religious figures, and statesmen, which insisted that there was no inevitable conflict between these spheres.[70]

In 1925, the issue of science's conflict with religion became even more prominent when William Jennings Bryan squared off against Clarence Darrow in Dayton, Tennessee, to argue whether the state had the right to ban the teaching of Darwin's theory of evolution in public schools. While the novel *Inherit the Wind* (1955) depicted Bryan as a simpleton, religious fanatic, and enemy of democracy, historian Edward Larson has pointed out that Bryan participated in the trial as a natural extension of his populist beliefs. To Bryan, scientific elites should not dictate unpalatable theory to the people, but rather, the people should have a say in the shaping of that knowledge. Larson also acknowledged that the trial opened a historic rift between Christian fundamentalists and Christian modernists who did not insist on biblical literalism.[71]

To this developing argument, Millikan offered a modernist perspective. Fundamentalists, he argued, were foolishly trying to turn the Bible into a science textbook. To rely on the Bible as a guide to nature was a losing proposition that Saint Augustine had decried as early as the sixth century. Any intricacy of the natural world was a "scientific matter with which religion as such has nothing whatever to do, and which should not have given it the slightest concern."[72] Descriptions of the natural world should be left to scientists, while religion must fulfill its function of creating a moral society. Religion, according to Millikan, was "the great dynamo for injecting into human society the sense of social responsibility, the spirit of altruism, of service, of brotherly love, of Christ-likeness, and of eliminating, as far as possible, the spirit of greed and self seeking."[73]

Millikan's model of the perfect relationship between science and religion was similar to the "separate spheres" model of gender relations in the nineteenth century. That model let men go out into the world of commerce and power and corruption, while women were to remain at home to inject virtue into their husbands, their families, and the body politic. Millikan's model substitutes "science" for "men" and "religion" for "women." Accordingly, science goes out questing into the world of nature and matter, while religion stays at home. Held to its proper sphere, the feminine religion was to avoid statements about the natural realm, and instead stand ready to purify and add moral fiber to society and the scientific project. Though there need be no conflict between these spheres, both could benefit from cooperation. "Science without religion obviously may become a curse, rather than a blessing to mankind, but science dominated by the spirit of religion is the key to progress and the hope of the future."[74] Millikan also decreased the tension between evolution and science with the dubious argument that evolution offered a mechanism for bringing into arrangements of matter "higher and higher qualities."[75] Willfully misreading Darwin's notion of random variations, Millikan insisted that nature worked by "law" not "chance."[76]

In his writings on the relationship between science and religion, Steinmetz generally followed Millikan's "no conflict" argument. One reporter paraphrased Steinmetz as saying that religion was a "very real power" that was not "measurable in watts or volts" and "beyond the reach of the chemist or biologist."[77] Steinmetz based this separation on epistemology, not moral imperative. Human consciousness with its subjective basis could not even flawlessly determine immutable physical laws. Nor could it pursue with precision spiritual "ultimates." Science and religion relied on "different and unrelated activities of the human mind."[78] That much could lie beyond the cognitive limits that bounded rationality was entirely acceptable to Steinmetz. However, he hinted that to attempt to step outside such limits would result in nonsense; he noted that science contented itself with studying the "finite" and a provisional version of the real while our minds yearned for an "infinite" that was ultimately unknowable.[79] When addressing the conflict between religion and science, Steinmetz offered philosophical rather than religious speculations.

To establish the limitations with which the human mind fashions a vision of the world, Steinmetz paraphrased from Kant's *Critique of Pure Reason*. Both "space" and "time" were categories not to be found in nature but in mind. To explain this, Steinmetz provided several fascinating thought experiments, which read like brief ventures into science fiction. For example, Steinmetz considered how human perceptions would be remolded if the human sense of time were one hundred thousand times faster or slower. In the world of slower sensations, the world of phenomena would appear to speed up one hundred thousand times. "Much of nature, all moving things, would be invisible to us. If I moved my arm, it would disappear to reappear again when I held it still. . . . The vanishing and the appearance of objects would be common occurrences in nature; and we should speak of 'vanishing' and 'appearing,' instead of 'moving' and 'stopping.' . . . Nature and its laws would appear to us very different."[80]

Many writers took the occasion of Steinmetz's death in 1923 as suitable material for moralizing. He was much eulogized, with poetic tributes treating him in a saintly manner. For example, one poem included the lines "We, whom he daily walked among, / Wondered that godlike head and majesty of brow / Were bound so meanly in flesh."[81] His death prompted another writer to criticize the still-current fad of eugenics, commenting that "strict eugenists would doubtless have cut off at birth the life of the deformed little German immigrant" and then went on to praise him for his genius and humanity.[82]

Novelist John Dos Passos relied on Steinmetz's death as an opportunity to critique big business. In *The Forty-second Parallel* (1937), Dos Passos described Steinmetz as a great man turned into a "pet" by General Electric. Dos Passos then shifted the "wizard" metaphor to make Steinmetz into GE's "parlor magician." Dos Passos wrote, "and the publicity department poured oily stories into the ears of the American public every Sunday and Steinmetz became the little parlor magician, / who made a toy thunderstorm in his laboratory." The engineer, Dos Passos concluded, "was the most valuable piece of apparatus General Electric had / until he wore out and died."[83]

Dos Passos was relying on more than poetic license when he insisted that GE sought to promote its research in terms of parlor magic. During the 1920s, GE's public relations department hired journalist Floyd Gibbons to give ten-minute radio broadcasts describing the research work being done in GE's research laboratory in Schenectady. Gibbons dubbed the research facility "The House of Magic."[84] After Steinmetz's death, for the 1933 Century of Progress Fair in Chicago, GE hired a magician to perform in its "House of Magic" in which they presented the results of their research laboratory's work. And for the 1939 World's Fair, GE placed a hundred-foot-tall stainless steel thunderbolt outside Steinmetz Hall. Inside, the exhibit included two generators, separated by thirty feet, which sent ten-million-volt bolts across the gap.

General Electric's use of Steinmetz as a figurehead makes a convincing case for corporate America appropriating the "science as magic" metaphor to amuse a passive, simpleminded public. The image of the parlor magician was not as threatening as that of a Promethean scientist who used all of nature as his or her workshop. Yet, this metaphor was not merely the concoction of public relations men but the end point of a longer cultural dialogue with many participants. William Hammer had been trying to impress a group of his friends when he created his "Electrical Diablerie" in Newark in 1883. The electrical industry had early on seized on the image of the Goddess of Electricity to suggest the awesome yet civilizing power of electricity. The float "Elektra" of 1892 suggested electricity was a heavenly force that had triumphed over and harnessed earthly dragons.

The appearance of the metaphor of the research scientist as parlor magician rather than as sorcerer also suggests that the issue of the heavenly or diabolical potential of technology had momentarily calmed in the 1920s and 1930s. Electricity, no longer a great source of wonder, had become normalized and domesticated. No longer need electricians be likened to Promethean figures who subjugated nature, but rather to stage magicians—technically capable performers who performed wonders using trick mechanisms.

The public fascination with scientists' pronouncements about religion in the 1920s suggests that if no longer regarded as "wizards," the public looked to scientists as custodians of new truths and as experts on final questions. Along with the controversy over evolution, interest in Spiritualism also flourished anew in the 1920s. While Tesla refused to address what he regarded as "superstition," both Edison and Steinmetz were willing to consider the resurgence of interest in Spiritualism. One of the primary goals of Spiritualism was to assure its adherents that the soul was immortal. While Tesla doggedly maintained his model of the human mind or consciousness as a simple mechanical arrangement, Edison and Steinmetz were eager to speculate about the nature of consciousness and the existence of the soul.

In his guise of philosopher, Edison's far from obvious conclusion was that it was foolish to talk of a single soul, as each person was a multiplicity. In 1910 he could write, "We are not individuals any more than a great city is an individual. If you cut your finger and it bleeds, you lose cells. They are the individuals."[85] Quite possibly, Edison was struck by this model because of his own experience as the corporate-entity "Edison"—he was both an individual and a group—since the productions and inventions that his entire team of researchers helped develop ultimately were labeled as his creations alone. To justify his vision of the multiplicity of the individual, Edison referred to medical researchers at Rockefeller Center who had managed to keep the organs of a chicken alive long after the animal itself had died, suggesting the life force could inhabit separate organs even when no longer in the organism as a whole.[86]

Edison later began to refine his theory to argue that millions of invisible "entities"—an entire swarm—gave each individual its life and shaped and directed life processes. He suspected that upon the death of the larger organism these indestructible atoms dispersed and entered new combinations, so destroying the former personality. By the 1920s, Edison reversed himself and thought there was some possibility that when, after the death of an organism, this swarm of minute "entities"—which he now believed were concentrated in the brain's Broca cells—"goes out into space . . . [it] keeps on, enters into another or last cycle of life and is immortal."[87]

Though unimpressed by what he derided as the "childish" methods of Spiritualists, with their rapping tables and Ouija boards, and even inclined to suspect they might be deluding themselves under a form of self-hypnosis, Edison thought there might be some basis to Spiritualist belief. A *Scientific American* article from 1920 confirmed that Edison, disgusted with the crude methods of the Spiritualists, was working on a sensitive device through which these entities, if they wished, could communicate with the living. He concluded the interview by saying, "I do hope that our personality survives. If it does, then my apparatus ought to be of some use."[88] Edison's interviewers returned to the notion that electricity was an etheric substance in between the realms of matter and spirit. Likewise, Edison suspected that if any spirits were to speak, they might be former "telegraphers or scientists, or others thoroughly understanding the use of delicate instruments and electric currents."[89]

Steinmetz, too, attempted to forge a scientific approach to the spiritual realm. He believed that consciousness defied a crude materialistic explanation. In a *Harper's* article, "Science and Religion," he toyed with the possibility that a form of mind, or what he called "entity *X*," might pervade the universe. Mind, or entity *X*, he argued, might be a quality or force quite separate from both matter or energy. Just as chemists had not included energy in equations until the late nineteenth century, Steinmetz queried whether entity *X* might one day be incorporated in equations along with matter and energy to explain thought processes. And if found there, then this same entity *X* might pervade all nature; the concentrations, however, would likely be much lower than in the human mind, making it undetectable to our crude senses.

Steinmetz insisted these speculations were not meant to encourage belief in "spiritism or other pseudoscience."[90] He was specifically disgusted by the vogue in séances for "materializations" in which mediums seemed to vomit up weird ectoplasm or other material as proof of visitation by spirits. Yet, his point of departure—an exploration of consciousness—was the launching point for many defenses of Spiritualism and spirituality. William James and other members of the Society for Psychical Research had long argued that transcendent states of mind might indeed bring some individuals in contact with spiritual realms. Mind, they argued, simply could not be reduced to matter, nor even properly bounded by the concepts of space or time; according to this view, for example, telepathy was a simple case of the transcendent mind—or subconscious mind—that all humanity shared making local links.

The question of the existence of the soul and the prompting to communicate with the spirit realm were far from trivial topics to citizens of the turn of the century. These concerns were in the mainstream. Nor did scientists and engineers shy away from speculations about such matters. These questions were continuously addressed in popular culture, not only in the parlors of Spiritualists and in popular magazines like *Harper's,* or in the addresses of inventors, but also in the tawdry wonder shows of "mystic vaudeville." There, hypnotists, mind readers, and anti-Spiritualist magicians introduced the public to mysteries and marvels, made promises that secrets would be revealed, and chased down Steinmetz's mysterious entity *X* while exploring hypnotic trances, telepathy, and other altered states of mind.

PART II

MYSTIC VAUDEVILLE

CHAPTER THREE

The Hypnotist

Before Professor Leonidas, a turn-of-the-twentieth-century performer, traveled to a new town, he would send his advance man ahead to install a coffin in a pharmacy window with a placard announcing an upcoming hypnotic show. Upon arrival, Leonidas would hypnotize his youthful assistant, sew his lips shut, set him in the coffin in the pharmacy window, and promise the crowd to revive the subject, or "window sleeper," on stage the following evening. During this same era, another stage hypnotist, Walford Bodie, M.D., offered a modernized gothic touch to his stage act: the electric chair. After describing the horrors of this new American form of execution, the dashing, caped hypnotist would find a volunteer in the audience, strap him in the chair, mesmerize him to "protect" him from the high voltage, and then throw a switch. The subject would twitch, scream, and otherwise suffer before the appalled yet fascinated crowd.

Bodie's use of the electric chair and Leonidas's use of the coffin as a prop—and the pharmacy window as a stage—suggest the stage hypnotist's conflicted agenda at the turn of the century. They presented themselves in "gothic" and "scientific" modes that often clashed. Leonidas's choice of the pharmacy window for his assistant's coffin display, however, reveals how the gothic and the scientific could be juxtaposed: at that time such establishments offered not only the "genuine" drugs that orthodox doctors might prescribe, but also the many flamboyant cures of the patent medicine industry.

Stage hypnotists, like patent medicine hawkers, promoted fantastic science and therapeutic wonders. In their promotions, hypnotists relied on two models of healing: one based on a scientific or psychological depiction of the mind, the second based on a religious vision of the mind's

—or soul's—capacities. They also added a gothic flourish, likening the hypnotic trance to death, and suggesting that their own abilities to "raise the dead" were a form of necromancy. This uneasy alliance of science with magic appealed to audience nostalgia, but hindered stage hypnotists' attempts to mesh with the late nineteenth-century interest in reform.

Early in the nineteenth century, mesmerists had highlighted the exalted state of consciousness reached by their somnambulists—or trance subjects. The somnambulist's trance suggested the perfectibility and spiritual potential of the individual. Mesmerism, it seemed, might be a tool for perfecting society. Yet by the turn of the twentieth century, diminished public belief in the marvels of the trance forced stage hypnotists like Bodie and Leonidas to make grotesque, humorous displays of their subjects, establishing the "operator's" own power and control. Paradoxically, such demonstrations intrigued many progressives as a tool for establishing order. But the stage hypnotists' dramas of power and enslavement also alarmed guardians of the public virtue, and turn-of-the-century stage hypnotists became a target of progressive reform when leading citizens sought to ban stage hypnotic shows.

Historians distinguish the Progressive Era from earlier reform movements by noting that the progressives rarely adhered to the utopian ideals and religious zealotry of the early nineteenth century. No longer concerned with ushering in a new millennium, the progressives believed that scientific methods and organization could moderate corruption in business and government. At the same time that progressives led important efforts to end corrupt machine politics, to curtail the excesses of monopoly capitalism, and to improve living conditions for the impoverished, they also campaigned against vice, and could target such breeding grounds for "white slavery" as dance halls and soda fountains, or condemn dime novels for inciting readers to acts of violence. The campaign against hypnotism was launched within the context of these other progressive battles—large and small, sublime and ridiculous.

Scholars such as Robert Fuller, Alan Gauld, and Alison Winter have examined the cultural significance of mesmerism in the early and mid–nineteenth century, but so far little attention has been given to the stage acts of the turn of the twentieth century.[1] Rather than share the progressives' distaste for these "degrading exhibitions" or disregard them as negligible, an examination of these performances can provide a fresh window into the era. As the rhetoric of stage hypnosis shifted from an early nineteenth-century perfectionist model to a turn-of-the-century progressive model, a campaign against stage hypnotism in America became inevitable. The strategies of the grotesque and "scientific occult," as well as the recourse to the vitalist worldview that variety and dime museum hypnotists employed at the turn of the century, while intriguing audiences, provoked hostile reactions and demands for reform.

Marvelous Somnambules

Mesmerism, derived from Anton Mesmer's theory of animal magnetism, evolved from a model based in physics and medicine to an early precursor to psychology. Anton Mesmer, an eighteenth-century Viennese physician, advanced the argument that the universe was suffused with "animal magnetism," a weightless, subtle fluid like electricity that need only be conducted into the suffering to cure them. Other mesmerists who followed explored the connection between mesmerism and electricity. One pamphleteer wrote in 1843, "Neurologists tell us that there is a fluid or *nerve-orer* which passes from all parts of the body more or less. . . . This *nerve orrer* [*sic*] is supposed by some to be a finer substance than Electricity. I think myself that probily [*sic*], it is on this ground, when an individual is charged with Electricity by a galvanic battery very hard, he feels some pains shooting through his system. It appears that it strains the small fibers through which it passes. . . . It is not so with the fluid, as *nerve orer* which passes from the Magnetisor, or the Magnetisee. It appears to harmonize with nature—giving no pain."[2] Such pamphleteers felt confident that "animal magnetism" was a genuine force, akin to electricity yet of finer, subtler stuff. Others attempted to deduce the properties and laws of the hypnotic medium. In 1838, for example, after proposing that hypnotism involved a "ray," English physician John Eliotson and science lecturer Dionysius Lardner successfully hypnotized subjects via mirrors to argue that hypnotism followed the laws of reflection.[3]

Largely because of the experiments of Mesmer's disciple, the Marquis de Puységur, however, interest shifted from the possible physical basis of mesmerism to its psychological implications. The debate over whether "animal magnetism" involved an actual "fluid" became less relevant than studies of mesmerism's effects. Puységur insisted that mesmerism provided a new model of the mind. Puységur believed that after establishing a magnetic rapport with some subjects, he could then transmit his thoughts and will to them. More significant from a medical point of view, Puységur also believed "that a somnambulist could see his own insides while being mesmerized, that he could diagnose his sickness [or that of others] and predict the day of his recovery, that he could even communicate with dead or distant persons."[4] Whatever its explanation, mesmerism had otherworldly implications.

In the mid-1830s Charles Poyen left France to give the first lectures about mesmerism in the United States. During his performances, Poyen magnetized volunteers and cured illnesses, and his "somnambule"—or trance subject—demonstrated the progressively more lucid—even clairvoyant—states of mind revealed under hypnosis. In these wonder shows, the somnambule often diagnosed the ailments of audience members and prescribed treatments. Though the mesmerist sought to impress the audience with his own grave powers, the somnambulist also was a featured player in the act. These subjects, like the mysterious "Veiled Lady" who traveled with a mesmerist in Nathaniel Hawthorne's 1852 novel *The Blithedale Romance,* could become objects of public interest and speculation.

Poyen's tour prompted an American fascination with animal magnetism and mesmerism. Soon dozens of visiting and home-grown mesmerists were traveling with their somnambules and giving demonstrations and offering cures in theaters, rented halls, and the homes of the wealthy. By 1843, according to one estimate, as many as two hundred storefront magnetizers worked in Boston.[5] Robert C. Fuller argued that by demonstrating hidden powers of the mind and suggesting a scientific basis for quasi-mystical experiences, including conversion experiences, this movement appealed to the revivalist climate in the United States and dovetailed with popular beliefs in the perfectibility of man and society.[6] Mesmerism also legitimized the beliefs of the followers of the Swedish mystic Emanuel Swedenborg and the beliefs of the Spiritualists—by offering an apparently scientific explanation for how humans could contact "higher" realms of spirit.

The religious revivals of the early nineteenth century encouraged hopes that a new millennium was dawning, which would see the perfecting of humanity. The works of John Bovee Dods provide a good example of how mesmerism could be adapted to the perfectionist mindset. Dods was a Massachusetts minister and an eloquent antebellum defender of mesmerism. He championed a perfectionist notion of mesmerism, employing ringing prose similar to that of the transcendentalists of his home state. Dods likened mesmerism to "mental electricity" and employed this tool in his investigations to uncover aspects of what he termed "electrical psychology." During a speech to the U.S. Senate in 1850, Dods remarked, "Man is intellectually a progressive being. Though confined to a narrow circumference of space, and chained to this earth, which is but a small part of the unbounded universe, yet as his mind wears the stamp of original greatness, he is nevertheless capable of extending his researches far beyond the boundaries of this globe. His mind is capable of ceaseless development of its powers."[7] Mesmerism, or electrical psychology, was one avenue by which Americans could seek to unfold the dynamic potential of mind.

The American medical community did not give mesmerism as hearty a welcome as the U.S. Senate offered the golden-tongued Dods. In 1838, physician David Meredith Reese wrote *Humbugs of New-York* and dedicated one chapter to mesmerism, which he called "the present reigning humbug in the United States." Of the somnambules' supposed clairvoyant abilities, Reese sarcastically remarked that one somnambule, deep in a trance, was asked to describe a stranger's house and said, "it was built of brick, that it had a front door, that there was a table and two chairs in the hall, a carpet on the floor, and on being asked if she saw anything else, she discovered a lamp, a back-door, or a staircase, with divers other similar wonders." [8]

Scare literature of the mid–nineteenth century also featured evil mesmerists who seduced innocents. Timothy Shay's *Agnes; or, the Possessed. A Revelation of Mesmerism* (1848) includes both a preface and afterword warning readers of the dangers of mesmerism, which has a "disorderly, and, therefore, evil origin." [9] In the novel a pretty and somewhat adventurous New England girl, Agnes, allows a visiting French mesmerist, Monsieur Florien, to magnetize her in order to have a tooth painlessly pulled. Soon she is under his spell and leaves her fiancé Ralph Percival to meet M. Florien in Boston for more treatments. There Florien and his accomplices Dr. T—— and the doctor's wife abduct Agnes because they feel she is an exceptional medium and wish to further their experiments in clairvoyance and extrasensory perception. After discovering that she was "still in the power of that arch-fiend of hell!" Percival tracks the knowledge-mad trio from Boston to New York, but they continually elude him. Agnes, however, finds her own way to salvation. One day, after Dr. T—— fails to magnetize her, she exclaims, "I am free. I had a friend of whom you knew nothing. I waited long, too long, before I called upon Him; but, when I did call, He came instantly to my aid. That friend is God! . . . Magician, I defy you in the name of God! You cannot stand against him!" [10] Percival then arrives, brandishing a pistol to complete the rescue, and Dr. T——'s wife informs Percival and Agnes that while a captive the girl remained "pure" thanks to her watchful eye. The last paragraph includes this admonition, "To all, the writer would say: Beware of mesmerism! Its origin is in perverted order, and it cannot, therefore, have a good influence." [11]

Despite such complaints from medical men and evangelical writers, it was not until the Progressive Era that concerns with the dangers of hypnotism led to legal efforts to ban the art. The orthodox medical establishment took periodic interest in hypnosis during the nineteenth century—first when it appeared to offer a promising form of anesthetic, and later when clinicians applied hypnotism as a treatment for mental illness. Beginning in the 1870s, the French neurologist Jean Martin Charcot used hypnotism to treat hysteric patients in his clinic outside Paris and theorized that the hypnotic trance represented a pathological state similar to hysteria. The work of Charcot's school provided a foundation for the psychoanalytic theory of the subconscious. It also was largely a result of Charcot's pathology theory of hypnotism that reformers became convinced that lay hypnotists could damage subjects. By 1900, however, most researchers of hypnotism had shifted allegiance from Charcot to his French rival Hippolyte Bernheim, who argued that the hypnotic trance represented a unique state of consciousness, but it was largely one of alert imagination, openness to suggestion, and desire to please the hypnotist.

With such redefinition, hypnotism became less of a symbolic threat to medical orthodoxy. Part of the fascination with mesmerism when it first arrived in America stemmed from the fact that mesmerism subverted the physician's traditional authority and heightened the patient's individualism and freedom. During the 1840s, colleagues of a leading British physician, John Elliotson, forced him to resign his post at University College Hospital when the working-class subjects of his popular mesmeric demonstrations in the medical theater began to present themselves in an uncontrolled manner, often deriding Elliotson during his lectures.[12] Such an inverted power relationship, which would clearly make modern physicians uncomfortable, also is demonstrated in the pamphlet of the Kennedy Brothers, late nineteenth-century performers who straddled the historical divide between the eras of mesmerism and hypnotism.[13] After a first session, they urged the practitioner to ask his subject how he felt, and then to ask, "First: Whether your manner of procedure agrees with him, and if he can point out a better; Second, whether he can think of anything that would be useful to say or advise . . . [and] whether he is able to look into your system, or his own, and say anything concerning them. . . . His answers to these questions will teach you how to interrogate or experiment with your subject, or whether you should at all."[14]

As hypnotism branched into the early twentieth-century fads for autosuggestion and positive thinking, a few researchers left open the issue of whether the hypnotic state itself might have the otherworldly attributes mesmerists once had claimed. In 1890, William James reported that the sensory powers of hypnotized subjects often increased greatly and argued, along with other psychic researchers, that the trance might indeed provide access to submerged streams of consciousness and, potentially, to the spirit realm.[15]

Dime Museum Scientists

Progressive Era stage hypnotists plied the marvels of hypnotism at a time when its marvelous nature, for the greater public, had diminished. Rather than insist on hypnotism's marvelous features, or the wonders of the trance, showmen instead made a spectacle of grotesquerie and mimicked the procedures of science. Hypnotists rarely made the better vaudeville circuits, and were usually featured instead in dime museums or traveled on circuits of their own making. Harry Houdini, for example, before gaining fame as an escape artist, performed briefly in circuses and dime museums as a hypnotist under the name Professor Murat.[16] Unlike Houdini, many turn-of-the-century hypnotists remained behind in the dime museum's curio hall—a room with small stages or platforms provided for the various attractions. There, a lecturer who used the title of "Professor" or "Doctor" introduced the hypnotist and other platform attractions: whether human oddities, entertainers, or educational artifacts. One week's curio attractions for Worth's Family Museum in New York City in 1891 included Mlle. Agnes Charcot, a female hypnotist (who borrowed her surname from the French neurologist); Professor Dufrane, "the anvil man," who allowed big stones to be broken on his chest; and Cunningham's Samoan Warriors.[17] And in 1892, a Philadelphia dime museum's advertisement could boast the "First Appearance in the country of THE HUNCHBACK PONY," "The WONDERFUL DE GRAY BROTHERS. Hypnotic Marvels," and "OTHER STRANGE CURIOSITIES."[18]

Robert Bogdan has argued that curio halls promoted freaks in both an "exotic" and an "aggrandized" mode. The exotic freak was presented as savage or degenerate, whereas the aggrandized oddity—often the same person—was introduced as a finely dressed and accomplished gentleman or lady whose talents had helped him or her overcome adversity. The aggrandized mode catered to the audience's sense of moral uplift.[19] In the curio hall, the Wonderful De Gray Brothers, hypnotic marvels, might outmuscle hunchbacked ponies—whether living samples or stuffed exhibits—but rank lower in status and pay than such aggrandized oddities as the Martin Sisters, the Beautiful Albino Twins—listed for the previous week at the same Philadelphia theater.

Like the other oddities of the curio hall, the hypnotist was promoting images of the "grotesque"—through the odd behavior he or she induced in trance subjects. The grotesque spectacle could generate reactions of both amusement and horror, as well as a moment of "wonder" as the spectator attempted to file the unexpected spectacle into comfortable cognitive categories. The hypnotic show created a temporary freak show, revealing the thin boundary that separated the norm from the bizarre. Such a connection was made explicit in the *How to Hypnotize* pamphlets of the era, which recommended that the hypnotist persuade a volunteer that he was running a sideshow, and this delusional stage barker would then regale the audience with his salty patter as he described the imaginary freaks around him.[20] In the dime museum milieu, the hypnotists' behavioral version of the "grotesque" was competing with the embodied grotesquerie of a bearded lady, a Wild Man, or a Samoan Warrior. The curio hall hypnotist might also be competing with "wholesome" displays of female beauty, beautiful baby pageants, or vigorous cowboys and lively acrobats in the main theater.[21] Having little choice, hypnotists accepted a middling rank even in the lowly curio hall, or created their own tours, visiting fraternal organizations and community clubs.

Hypnotists attempted to boost their appeal—or in Bogdan's terms, make themselves into "aggrandized" freaks—by presenting themselves as gentlemen or gentlewomen of science. Itinerant mesmerists had first employed this strategy. For example, the narrator of Hawthorne's *The Blithedale Romance* (1852) remarks, "Now-a-days, in the management of his 'subject,' 'clairvoyant,' or 'medium,' the exhibitor affects the simplicity and openness of scientific experiment. . . . Twelve or fifteen years ago, on the contrary, all the arts of mysterious arrangement . . . were made available in order to set the apparent miracle in the strongest attitude of opposition to ordinary facts."[22] A pamphlet from 1900 even argued that hypnotism was superior to other sciences, as it dealt with subtleties of the mind rather than the laws of crude matter. The pamphleteer provided a sample speech for exhibitors that argued, "I call it a science. . . . It deals with the invisible but living mind, the thinking part of our nature, while the other sciences have their application only to lifeless matter. For that reason alone we must believe its applications are boundless."[23]

Stage hypnotists maintained this strategy well into the twentieth century. According to vaudeville chronicler Joe Laurie, hypnotists of the 1920s and 1930s continued to stress science, noting they "never referred to it [hypnotism] as an 'act,' but always billed it as a 'scientific demonstration.'" Of a favorite showman, Laurie wrote, "Pauline was the tops; he had a fine personality, spoke like an actor–doctor, always referred to 'this experiment, which I performed before the world's greatest scientists.'"[24] Another early twentieth-century performer, the Great Newmann, often advertised his act as "Two Hours of Clean, Psychological Amusement, Scientific Entertainment and Mysterious Fun."[25]

In insisting on a quasi-scientific presentation, hypnotists were neatly accommodating their acts to the worldview the dime museum promoted. By the late 1800s, promoters of dime museum and sideshow acts chose to introduce their human oddities less as extravagant creations and more as medical case studies. This strategy changed the human oddity from the level of simple spectacle to that of an educational exhibit. The dime museum, a showcase for vulgar wonder, offered a funhouse mirror reflection of the scientific and medical world—mimicking rituals such as the experiment and the medical demonstration to gain credibility.

Enter the Master of "Bodic Forces"

In the 1890s British hypnotist Walford Bodie, M.D., pushed the dime museum conventions to their limit. He created a stage spectacle that combined elements of the grotesque, magic, and science by splicing together electrical displays, electrical executions, and the medical marvel of hypnotic healing. Bodie was born Samuel Murphy Brodie in Scotland in 1869. As a child he practiced magic and ventriloquism, and as a young man he worked for the Scottish National Telephone Company and so gained expertise in the then-pioneering field of electricity. His sister married into a theatrical family, and soon after, Brodie helped manage a variety theater. One year later he took on the stage name "Doctor Walford Bodie, M.D." (as he explained to one judge, the initials "M.D." stood for "Merry Devil") and initiated an act that was to make him a frequently copied showman in Europe and America.

To the standard gentleman's garb and demeanor, Bodie added a moustache that would have made Salvador Dali envious: a neat rectangle flanked at each side with two thin, tweaked daggers, upended at forty-five-degree angles. This moustache was striking enough for Charlie Chaplin to imitate on the English variety stage in 1906 when he burlesqued the great hypnotist.[26] Performer J. F. Burrows, who used the stage name Karlyn, in 1912 wrote a book "unmasking" Bodie's act; the frontispiece photo of Burrows, Bodie's "unmasker," shows him with a Bodie moustache and a similar evening coat and hairstyle. Like the era's muckraking journalists, turn-of-the century vaudeville performers often "unmasked" one another—indicative of an ongoing cultural obsession with authenticity.[27]

Walford Bodie, M.D., the "stage electrician" and "bloodless surgeon," was a frequently imitated performer through the 1920s. Rivals even replicated his one-of-a-kind moustache. Performing Arts Collection, Harry Ransom Humanities Research Center, University of Texas at Austin.

Bodie's act relied on spectacles both magical and scientific—on faith healing and on displays of electrical effects. Bodie melded these realms with his pronouncements about the healing force that had once been called animal magnetism but which he preferred to call the "Bodic Force"—also at the heart of the new philosophy of "Bodieism."[28] As he noted in the preface to the *Bodie Book* (1905), "My method of cure, being connected with electricity on the one side and with the mysteries of occult science on the other, could not be explained in a few words." But, sounding much like the earliest mesmerists, he insisted that "the sorcery and supernatural agency of the dark ages have become the scientific facts of today."[29] In step with the Progressive Age, his was a confused gospel of scientific—and religious—miracles, expressed on stage with Leyden jars, condensers, and machines that sparked and caused strong men to quiver, and cripples to throw aside their crutches and promenade.

Bodie's use of electrical devices helped to reinvigorate the somewhat tired hypnotic act and to promote public faith in scientific progress. Bodie's act clearly imitated inventor Nikola Tesla's penchant for bathing himself in 250,000 volts or more of high-frequency (but low-amplitude) current, which made his entire body glow with an aura of electrical flames.[30] Bodie was among the more prominent of the entertainers who brought lesser electrical effects to the variety stage. According to Odell's *Annals of the New York Stage,* dime museums in 1890 also featured such acts as Mattie Lee Price, the Electric Girl; the Electric Three; Barella and the Electric Chair; and La Pierre's Electric Exhibition.

Bodie's staging extended to the theater lobby. He transformed it into a simulacrum of the entry room of a holy shrine where miracles take place, filling it with framed testimonials to cures, as well as displays of crutches and "irons" that the formerly paralyzed or disabled had thrown aside. Such trappings suggest Bodie's insight into what anthropologist Victor Turner later termed the "liminal" aspects of theater—its status as a separate place where stagings might evoke in the spectator a sense of transformation akin to that of a tribal rite of passage.[31] Bodie's electrical apparatus could feed the hopes of spectators who had walked through the lobby and awaited his appearance in the darkened theater.

Bodie's use of the stage as a setting for miracles extended the sense of awe first evoked in the spectators when they passed through the lobby. When the curtain opened for the act, the stage was revealed with Bodie's somber assistants standing near magnificent electrical apparatus, throbbing with lights and sparking to dramatic effect. Bodie said that when he strode out into the limelight and began his explanation of his art and healing abilities and commented that the newspapers liked to attack him for his "modern miracles," a heckler might shout, "Miracles! Oh, oh! D'you work miracles then?"

To such hecklers, itching for a fight, Bodie would respond, calmly, "I said what they CALL miracles."

"Right! Take it! But do you call them miracles?"

"I do not claim to work miracles. . . . But if I did, the very first one I should attempt would be to instill a grain of sense into your head."[32]

As Burrows suggested, since overcoming hecklers helped establish the hypnotist's authority—and in this case prepared the audience for possible miracles—such useful souls were often planted in the audience. The first part of his act was designed to establish the hypnotist's mastery of mysterious forces and convince the audience that miracles would indeed occur. To this end, Bodie would explain the electrical equipment, invite a "committee" on the stage, cause some of them to go in convulsions while gripping the same handles he could hold with no effect (because he was standing on an insulated pad on the stage), make sparks fly between subjects' fingers and the apparatus, and so on.

At the turn of the century, there was much discussion that middle-class males were growing weak and effeminate because of their white-collar work. The "cult of the strenuous life" developed to encourage physical fitness and toughness. Many young men of the patrician class went west to "toughen up" and prove themselves, including Theodore Roosevelt, Frederick Remington, and Owen Wister, who penned numerous articles about the West, and several novels, including the influential *The Virginian* (1911). Social reformers encouraged participation in organized sports such as baseball, football, and gymnastics. Bodybuilding also became an emerging obsession. Recent scholars also have placed the strenuous stage escapes of Harry Houdini within this cult.[33]

In keeping with this turn-of-the-century concern, Bodie, a friend and admirer of Houdini's, boasted he could withstand electrical forces that would stun or kill the average man. In his earlier days he insisted that four thousand to six thousand volts passed through him. Perhaps after learning that Tesla submitted himself to three hundred thousand volts and more of electricity, Bodie continued to up his voltage count while regulating the amperage (or current) to keep the power output at a safely low level.

These demonstrations of Bodie's power followed from romantic notions that genuine electricity and "animal magnetism" were analogous—a theory that had convinced many to submit themselves to electrification as a therapeutic method. The writing of mid–nineteenth century mesmerists, like Dods, who preferred to call their field "electrical biology," reinforced such associations. Even in the late nineteenth century, the Scottish physicist Lord Kelvin was still positing that there was a connection between electricity and the life force, giving credence to electrical therapy.[34] New electrical healing devices abounded. Tesla electrified himself daily as part of a therapeutic regime, while the Sears catalogue peddled electric nerve toners to the public.[35]

At a time when the "body electric" was becoming a reality, Bodie's act and persona made sense. His dazzling stage displays and his apparent ability to cure people with hypnotism and electricity helped to reinforce faith in electricity and progress. Bodie then tested that faith in progress —following his opening sequence of electrical effects, Bodie would dismiss his bedazzled and traumatized stage committee, and his assistants would bring out a replica of the American electric chair. This, too, had been displayed in the lobby. In the 1920s, Bodie began to display the original electric chair used at Auburn Prison in New York to execute Kemmler. His friend Harry Houdini had purchased the chair and sent it as a gift after receiving several pleading letters from Bodie.[36] With the electric chair beside him, Bodie would make a reform-minded speech describing the horrors of this method of execution.

According to Burrows, during this part of the show, the stage hypnotist "thrills the audiences with accounts, more or less imaginary. . . . He tells how . . . one was tortured for half an hour before death released him, and another was practically burned alive."[37] Despite Burrows's sarcasm, early accounts of electrocutions suggest that Bodie did not need to stray far from the truth in describing the horrors of this form of execution. After his speech Bodie would strap an assistant, posing as an audience volunteer, into the chair. Next Bodie hypnotized the subject to "prevent electrocution." After further demonstrations of the chair's functions, Bodie or his assistants turned on the full current, and the subject began to tremble and shake. According to the *Aberdeen Journal,* during one such performance, "the awfulness of execution was borne to the audience. . . . Dr. Bodie watched closely and when the [subject's] face became black the current was switched off. After vigorous slapping, the subject was restored to consciousness."[38]

In Bodie's cunning show, death was followed with symbolic resurrections. To top the thrills of the electric chair mock execution, Bodie moved on to his cures. The hypnotist and his assistants had earlier auditioned the town's disabled population and chosen the best candidates for a stage cure. During the performance, his assistants carried the disabled people on stage. Once seated, the patients were "hypnotized" by Bodie, who made a few passes in front of them. Next he manipulated and massaged their damaged limbs to "remove adhesions"—or old scar tissue that blocked motion. To complete this "bloodless surgery," Bodie applied electrical current. The patients then were revived from their slumber, and to the tears and astonishment of the audience and appropriate music from the pit orchestra, they walked off stage unassisted.

Bodie, like Katterfelto and Charles Came before him, was running a medicine show. He auditioned the disabled population in towns, chose some for his acts, and sold those he rejected useless liniments. His act made the most of theater's "liminal" aspect. His hypnotic and technological miracles prepared his audience for the cures, just as tribal

THE ERA.

I LEAD. FOLLOW WHO CAN.

THE MONEY MAGNET. THE RECORD BREAKER. THE GREATEST NOVELTY ON EARTH.

THE RAGE OF LONDON.

Cheered to the Echo Nightly. Greeted with Thunders of Applause. Doctors Amazed, Scientists Staggered.
The Skill of the Scientific World Baffled.

DR. WALFORD BODIE AND HIS PATIENTS.

DR. WALFORD BODIE, M.D.,

THE MODERN MIRACLE WORKER, THE PROTECTOR OF SUFFERING HUMANITY,

who has caused the Greatest Sensation ever known in London by any Artiste. Audiences spellbound, and struck with Awe, Amazement, and Wonder at the doings of this Great and Wonderful Man. Undoubtedly,

THE SENSATION OF THE AGE,

whose will is indomitable and whose power is inexplicable and resistless. As predicted, all London has flocked to see Dr. Bodie, and, owing to his unprecedented and phenomenal success, he has been rebooked for Two Dates at the Britannia Theatre, London. Also the following Halls:—Camberwell Palace, Two Weeks; Hammersmith Varieties, Two Weeks; Canterbury, Two Weeks; Empire, Islington, Two Weeks; Empress, Brixton, Two Weeks; Duchess, Balham, Two Weeks; Euston Palace, Two Weeks; also the London and Collins's, at the Biggest Salary ever paid to any Artiste in London. Surely a record to be proud of. Also booked with my own Combination the whole of the Barrasford Tour, Two Weeks each Town. No Vacant Dates this Year or Next. Now fast booking 1906. Thanks to Managers for offers. To save time and trouble, my actual Lowest Terms are meantime £200 per week for my own Act alone. With Combination in Provinces, Sharing Terms only.

London Agent, Mr GEO. FOSTER.

All letters address to Dr. WALFORD BODIE, M.D., the British Barnum, London's Pride,

THEATRE ROYAL, WOOLWICH.

LOOK OUT. MORE STARTLING REVELATIONS SHORTLY.

This advertisement for Bodie, circa 1910, indicates his miraculous abilities to heal the lame and demonstrates the grandiose claims that led to his court appearances. Performing Arts Collection, Harry Ransom Humanities Research Center, University of Texas at Austin.

shamans offered demonstrations of power during healing ceremonies to build up their audience's expectations. Bodie's apparent ability to kill and then rejuvenate a subject with the electric chair mimicked the death and resurrection patterns common to the shaman's ritual.

Bodie's decision to exhibit himself as a miracle worker led to court challenges and a public humiliation from which he never entirely recovered. The established medical community saw him as a nuisance at best. In 1905 the Medical Defense Union in London sued him for fraudulent claims on his posters and the court gave Bodie a small fine. In 1909, a gullible assistant who had paid Bodie one thousand pounds to learn the medical hypnosis trade sued Bodie because he had only been taught how to fake tricks on stage. The London papers covered the colorful trial and ran articles with headlines such as "Secrets of a Hypnotist—Extraordinary Disclosure of Imposture in Music Hall Entertainments." While his medical degrees and most of his stage work, including cures, were shown to be phony, Bodie did bring in several witnesses who testified to successful cures. But Bodie was found guilty of breach of contract and his assistant awarded one thousand pounds in damages.

The week after the 1909 trial, students rioted at a Bodie performance in Glasgow. No sooner had he stepped on stage than "the fusillade began." Bags of flour, eggs, and red and yellow ochre were thrown at him. A second riot broke out during the same theater run, and Bodie was forced to come out and apologize to the students for calling them "no gentlemen, and a disgrace to the university." After his apologies the students marched, chanting, "Victory is Ours!" And "Bodie, Bodie, quack, quack, quack!"[39]

Bodie's humiliation was an outgrowth of his risky strategy of taunting the medical establishment. Four years prior to the trial and riots, Bodie had dedicated his *Bodie Book* to "British Medical Men" in the hopes that it might "enable them to perform even more efficiently than at present their duty to the millions who turn to them in times of pain and sickness." He also insisted he could always spot a medical man in the audience, diagnosing specimens of this type as "suffering from acute bigotry, brought on by excessive dogmatism."[40]

Following the miraculous cures, the final part of an evening with Bodie would involve a more typical display of stage hypnotism, played out to humiliate subjects and amuse audiences. Electricity was added to this part of the night as when two "lovers" were told to kiss and actual sparks flew between their mouths—a hoary trick first developed in the 1700s. At the climax of this section of the act—true for most hypnotic shows—the hypnotist would take the subjects out of their trances, but one would not respond.

The hypnotist would gravely tell the audience that this subject had reached the deepest stage of hypnosis, catalepsy. He would then walk up to the unfortunate sleeper, make a pass, and say, "Go rigid!" This subject, commonly called a "plank," would then be set across the backs of two chairs and heavy anvils would be placed on his chest and struck with hammers, or the hypnotist would put a boulder on the sleeper's chest and ask a strong volunteer to smash it to pieces with a sledgehammer. Several volunteers might also be encouraged to climb up and stand on the "plank." While standing on top of the subject, the hypnotist might command the sleeper to relax and then go rigid again, riding him down and up like a wave. Ultimately, the "plank" would be revived. The miracle here no longer focused on the subject's powers of mind, but on the subject's body—more specifically the powers of the hypnotist's mind over his subject's body.

The Regulation of "Degrading Exhibitions"

Displays of the hypnotist's power led to public concern. The control of the hypnotist over his subject was explored in George du Maurier's 1894 novel, *Trilby,* which introduced to the world the fictional mesmerist Svengali. Adapted to the stage in America in 1895, it was enormously popular. By 1896 numerous productions were running simultaneously in American theaters. The play describes a Bohemian artists' model with no musical talent who falls under the hypnotic spell of Svengali, a music instructor. When placed in trances she becomes a celebrated concert singer. But when the mesmerist dies, her gift vanishes. While *Trilby* may have helped drum up audiences for stage hypnotists—frequent references to *Trilby* began to appear in their publicity[41]—the play also highlighted a troubling side of the hypnotist's alleged powers and added to the age's concerns about hypnotism's dangers.

While many had once hailed mesmerism as a metaphor for the liberation of spiritual potential, many now easily could criticize hypnotism as a metaphor for exploitative control. Borrowing from the "Svengali" mold, the covers of nineteenth- and twentieth-century pamphlets for hypnotism tend to show a male hypnotist exerting electrical power over a weakening female subject, supporting a popular notion that the hypnotist was imposing his will—and might make a sexual slave of a weaker subject.

According to this line of reasoning, the more feminine and passive the subject, the better. Hypnotists' pamphlets often argue that not only women but also factory workers and soldiers—trained to obey—made

The zigzag rays emanating from the male hypnotist's eyes on this Depression-era pamphlet liken hypnotism to an electrical force that only the gifted could transmit. Women and "the fair" were particularly susceptible.

ideal subjects. One of mesmerism's earliest critics, New York physician David Reese, in 1838 noted that somnambules tended to be "factory girls"—at that time manufacturers also preferred young women as workers, as they were assumed to be more tractable than men. A pamphleteer of the 1880s, curiously attributing a similar docility to Russians, insisted that of all nationalities they were the most easily mesmerized. Natives of tropical climates were also favored. The Kennedy Brothers urged mesmerists to "seek those of lighter eyes and complexion than yourself; it is found exceedingly hard to affect those of darker eyes. . . . Blacks, nevertheless, make capital subjects for exhibiting the physical phenomena."[42] The Kennedy Brothers preferred lighter-skinned subjects to exhibit the "higher" phenomena such as clairvoyance. But an African-American was fine for behavior modification displays. Such notions fit the racist stereotype of the plantation tradition that viewed African-Americans as childlike and simple. Such racist stereotypes continued so that a hypnotist in the mid–twentieth century could still argue that "members of the black race are easiest to hypnotize, probably because their origin is in the torrid zones."[43]

There were at least two African-American hypnotists. For example, Henry "Box" Brown, an abolitionist speaker who escaped from slavery by mailing himself via an express package service from the South to Philadelphia, took up a public speaking career in England in the 1850s that included abolitionism, and demonstrations of mesmerism and electro-biology.[44] His contemporary, H. E. Lewis, also African-American, likewise lectured on electro-biology in 1851 in Britain to great acclaim. Lewis clearly had abolitionist tendencies; during one performance, he induced a Scottish woman to visualize an imaginary tour of America, prompting descriptions of Niagara, Buffalo's backlots, and also her "horrified" description of a slave market in Louisville, Kentucky.[45]

Most pamphleteers did not even propose that a "member of the black race" could be a successful "operator" or hypnotist; however, they often did give grudging room to female hypnotists like Mlle. Agnes Charcot who performed at Worth's Family Museum in New York City. One pamphleteer, Albert Cavendish, set the record straight by noting, "There has been several ladies who have been expert and powerful operators, getting even very strong men very quickly into mesmeric coma."[46] In his 1901 primer, Professor Leonidas also stated that women could be fine hypnotists but cautioned that a "lady of genteel bearing is the one for the hypnotic stage. She must never assume the masculine attitude."[47] He also insisted that boys were preferable as subjects, primarily because they were more suited to the rough-and-tumble of life on the road.

Leonidas and other hypnotists also played with gender and identity roles on stage, as when he recruited two young men from the audience, hypnotized them, and assigned them the roles of Romeo and Juliet.[48] If Bodie's ability to withstand electricity embodied the cult of the rigorous life, then Leonidas's stagings explored the fears of "feminization" on which that movement was founded. Such acts amused audiences, discharged their anxieties, and established the operator's power at the expense of humiliated subjects.

The hypnotist's reliance on a variety of grotesque performance modes added to public fears of the hypnotist's conceivable misuse of power. Leonidas defended hypnotism as a "science that has been much abused," but he also didn't apologize for promoting a gothic vision of hypnotism. He carefully chose his stage name and recommended that other performers also "choose an old world name; something that savors of the pyramids, ancient, antique."[49] And, as indicated at the beginning of this chapter, after making a speech before a pharmacy window and hypnotizing his nightgowned assistant, Leonidas would sew the assistant's lips shut, lay him down in a coffin, and promise the crowd to revive the "window sleeper" twenty-four hours later on stage.[50]

Other hypnotists increased the macabre element of such promotions. Some hypnotists, desperate to drum up public interest, promised to bury their subjects in graves for days at a time. Hypnotist George Newmann kept a clipping that described a hypnotist who had buried his sixteen-year-old assistant for three days in the Woodlawn Cemetery in Lexington, Kentucky. A crowd of one thousand came to see the assistant wake up and then proceed to drink water and eat graham crackers.[51] Such displays encouraged a vision of the hypnotist as "vampire"—feeding off the lifeblood or animal magnetism of his subjects. In Henry James's *The Bostonians* (1886), the father of the heroine is the seedy magnetizer Selenah Tarrant, whom James constantly likens to a vampire.

Hypnotists also resorted to grotesque antics in a more "scientific" mode for dramatic effect. During his act, Leonidas would call local physicians to the stage and have them monitor the blood pressure and pulse of hypnotized subjects that he would then jab with needles. He would explain to the audience that in many cases major surgery had been performed with hypnosis serving as the "anesthetic."[52] As a special shock, he might sew the lips of two youths together and command them to laugh. Such acts were great hits. Defending his tactics, Leonidas remarked, "The best entertainment in hypnotism is that which possesses the funny side, presents the grotesque and at the same time does not give anything that is really injurious to the subjects."[53]

In the 1890s, various cities in Europe and the United States began to consider laws outlawing stage hypnotism. In addition to concerns about

In 1947 hypnotist Harry Arons demonstrates the ever-popular "human plank" trick that was the focus of turn-of-the-century medical concerns about hypnotism's dangers. Performing Arts Collection, Harry Ransom Humanities Research Center, University of Texas at Austin.

criminal or sexual abuse via hypnotism, physicians added that hypnotism could damage the health of subjects. Individual performers occasionally were brought to court for damaging or failing to cure volunteers. Medical experts who accepted French neurologist Charcot's influential studies connecting hysteria and hypnosis concluded that "a dormant hysterical tendency could be awakened by a non-professional hypnotist."[54] Physicians were concerned about the physical and psychological side effects of hypnotism on trance subjects, both amateurs and professional "human planks."

To claim that physicians were chiefly responsible for the turn-of-the-century push to ban stage hypnotism in America would be an exaggeration. As of 1900, the American Medical Association had little national power and only eight thousand national members.[55] In this period, national and local chapters of medical associations had larger targets. For example, they sought to reform medical education—aiming at the schooling of unorthodox groups such as osteopaths, homeopaths, and the herbal remedy–based eclectics. Orthodox physicians—or "regulars" —also made efforts to regulate the lucrative patent medicine industry and with it the power of pharmacists to prescribe. Concurrently, the medical orthodoxy sought to outlaw midwifery in order to gain a larger share of the valuable obstetrics market.[56] Though the orthodox medical community did not mount an organized attack on stage hypnotism, physicians gladly did testify against itinerant hypnotists when the opportunity arose.

In 1889, Clark Bell, president of the New York chapter of the Medico-Legal Society, who also had overseen a study recommending the electric chair as an execution method, canvassed psychologists and physicians about cases of stage hypnotists damaging subjects. One asserted that hypnosis was dangerous both physically and morally and inevitably would lead "to imbecility or insanity."[57] A *New York Times* editorial of 1890 concluded that despite its good uses, hypnotism seemed of dubious value overall and urged the banning of stage exhibitions. The editorial characterized such stage exhibitions as "degrading" and asserted that "the exhibitor might as well be allowed to chloroform people in public in order to amuse a mixed audience with the phenomena of their narcotization."[58] At least one American city, Cincinnati, made stage hypnotism a misdemeanor offense in 1891. This law followed the much-publicized conclusion of an autopsy panel in New York state that Spurgeon Young, a young African-American who worked as a hypnotist's assistant, had died of diabetes aggravated by his work as a "human plank," which involved holding enormous weights on his torso.

Efforts to reform hypnotism came less from the organized force of medical associations and more from those medical practitioners who relied on hypnotism in their practice. One of the more powerful and idiosyncratic voices in this battle belonged to Sydney Flower, owner of the Psychic Publishing Company of Chicago and publisher of *Hypnotic Magazine,* a monthly that ran from 1896 through 1898. The magazine was a curious amalgam—its tone at times comical, at other times high-minded and focused on the public good. It included articles from contributors such as Clark Bell about the ethics of hypnotism, reviewed the Spurgeon Young case, ran advertisements for the *Medico-Legal Journal* as well as for Spiritualist journals such as *Light* and the *Christian Metaphysician,* and ran advertisements for various books on telepathy and psychic arts. Flower would gently advocate the reality of psychic matters, yet poke mild fun at the fringe sector of his readership with satirical poetry about vegetarianism, or in such an article as "Baldness versus Mental Treatment," which questioned why, if so powerful, positive thinking had never cured this ailment. Yet Flower attempted to extend his readership to the professional classes and to gain support for his advocacy of hypnosis as a powerful therapeutic tool. Flower had the respect both of professionals such as Clark Bell and of stage hypnotists such as X. LaMotte Sage, who remarked that Flower was "undoubtedly possessed of more than ordinary erudition and genius."[59]

Hypnotic Magazine was affiliated with a small clinic with the grand name of the Chicago School of Psychology. Flower was the school's acting secretary, and in each issue of his journal he published an account of the hypnotic treatments administered in that clinic by Dr. Herbert A. Parkyn. Parkyn would take on cases of insomnia, rheumatism, kidney disorder, deafness, bronchitis, stammering, loss of appetite, and sexual dysfunction, among others. Most frequently, the published proceedings indicated successful treatments, or remarked that not enough long-term data was available to pronounce the treatment a success or failure.

Besides providing publicity for Parkyn's clinic and his own publishing house, Flower's *Hypnotic Magazine* had a mission—to evangelize for hypnotism, to demystify it, and to insist that it be regulated and legally limited to physicians who could do the most good with it. In the introduction to the first issue Flower asked, "Why should the whole field of mental therapeutics be left in the hands of pseudo 'professors,' mental healers and charlatans? Surely it is the province of the duly qualified M.D. to possess himself of the facts."[60] Hypnotism, he argued, had acquired a bad reputation because of such unfortunate affiliations. It also had a bad reputation because of aggrandizing performers who insisted that they had "hypnotic powers" unavailable to others. If physicians and businessmen did not have the time to investigate the supposed dangers of "hypnotic influence," Flower and his magazine would do so.

Meshing with his interest in things psychic, Flower argued in the magazine that there was indeed a close link between mind and body, and that hypnotism or "suggestion" was the best way to trigger healing through this link. Hypnotism's "value to the physician and to the psychologist cannot be estimated. It affords a means by which the power of the mind to heal the body may be manifested."[61] The good healer evoked the patient's own ability to heal himself or herself.

Correlating with his distaste for stage hypnotism, Flower thoroughly covered the death of hypnotic assistant Spurgeon Young and published many documents related to the 1896 case. Among them were letters that

Clark Bell, acting on behalf of the New York State Department of Health, had requested of experts in hypnosis, medicine, and neurology to determine if Young's diabetes could have been aggravated by his work as a hypnotic subject. Bell's list of correspondents, which included Sidney Flowers and several other *Hypnotic Magazine* contributors, suggests that this magazine was not a mere fringe effort but part of that era's network of experts on hypnotism.

The answers to Bell's request covered a great range—most of the experts queried thought it unlikely that hypnosis could trigger diabetes, but they did believe that Young's work could have led to "malaise and physical prostration."[62] Thomson Jay Hudson, a frequent contributor to *Hypnotic Magazine,* replied that Young's profession and its toils would cause "but one inevitable result, namely, a shattered, nervous organism, leading, eventually, if life is prolonged, to imbecility or insanity."[63] Other experts said it was conceivable that holding large weights on his abdomen could have contributed to the stage assistant's diabetes, but noted such a line of reasoning was quite speculative.[64]

Bell appeared at the autopsy trial and requested more time for responses to his survey. Instead, the jury accepted the expert opinion of J. D. Buck, a professor of medicine and nervous and mental disease in Cincinnati. Buck answered that "cerebral softening and diabetes might result from repeated hypnosis."[65] A newspaper reporter also characterized Buck's letter to the coroner as including the opinion that "every hypnotist who comes to a town should be knocked out."[66] The jury concluded that Young had died of "diabetes and nervous exhaustion caused by hypnotic practices."[67] It is likely that Buck's hatred of stage hypnotists and campaign against them led to the statute in Cincinnati banning stage hypnotism.

In every issue of his magazine, Flower published scornful accounts of stage performers, whether phony psychics or hypnotists. He found exhibitions of "window sleepers"—like those of Leonidas—distasteful, as was stage hypnotism in general. Flowers quoted an issue of the *Baltimore Citizen* that described a hypnotist whose subject had been sleeping in a storefront theater window for more than a week, gathering large crowds of spectators on the street. Concerning such spectacles, Flower's general conclusion was, "I wish to express my intense dislike of the induction of hypnosis for the purposes of amusement. There is no good end to be gained by the public exhibition of somnambulistic feats. . . . It is not possible to restrict the use of hypnotism as a therapeutic agent to the medical profession, but it is possible to bar the hypnotic 'entertainment.'"[68]

The stage hypnotists had varying responses to such calls. Many stage hypnotists, particularly those who affected cures in their offstage practice, sought the moral high ground of the perfectionist model. The influential, often-plagiarized stage hypnotist, mind reader, and healer P. H. McEwen argued against medical control of hypnotism in *Hypnotism Made Plain* (1897). McEwen, who was a lay healer, insisted, "Not until doctors have proven themselves more intellectual and virtuous than their fellow men, should they be given the monopoly of one of the greatest God-given benefits to mankind."[69]

McEwen defended hypnosis in general with the argument that the subject underwent a valuable spiritual transformation when hypnotized. "When one has thus learned to control himself, or, in other words, has learned how to overcome the material body by asserting the rights of the true ego, he has accomplished much towards the development of the soul, giving to it the place to which it rightly belongs."[70] Other stage hypnotists borrowed McEwen's defense that hypnotism revealed the soul. For example, L. A. Harraden suggested the following be worked into opening speeches: "Our object tonight is to cause the flesh and its power, the intellect and reasoning faculties, to slumber; while we thereby temporarily set free the invisible spirit which we call the ego, or soul, or the subjective mind."[71] This rhetoric might be appropriate to acts in which the hypnotist was revealing some greater power of the subject's mind or exhibiting the subject's apparent ability to self-heal; however, such a defense could only uneasily apply to stage performances in which hypnotic subjects followed humiliating commands.

Many hypnotists staged their entire acts and never placed their subjects in trances. These "fake operators" must have been amused at reformers' calls to ban stage hypnosis on the basis that the hypnotic trance damaged subjects' health. Newspaper reports in 1885 indicated how a mesmerist performing in Chicago managed such stage fakery, and also suggested that orthodox physicians of the era viewed stage hypnotists as a "problem" to solve, often using the "unmasking" of fraudulence as their chief weapon. A group of Chicago physicians had been attending hypnotist Dr. Townsend's performances at Grenier's Theatre nightly, sitting in the twenty-cent seats, fascinated. One evening, H. R. Robinson, one of the assistants—or "horses"—of Dr. Townsend, angered about not being paid, disrupted a performance, shouting, "This thing is a fraud, and I can prove it. I've been a subject here and I can stand any kind of test." Robinson later arranged for several physicians who had been attending Townsend's "séances" to test him and other assistants, showing that—when completely awake—they could be jabbed with needles under their fingernails and through their tongues, burned with lit cigars, have cayenne pepper thrown in their eyes, and made to swallow "the bitterest drugs" without reacting.[72]

Professor Leonidas insisted such fraudulence was not common. He asserted that while hypnotists often did use one or two professional subjects, or "horses," these were people who were easily hypnotized and extremely valuable. He urged the would-be performer to "avoid all fake work" and to get a genuine subject, "a good subject; one who can be put into catalepsy or made to eat the delusive strawberry. . . and he is the boy to purchase, hire or kidnap!"[73] Another hypnotist, X. LaMotte Sage, insisted that such frauds were generally only found in dime museums and were inferior to genuine subjects, adding that "no high class performer could possibly use them."[74]

Stage hypnotists also often agreed that hypnotism could damage subjects, but insisted that they were as expert at avoiding such dangers as were those with more than mock credentials in psychology or medicine. Savvy stage hypnotists could regulate their own profession through study and practice. One hypnotist's 1896 handbook encouraged operators to combine the study of hypnotism with the then mildly credible science of phrenology, in order to work "phreno-manipulations" on subjects. If an operator touched the area of the skull corresponding to the organ of "Imitativeness," they could get some wonderful impressions of parrots out of the subject. But operators were warned to steer clear of the "organs of our lower nature, such as 'Amativeness,' 'Destructiveness,' 'Combativeness,' 'Fear,' & c., as their manifestations are not always of the most agreeable character."[75] And in a 1907 pamphlet, the English stage hypnotist George White suggested that assistants should be put into "cataleptic" trances sparingly. The human plank trick was especially dangerous and might lead to subjects whose "nervous systems have been completely shattered."[76]

Leonidas also warned his readers about "bad" subjects, people with constitutional weaknesses who are "usually pale" and "wabble [*sic*] perceptibly" when going into a trance. "Here," he remarked, "is a case in which the inexperienced operator will feel his heart growing weak. These cases are rather frequent and must be treated 'heroically.' That is, the subject must be brought to the waking state without delay, or—well, there might be a bit of a sensation. . . . If left alone they might come out of the sleep in half an hour or they might sleep a week."[77] The able hypnotist had to steer a careful course and rely on his or her own wits. Like McEwen, Leonidas thought himself more of an expert than many of the era's physicians then experimenting with hypnosis.

Leonidas's defense of hypnotism avoided the spirituality-based argument of McEwen; instead he insisted that hypnotism was justified because it was both entertaining and educational. Commenting: "P. T. Barnum once said that the public wants to be humbugged." Leonidas added,

"They do to a certain extent. That is, they—and especially Americans—want to be entertained. They look for variety and not reform."[78] He saw education as part of his mission, however, and comes off as a somewhat vague apostle of positive thinking—an offshoot of mesmerism still powerful today in American culture. "True," he remarked, "the work that is seen in the average hypnotic show is not illustrative of the highest type of psychology. But it has its mission and always will have—or until people have been educated to that point wherein they can utilize the mental forces in every-day life."[79] Leonidas, like McEwen, Sage, and others, sought to emphasize hypnotism as a model for "emancipation" and not one of "enslavement." Though his primer may still persuade readers of his integrity, Leonidas's career as a hypnotist languished. In 1903, the American Mutoscope and Biograph Company released the short film *Stealing a Dinner*—designed for mutoscopes with their flapping photographic cards—that featured Professor Leonidas, along with his troupe of trained dogs, involved in high jinks at mealtime.[80] Clearly, this career shift suggests he never became a headliner as a hypnotist.

Progressive Entertainment or Grotesque Performance?

Stage hypnotism foundered as an entertainment form when it entered the cultural currents of the Progressive Era. Regulation attempts imply a rising American middle class that had begun to view hypnotism more as a metaphor for enslavement than liberation. In this same era, critics frequently used the word "hypnotizing" to describe the allures of the city with its electric lighting, window displays, and consumer abundance. In this melodramatic formulation—often found in "white slavery" tracts—innocents encounter modernity and become charmed, and then their virtue is severely challenged. The metaphoric implications of hypnosis were all the more reason that "degrading spectacles" be stopped.

While reformers of that era hounded stage hypnotists, some new adepts within the ranks of the progressives seized on hypnosis as an ideal tool for social reform. In 1888, a reviewer of a book titled *Animal Magnetism* by A. Binet and C. Fere of the Charcot school in France commented, "Much curious information is given as to the production and effects of hypnotism. . . . The science of magneto-therapeutics is certainly still in its infancy, but if it can give us new moral agents and effect the reform of every criminal, let it be developed by all means."[81] This formula was revisited a decade later when a *New York Times* article from 1899, titled "Hypnotism the Cure-All," highlighted the work of a Columbia professor who claimed, "Hypnotism, as a means of reforming criminals and of removing crime and moral obliquity . . . is the latest theory which advanced science has to offer."[82] The article went on to discuss how hypnotism could end drug addiction and moral perversions, and turn thieves into upright citizens. A volume cowritten by a minister and two physicians in 1908 continued this line of reasoning. They insisted that hypnotism and autosuggestion could be useful in treating nonorganic mental disorders and to reform moral habits. Hypnotism could help reeducate and reform prostitutes, as well as treat alcoholism, drug addictions, neurasthenia, sexual aberrations, bed-wetting, and "incorrigible children with vicious habits."[83]

Capitalizing as best they could on their notoriety, hypnotic showmen navigated the tricky currents of reform. Those with the greatest pride in their art likened themselves not to medicine showmen but to genuine healers. McEwen, for example, wrote of his own successful efforts to cure patients of illnesses and to alleviate mental and moral disorders. He also argued that hypnotism revealed the human soul, relying on an argument that would have had more appeal during the earlier perfectionist era.

While Charles Came added electrical exhibitions to the possibilities of the medicine show, Dr. Bodie, M.D., added both electricity and hypnotism. He positioned himself as a man of science, complete with a string of honorifics following his name, and as someone who happened to be in touch with mystic forces. While his electrical displays appeared to affirm modernity, his use of the electric chair critiqued its products. Likewise, if his stage work made his subjects into grotesques, his stage cures symbolically redeemed paralytics from their own status as grotesque displays. Though Bodie continued performing into the 1920s, his popularity was greatest before the Glasgow theater riots of 1909.[84]

Professor Leonidas traveled lighter than Bodie and gained fewer enemies, but ultimately abandoned the art to become the leader of a trained dog troupe—a move that clearly signals the decline of stage hypnotism. Leonidas's act, like that of most hypnotists, could be said to provide greater sociological insights than the psychological insights he promoted: his stagecraft revealed the grotesque product that ensues when social roles are violated—i.e., when a middle-aged gentleman assumes that he has been transformed into an international opera diva named Madame Squeeba.[85]

Regarded in a positive light, the progressives' battle against stage hypnotism, led by advocates of medical hypnosis such as Sydney Flower, illustrates that by this era the middle class would no longer tolerate exploitation, whether of child laborers, sweatshop workers, or hypnotic subjects. In a more critical light, the growing distaste for hypnotism reveals a middle class that preferred a sanitized culture in which older mysteries and marvels no longer had a place. Yet, not only middle-class guardians of public virtue were to blame for stage hypnotism's demise, but also the new mass culture forms such as vaudeville, cinema, and radio. While stage magicians were far more successful at adapting to the demands of the new era, seedy performers such as stage hypnotists and such venues as dime museums were quickly vanishing. Hypnosis "via radio" stunts could only amuse the public so long.

Hypnotists were up against more than the outrage of crusaders like Sidney Flower or Dr. Buck of Cincinnati. In the 1890s, with the advent of the Keith Circuit on the East Coast and the Orpheum Circuit in the West, variety entertainments were becoming less bawdy and offensive. As the vaudeville network expanded and became standardized, acts that relied on the grotesque, like those of the hypnotists, were pushed even further to the margins. Vaudeville, the leading edge of the mass entertainment industry, itself soon displaced by the film industry, was then helping to shape America's middle class.[86]

When Leonidas encouraged would-be hypnotists to use a "window sleeper" on the small-town circuits—that is, to place an entranced assistant in a coffin—he may have been unconsciously announcing the slow death of stage hypnotism as a viable trade. Turn-of-the-century hypnotists had indeed symbolically "killed" their assistants—the modern hypnotist's "horse" no longer displayed the amazing powers of mind that mesmerists had once elicited from their somnambules. By exalting his own role as an "operator," Leonidas admitted that for subjects he might as well "go to the dogs."

CHAPTER FOUR

The Magician

In 1873, in Genoa, New York, townspeople saw the following poster announcing an upcoming performance: "IF NOT SPIRITS WHAT IS IT. THE MYSTERIOUS MAN WILL PERFORM THE WONDERFUL MANIFESTATIONS PRODUCED BY ALL THE NOTED MEDIUMS OF THE DAY." The poster went on to describe the performer's abilities, announcing, for example, that he "plays on several musical instruments while firmly bound with ropes" and that he "is released after being bound by a committee in less time than is taken in binding him." Using rhetoric common to séance invitations, the poster urged the audience to "Come and Investigate." Men were charged twenty-five cents and women and children fifteen cents.[1]

The poster did not make clear whether the "Mysterious Man" was one of the many Spiritualist performers of the era or that breed's prime enemy: the stage magician or "anti-Spiritualist" whose goal was to prove the Spiritualist a fraud. The Mysterious Man occupied a low tier of the show business world where such ambiguity could widen his appeal. Prominent magicians chose sides more precisely. To defend their prior claim to the mystery stage, the magicians employed a two-fold strategy against newcomers in the trade that did not present themselves as kindred performers. First, they deflated the pretensions of mesmerists, hypnotists, and Spiritualists to otherworldly powers and scientific status, and second, they absorbed their competitors' marvels into their own magic acts.

The ensuing turf battles between the era's anti-Spiritualist magicians and mystic performers mirrored cultural tensions between scientifically minded skeptics and followers of Spiritualism. The stage magicians' decision to police the line between "genuine" and "fraudulent" stage presentations underlined their alliance with what they took to be progres-

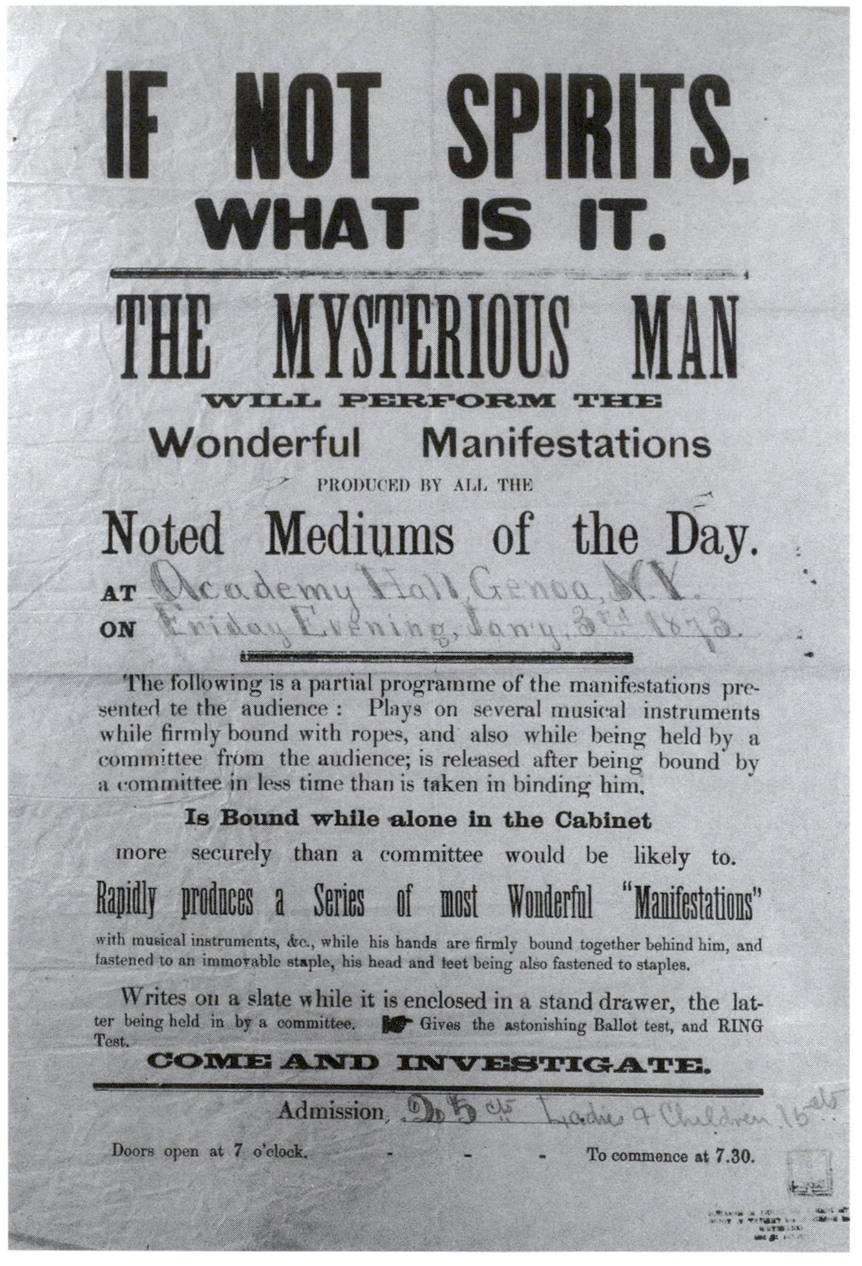

This 1873 poster is for the "Mysterious Man," a performer who offered "cabinet effects" like the Davenport Brothers and who also performed slate tricks and escapes. Although he does not identify himself as a Spiritualist performer, his invitation to "come and investigate" is similar to the phrasing on many séance invitations. Library of Congress, Rare Book and Special Collections Division.

sive scientific forces. Sociologists of science insist that working scientists inevitably must patrol the boundaries of science and expose "deviant" science or pseudoscience to maintain the integrity of their discipline.[2] A prominent nineteenth-century scientist such as Michael Faraday, for example, only hesitantly "exposed" table tipping. Although he argued that the lifting of tables was an outcome of the unconscious muscular action of séance attendees, Faraday was unsure whether his scientific attention to the topic would only further the cause of Spiritualists. With prominent scientists hesitant, in the mid–nineteenth century stage magicians jumped to the aid of the rationalist cause. Anti-Spiritualism was one of the first strategies stage magicians adopted to secure a symbiotic relationship with the scientific project and to promote themselves as exemplars of the "modern."

Yet the magicians' attempt to embrace "science" and reject "superstition" hid the deeper ambiguities in such performances. Magicians wished to offer "shows of wonder" that fulfilled the audience's nostalgia while yet assuring audiences that miracles, ultimately, had no place in the modern age. Further, anti-Spiritualism could be an extravagance that only the more successful performers could afford. Just as the Mysterious Man's poster makes it unclear whether he was a Spiritualist or anti-Spiritualist, many magicians either started their careers as mystic performers or occasionally turned to such venues when desperate.

One graphic example is the career of turn-of-the-century performer Harry Houdini. His professional life recapitulated the curious kinship and hostilities that informed the relationship between Spiritualists and anti-Spiritualists. Though now perhaps the best known of anti-Spiritualist performers, early in his career Houdini moonlighted as a fraudulent Spiritualist performer. And even when he later battled Spiritualism, taking on the mantle of science and progress, Houdini never became a comfortable symbol of the status quo. Instead, like other artists of the modernist era, he positioned himself as a rebel—an individualist whose "natural" humanity freed him from most forms of authority: whether the encroaching regimentation and "feminization" of daily life; the powers of police and their jails; the restraints of straitjackets and psychiatry; or the charlatans of the religious or occult worlds.

To further complicate matters, Houdini's beliefs about Spiritualism were as ambiguous as his attitude regarding authority. Both a seeker and a skeptic, Houdini's escape act indicated a fascination with the borderland between life and death that placed his interests close to those he mocked in his anti-Spiritualist lectures and exposés. Houdini reflected his era—a time of modernist skepticism of authority, but also a time when audiences thirsted for spirituality and illusions in any form.

If Not Spirits What Is It?

The Spiritualist movement created a context in which stage illusionists of all sorts could flourish. Spiritualism was launched in 1848, when the Fox sisters in upstate New York began to hold séances in which spirits "rapped" responses to questions. Soon other houses in the area were subject to ghostly "rappings." Spiritualist societies blossomed, and séances were held, both as a form of worship and as a moneymaking method for mediums to help supplicants communicate with spirits of the dead. Andrew Jackson Davis, who began as a mesmerist's subject in New York State, experienced a trance-based illumination in 1843. His published works that followed, describing his "harmonialist" philosophy, met the philosophical needs of the developing Spiritualist community.[3] Within a few years, thousands of mediums set up shop in the United States. The advent of the American Civil War, with its high casualty rates, also aided business, just as the aftermath of World War I was later to revive fascination in Spiritualism.

Spiritualism also soon sprawled into show business. Soon after their initial séances, the Fox sisters presented a public séance and charged admission in Rochester; in the summer of 1850 they stayed at Barnum's Hotel and offered séances to genteel New Yorkers, impressing, among others, Horace Greeley, publisher of the *New York Tribune*. In the 1850s and following decades, young and pretty Spiritualist "trance speakers" such as Cora Hatch and Achsa White Sprague became the equivalent of today's rock stars when they lectured in an inspired and purportedly unconscious state about politics and women's rights.[4]

Shortly after the launching of Spiritualism, the Davenport Brothers, also from upstate New York, became the most famous of all performing Spiritualists. Though they never publicly stated that their mediumship was genuine, the Davenports' act usually began with a minister solemnly explaining to the audience the value of Spiritualism to the renewal of Christian faith. In the Davenports' cabinet act, which they launched in the 1850s and performed before Spiritualist societies, variety audiences, and royalty, a "committee" from the audience tied the brothers' hands behind their backs and fastened their ankles. The committee bound them to a bench within a large cabinet—essentially a large wardrobe—that held suspended musical instruments. Assistants lowered the stage lights and closed the doors of the cabinet. Soon after, the audience could hear guitars, violins, bells, and tambourines playing inside. But when the doors of the cabinet were opened, the Davenport Brothers sat calmly, hands and feet tied in place.

To make their acts credible, Spiritualists often appealed to the scientific beliefs of the age. The result, for believers, was a show of wonders with a scientific aura that assured audiences that human powers could outpace scientific progress. As mesmerism had some scientific cachet, prior to a séance, assistants or handlers might make mesmerizing passes before the mediums to help them reach their trance state. Spiritualists also liked to insist that they were conducting scientific experiments and asked observers, as did the Mysterious Man, to "come and investigate" and draw their own conclusions from the empirical evidence. Likewise, both mesmerists and Spiritualists insisted their wonders had some basis in electrical phenomena. At séances, for example, men and women were alternated around the table in order to balance out "negative" (female) and "positive" (male) forces. Mediums likewise urged séance goers never to "cross themselves" as this might create a short circuit. The telegraph, which was a popular sensation when the Fox sisters began to receive their rapped messages, also provided a technological metaphor for Spiritualist practice. Spiritualists insisted they were simply employing the "spiritual telegraph," giving a contemporary twist to the archaic practice of communicating with spirits.

Stage magicians, too, long had made pretenses to scientific status to deflect charges of frivolity and to promote their acts as "moral entertainment." As early as 1787, a contemporary of Katterfelto's named Falconi described his show in Baltimore as a series of "Natural Philosophical Experiments." Falconi emphasized the word "experiment" rather than "trick" or "illusion," and his playbill emulated the prose of a "natural philosopher"—or man of science.[5] A Mr. Charles performing in Boston in 1819 billed himself as a ventriloquist and "professor of mechanical sciences to his Majesty the King of Prussia."[6] And a Mr. Stanislaus who performed in Boston in 1823 described himself as a member of the Academy of Arts and Sciences in Paris and a professor of "Natural and Experimental Philosophy." His show included sleight-of-hand and philosophical experiments with names like "No.1 the gallant Mercury."[7] An 1837 broadside for Mr. Baldwin referred to him as "The Unrivalled American Magician, Professor of Magical Hydraulics, Metamorphoses, Scientific Ledgerdemain [*sic*], &c."[8] Such presentations, which were common from the late eighteenth century through the first half of the nineteenth century, seemed appropriate to an age when a scientific experiment generally referred to a public demonstration. Natural philosophers often sought to intrigue and entertain, as when Humphrey Davy demonstrated the effects of laughing gas at the Royal Institution.

In the 1860s, New York State required magicians or "jugglers" to take out licenses to perform, and some entertainers may have posed as scientific lecturers to spare this extra expense. John H. Anderson, the son of the well-known magician Professor Anderson, described his occupation as "giving scientific lectures with mechanical experiments" when he appeared as a witness at a trial prosecuting a fraudulent Spiritualist in 1865.[9] Likewise, nineteenth-century magic catalogues often included "scientific apparatus" or "philosophical apparatus" among their offerings.[10] One prominent catalogue from 1876, for example, was titled "A New and Descriptive Catalogue of Magical Apparatus and Scientific and Mechanical Novelties."[11]

Stage magicians gained additional insights from nineteenth-century science popularizers. These writers or lecturers often defined themselves by taking a strong stance against "superstition." Such popularizers frequently relied on the device of debunking superstition or "correcting error" as a prelude to their own explanations of scientific phenomena.[12] In their rationalist stage performances, the magicians demanded the audience decode their acts and search for the physical explanation, much as a scientist like Faraday might solve the puzzling problem of table tipping. Popular science magazines also recognized the affinity between science and stage magic and often published explanations of the stage illusions of magicians and Spiritualists. Magicians found that adopting a similar "anti-superstition" stance, as well as teaching audiences how to avoid cardsharps and confidence men, could align their craft with progressive forces while releasing them from the strain of directly imitating a scientist or natural philosopher while on stage.

The stage magicians' anti-superstition efforts pitted them against their stage rivals—the era's mesmerists and Spiritualists. To debunk these popular movements, magicians duplicated the effects that Spiritualists and mesmerists supposedly achieved through occult powers at séances. Magicians launched "Second Sight" acts as early as the 1830s to imitate the performances of mesmerized subjects who exhibited clairvoyance and described objects or places apparently out of view. In the 1840s, the older Professor Anderson mocked the Fox sisters, calling them "conjurers in disguise," and soon added spirit rappings to his act.[13] Prominent stage magicians Harry Kellar and John Nevil Maskelyne both debuted as debunkers of the Davenports' cabinet act. Maskelyne and his partner George Cooke launched their duplication of the Davenports in 1865, shortly after the Davenports first visited England. The entertainers added a few humorous touches, as when the cabinet doors were opened and the shackled magicians had transformed themselves into an ape and a lady. Maskelyne and Cooke also introduced Spiritualist stage farces at their London theater, Egyptian Hall. Similar farces soon appeared in magic catalogues, with titles such as "'Spooks' or the Spiritual Cook,"[14] or "Lady Daffodil Downy's Séance," an "excellent anti-Spiritualist farce

In this 1893 promotion, magician and anti-Spiritualist John Nevil Maskelyne manhandles the serpent "Humbug." Maskelyne came to prominence with a rationalist reconstruction of the Davenport Brothers cabinet act. From Milbourne and Maurine Christopher's *Illustrated History of Magic*.

for three or more persons, as introduced in London by Maskelyne and Cook."[15]

Yet the conflict between magicians and Spiritualists hid deeper ambiguities. The poster of a performer like the Mysterious Man of the 1870s obscured whether he was a Spiritualist or an anti-Spiritualist. A similar performer named George Everett, offering Spiritualist-styled escapes from ropes, was denounced as a fraud in 1878 and responded that he made no claim to Spiritualism or any other "ism," but simply "gave what he had acquired in his own investigations and the public must judge as it saw fit."[16] Meanwhile a Spiritualist performer such as Professor Laroy Sunderland offered a show that included "original experiments in Mental Magic, Musical, Wonderful, Mirthful, demonstrating new discoveries in psychology and other 'ologies.'"[17] The decision to define oneself as Spiritualist, anti-Spiritualist, or in between often had less to do with ethics than with box office concerns.

Despite their public conflict, throughout this period magicians, Spiritualist performers, and anti-Spiritualists purchased their tricks and cabinets from the same catalogues. From the mid-1800s, the catalogues of shops that sold magical devices to performers frequently devoted pages to "anti-Spiritualistic devices," such as the "Spirit Bell," "Rising Tables," "Rapping Hand," "Luminous Materialistic Ghosts and Forms," and "Magic Slates." The catalogues also offered entire acts such as "Etherialization"—which enabled a medium "to produce any number of spirit forms, in the perfect dark, which have the appearance of a fine, misty, luminous vapor . . . fading away, producing a weird and wonderful effect."[18] Hands that could rap out messages were also available, as well as the "New Flying Music Box," "Slate Tricks," and "Spirit Lectures," "meant to [be] use[d] in combating spiritualism, or by anti-spiritualists. Are suitable for delivery from the stage, parlor or pulpit."[19]

Some catalogues catered to fraudulent performers, and made no demand that the tricks be used in an anti-Spiritualistic context. To take a late example, the copy of a Depression-era magic catalogue entry for "Distant Hypnotic and Magnetic Force" ran, "Actually convinces the suckers that you do possess some strange force."[20] More reputable dealers avoided such direct appeals to fraud and insisted that Spiritualist effects and devices were intended for the entertainment of friends in the parlor only. These contrasting marketing strategies underline the ambiguities of using "Spiritualist" and "anti-Spiritualist" as categories. With the aid of the same technology, magicians and Spiritualist frauds were able to carry on their ongoing cat and mouse game. Self-proclaimed Spiritualists used the equipment to defraud audiences, while magicians used the same equipment to become champions of rationality.

Howard Thurston, one of the top early twentieth-century stage magicians, was sympathetic to the Spiritualist cause, but this poster advertises an anti-Spiritualist cabinet act. Thurston toured with his "Wonder Show of the Universe" into the 1930s. From Milbourne and Maurine Christopher's *Illustrated History of Magic*.

The Strenuous Life of Harry Houdini

Although Harry Houdini came to prominence as an escape artist and later as an anti-Spiritualist crusader, he began his career in the more ambiguous realm of the mystic performer. In the 1880s and 1890s, young Houdini attended many séances in New York City, eager for marvels, but reported constant disappointment at the obvious trickery and greed he and his friends uncovered. Nevertheless, before establishing himself as an escape artist, Houdini occasionally relied on the ambiguities of "anti-Spiritualistic devices" and routines. Between bookings as a magician, Houdini and his wife conducted phony séances in the Midwest. His séances included such phenomena as floating tables, self-playing accordions, and the appearance of spirit faces. He also gave advice and messages to the bereaved. He relied, in short, on the products of what he and others later would call "spook racketeers."

Houdini was able to leave these days behind when he forged a new career as an escape specialist and became billed the "Handcuff King" as early as 1899. Yet many of Houdini's escapes were deftly related to Spiritualistic practices.[21] His escapes almost always took place in small curtained areas on the stage or in wooden cabinets like those of the Spiritualists. Spiritualist mediums had first introduced cabinets as a way to isolate themselves and prove that they were not working with confederates in pitch-dark séance parlors. Mediums also claimed that the dark cabinet helped them build up the spirit forces necessary for manifestations. A Spiritualist character in one of believer Arthur Conan Doyle's novels explained that the cabinet "serves as a reservoir and condensing place for the ectoplasmic vapour from the medium, which would otherwise diffuse over the room."[22] But the cabinet, as even Conan Doyle acknowledged, also opened up new possibilities for trickery.

Like the Davenports, Houdini relied on the Spiritualist's cabinet as his basic stage property. In the feat that first got him started, "Metamorphosis," Houdini had his hands tied behind his back, and then was draped in a coat borrowed from a spectator. He then climbed into a sack in a trunk. The sack was tied, its knots sealed, and the trunk shut and also bound in ropes. Then the trunk was placed in a cabinet. His assistant—usually his wife, Bess—rushed into the cabinet, clapped three times, and Houdini sprang out on stage. His wife was then found bound in the trunk, with her hands tied behind her back and the borrowed coat on her. This lightning exchange amazed audiences. Houdini's other escape acts were also inspired by the Davenports and other Spiritualists—or "anti-Spiritualists"—who insisted on having their hands and feet tied or

Houdini, the Handcuff King, in an early publicity photograph, preparing to break free not only of the leg irons and handcuffs, but also from the gentlemanly restraints of the stage magic act. Library of Congress.

locked or who had themselves bound in sacks and nailed to floors to convince audiences that no trickery was involved.

Though not the originator of handcuff escapes, Houdini certainly brought this act to prominence. Magic catalogues of the late nineteenth century—when Houdini was just beginning his career—were full of devices in which anti-Spiritualist performers could be tied up, chained, or bound. These catalogues sold the secrets to trick knots, as well as handcuffs, "spirit collars" that came with padlocks, "spirit benches" that performers could be locked to, and medieval stocks in which a performer's head and hands could be secured. After Houdini rose to prominence, catalogues added numerous listings specifically for handcuffs.

Houdini's mastery of Spiritualist handcuff escapes—or his "Handcuff Challenge"—helped edge him onto the vaudeville circuit. To prove that the handcuffs he escaped from were genuine, Houdini published invitations for spectators to bring handcuffs—as long as they weren't doctored—to theaters where he was appearing and lock him up. Legend has it that after vaudeville booker Martin Beck saw the Houdinis perform in a Midwest dime museum, he arranged a challenge and was impressed at Houdini's easy escape. Beck then booked the young escape artist on the Orpheum circuit of western vaudeville theaters, giving Houdini his first taste of success. Soon he was not only freeing himself from handcuffs, but also liberating himself when sealed in a milk can full of milk, when manacled, boxed, and thrown in a river, when chained inside the body of a "sea monster" dredged up by Cape Cod fishermen, or when hanging upside down in a straitjacket from the top of a building.

Houdini's escape acts, particularly the simple Handcuff Challenge, had subversive potential, as it identified him with criminals and lawbreakers. Houdini often publicized his Handcuff Challenge by visiting the police stations of local towns and challenging jailers to keep him locked up. They usually complied, and handcuffed him and locked him in a cell. He would soon walk out free, to the delight of reporters. A typed 1905 testimonial from the Rochester New York Chief of Police is typical: "We, the undersigned, certify that we saw Harry Houdini, the bearer of this note, stripped naked, searched, locked in one of the cells . . . handcuffed with three paris [*sic*] of cuffs; also strapped with a strap extending from pari [*sic*] of cuffs and buckled at the back."[23] Such escapes appealed to a public disgusted at the corruption represented in civic authority at the century's turn. Endless muckraking exposés of the era revealed police graft, brutality, and complicity in prostitution and racketeering.

This was not news to Houdini. In a file titled "Police," he compiled clippings from English and American newspapers that described acts of police crookedness, misconduct, false imprisonment, and brutality. Cops as robbers was a favorite theme. For example his 1912 clipping from the *New York Evening Telegram* is headlined "Held, Accused of Robbing Garage While Policeman." Clippings from 1913 include such articles as "Action against a Police Inspector—Damages Awarded" and "Two Policemen Accused of Night Robberies." Some of the clippings are lighthearted—for example, a story that describes how "two members of the [Bristol] city police force . . . [were] charged with breaking into a bakery, and stealing a sponge cake, value one penny." More ominous is a

1912 front-page cartoon from the *New York Evening Journal,* showing a line of huge, headless policemen holding clubs with the word BLACKMAIL over their heads. A body lies on the ground behind them, with the sign "A Dead Man Tells No Tales" on it, while a small figure of Justice before them is ignored.[24]

Houdini's ability to escape prison and stroll out to the street fully clothed, a free man who even had spared himself lawyer's fees, had great emotional value.[25] In identifying himself with lock pickers, jailbreakers, and other thieves, Houdini took on the aura of the heroic antihero appropriate to the age of the muckrakers. He confronted not only the police and the apparatus of the state but also the psychiatric profession when he began to perform straitjacket escapes. His brother, Hardeen, also an escape artist, had discovered that this act was more effective when performed writhing on the stage—that is, in real time—rather than when hidden in his cabinet.

Historians generally suggest that the nineteenth-century cult of the strenuous life was based on white male fears that modern life was overly regimented, feminizing men and making them into slaves of technology and bureaucracy. This cult stressed exercise, sports, fitness, and asceticism. Self-liberator Houdini was one of the cult's exemplars.[26] Houdini showed off his physique, conditioning, ingenuity, and bravery in his escapes on stage, and these qualities were highlighted in his many seminude publicity photos and in dime novel accounts of his exploits. Such tricks as his straitjacket escape made it clear that often there was no "trick" coming to his aid besides his strength, dexterity, and cunning. This emphasis set Houdini more firmly in the currents of his age and further guaranteed that his escapades would take on the aura of myth.

Houdini's advocacy of manly fitness and his desire for publicity made for a playful confrontation with London feminists. In 1908 Houdini responded to—or, more likely, choreographed—a public dare from London's suffragettes, whose printed "challenge" complained that "so far, only men have tried to fasten you." Relying on tools of the domestic sphere, his six female challengers promised to bind him "to a mattress with sheets and bandages." An early biographer solemnly remarked, "He regarded this as one of his most difficult escapes."[27]

If There Is Anything in This Belief in Spiritism

The thin boundary separating the anti-Spiritualist from the Spiritualist performer bewildered many. Despite Houdini's insistence to the contrary, Spiritualists tended to believe the magician was one of them. How else to explain his ability, demonstrated on a New York stage, to walk through a freshly built brick wall? Though the trick involved a trap door, many Spiritualists suspected he had the superhuman ability to "dematerialize" and reappear. His escapes also seemed to involve similar supernatural powers, and his frequent attendance at séances implied some unclear fascination with the spirit realm. Many Spiritualist believers claimed Houdini as one of their own.

In his autobiographical account, *A Magician among the Spirits* (1924), Houdini toyed with this interpretation when he admitted that on a boy's prompting he once had made it rain and then stop on command. Houdini's wife also ambiguously raised the possibility that psychic powers aided Houdini's escapes in a letter she penned to his former friend, Spiritualist Arthur Conan Doyle. She wrote: "As I often told Lady Doyle, often he would get a difficult lock, I stood by the cabinet and would hear him say, 'this is beyond me' and after many minutes when the audience became restless, I nervously would say, 'Harry, if there is anything in this belief in Spiritism, —why don't you call on them to assist you' and before many minutes had passed Houdini had mastered the lock. We never attributed this to psychic help."[28]

Houdini's reliance on the Spiritualist cabinet as an escape prop also might have resonated with the Jewish liturgy he witnessed as a child. Houdini's father, Mayer Samuel Weiss, had been a rabbi, and as a rabbi's son, Houdini would have sat front row at services watching his father open a wooden ark to carry the Torah through the congregation, then remove the Torah's covering and ornamentation before unrolling it and chanting prayers. Arguably, Houdini's act had religious reverberations for himself and his audiences. In such a reading, Houdini would substitute his own body for the holy Torah, and his magic cabinet then would become a place of true miracles. And, although his father, with his Old World ways, lost his congregation and stature in America and eventually was reduced to the role of an impoverished garment worker in New York City, Houdini was able to reclaim center stage with his own career and enactment of miracles.

Rogan P. Taylor has argued that Houdini's popularity resulted from his shamanistic aura.[29] Taylor identified the stage magic show as a descendent of the shaman's healing ritual. In the context of shamanism, the nineteenth-century stage magician's popular trick of decapitating a subject's head and restoring it, or the twentieth-century variant of sawing a woman in half and then restoring her to life, takes on a new resonance. Houdini's escapes from chains and ropes are also suggestive of the shaman's ability to escape this world and travel in another. Taylor argued, somewhat vaguely, that Houdini's performances, like that of a shaman, unleashed healing forces in his audiences. Other writers have argued that Houdini's strong following among the working class suggests his act served as a metaphor for freedom from exploitative labor. However, one does not have to insist that Houdini was functioning as a shaman, rabbi, or Marxist educator to agree that his performances with their symbols of liberation had mythic resonance.

The theme of death and resurrection, or survival of the soul, crucial to Spiritualism, more sharply emerged in Houdini's act after his mother's death. Even before his father's death when Houdini was eighteen, Houdini's mother, Cecilia Weiss, was at the family's center. Houdini in particular doted on her. Throughout his life, Houdini liked to rest his head against his mother's breast to listen to her calming heartbeat. When news of her death arrived as he was performing in Copenhagen in 1913, Houdini fainted. After returning to America, Houdini suffered a breakdown and spent days and nights lying on his mother's grave, his thoughts turning to death and suicide. When he resumed touring, he would spend his off-hours in cemeteries, fascinated by the graves of suicides. He also toured lunatic asylums, as one biographer put it, "morbidly convinced that he would end his days in one."[30]

After his mother's death, coffins appeared in his act—one of which he was later buried in. In 1916, a few years after his mother's death, as an experiment, in a field outside Santa Ana, California, he had his assistants dig a grave. Houdini climbed in and lay down, slightly hunched to give him room to maneuver. His assistants then shoveled dirt down until he was completely covered. He struggled with the weight of the earth and nearly suffocated before his assistants saw his hands break the surface, frantically clawing, and so pulled him out.

Houdini's interest in coffins and graves was rekindled in 1925 when Rahman Bey, "the Egyptian Miracle Worker," came to New York and garnered headlines with his stunts. After going into a trance, Bey was sealed in an airtight coffin for ten minutes. Doctors claimed the coffin contained only enough air to sustain a person for three minutes. Next Bey arranged to have a coffin lowered into a swimming pool and remained inside it for an hour. These miracles were credited to Bey's trance powers. Houdini, then fifty-one years old, couldn't stand the headline competition. He arranged to duplicate the stunt. Houdini stayed submerged in a coffin for an hour and a half in a swimming pool and showed that conditioning and ingenuity could surpass Bey's alleged mystical powers.

Following his mother's death, Houdini again began to consult Spiritualists, hoping to receive word from her. Bess Houdini remarked, "Often in the night I would waken and hear him say, 'Mama, are you here?' and how sadly he would fall back on the pillow and sigh with disappointment."[31] Houdini's interest in Spiritualism, however personal its basis, was also well-timed to keep the aging escape artist's name in the public realm. The aftermath of World War I led to a resurgence of interest in séances and attempts to contact the dead. In the 1920s, during a long performance run in England, Houdini befriended one of Spiritualism's great champions, Sherlock Holmes's creator Sir Arthur Conan Doyle. According to most accounts, Conan Doyle, led to Spiritualism after the death of his son in World War I, was a true believer, whom fraudulent mediums easily fooled. Claiming he had an open mind on the subject, Houdini attended his first séances in England with the help of Conan Doyle. Houdini's fourth venture into filmmaking, *The Man from Beyond* (1922), included some Spiritualist influences and a nod to the writings of his friend.

In *The Man from Beyond,* Houdini plays a seal hunter lost at sea and frozen into the Arctic ice in 1820, whose body is found and revived in 1920 by the scientist Dr. Strange. The explorer breaks up the wedding of the scientist's daughter and is desperate to marry her because she looks exactly like his fiancée of a century earlier. Her enraged father locks him up in a lunatic asylum. After grappling with mad scientists and a variety of restraints, the film ends with the hero and his young love at peace, while a "ghostly" superimposed image of the sealer's nineteenth-century fiancée eases into Felice Strange's body. As this miracle occurs, the camera cuts to a book Felice is reading, Conan Doyle's *The Vital Message,* and the quote, "The great teachers of the earth—Zoroaster down to Moses and Christ . . . have taught the immortality and progression of the soul."[32]

A séance was at the heart of Houdini's eventual split from Conan Doyle and from the Spiritualist community. Houdini and his wife joined Conan Doyle and his family in Atlantic City in the summer of 1922, and during a séance in the writer's hotel room, Conan Doyle's wife contacted Houdini's beloved and dead mother, Cecilia Weiss, and recorded her pronouncements in a bout of automatic writing. The fifteen-page transcript included, "God bless you, too, Sir Arthur, for what you are doing for us—for us, over here—who so need to get in touch with our beloved ones on the earth plane."[33] The Doyles were quite pleased with the results, and Sir Arthur later noted that Houdini had been visibly shaken and moved. Conan Doyle's wife surely meant well, but Houdini seethed. His "mother's" elocution seemed oddly formal to him; he also claimed she should have dictated in German, not English, which she didn't know; likewise, the content of her message didn't include any revealing personal references; further, Houdini had chosen his mother's birthday for the séance, and he felt that "if it had been my dear mother's Spirit communicating a message, she, knowing her birthday was my most holy holiday, surely would have commented on it."[34]

When Conan Doyle returned for his second lecture tour of America in 1923, Houdini finally began to air his skepticism about Spiritualism and about the Atlantic City séance as well. Soon the two friends were exchanging angry retorts via the *New York Times* letters page, at turns denouncing one another and their respective beliefs. The newspaper war continued throughout Conan Doyle's lecture tour, aiding their mutual needs for publicity but ending any semblance of a friendship. Houdini's book, *A Magician among the Spirits,* a long exposé of Spiritualist frauds, continually pointed to Conan Doyle's credulity. The title page of Conan Doyle's copy of Houdini's book has this comment from Conan Doyle: "A malicious book, full of every sort of misrepresentation." In his marginal comments, Conan Doyle frequently used the words "bosh!" and "rubbish!"[35]

Conan Doyle was particularly disturbed that Houdini's skepticism was limitless. He noted that Houdini never explained what would be credible evidence. In his autobiography, Houdini wrote, "Were I at a séance and not able to explain what transpired it would not necessarily be an acknowledgment that I believed it to be genuine Spiritualism."[36] Conan Doyle added the exasperated marginal note, "This really means that nothing could convince him."

The Medium Stripped Bare by Her Bachelors, Even

If Houdini briefly had lapsed into the role of earnest seeker, after the break with Conan Doyle, he became an enemy of all Spiritualists. Houdini commented in a 1925 article in the *New York American,* "There's a regular tidal wave going around the world. There should be a law passed that anyone pretending to be able to communicate with the spirits ought to prove it before a qualified committee."[37] In fact, Boston, Chicago, and several other cities did pass anti-fortune-telling laws, and officials often included séances within the jurisdiction of such laws. Houdini pushed New York congressman Sol Bloom to propose a similar law for Washington, D.C. When the law was considered in 1926, Houdini testified before a congressional subcommittee. Despite Houdini's colorful confrontations with the Spiritualists in the audience, the bill never went beyond draft form.

Arguing, as had dozens of magicians before him, that it took a skilled trickster to spot another skilled trickster, Houdini insinuated himself into the public eye as a writer of articles denouncing Spiritualists in *Pop-*

ular Science Monthly, Scientific American, and daily newspapers. He also gave lecture tours debunking fraudulent Spiritualists and served on committees investigating—and ultimately rejecting as phonies—Spiritualists who wished to claim prizes for their genuine abilities—often mediums who had previously been approved by more gullible men of science and business. Houdini also incorporated medium busting in his stage acts. As with the nineteenth-century efforts of Robert-Houdin, Kellar, and Maskelyne to reproduce occult effects by natural means, Houdini helped reestablish the magician's critique of Spiritualist fervor. And in a parallel to his earlier escape acts, Houdini was now metaphorically freeing the public from the bondage of superstition.

The efforts of Houdini and other stage magicians to either replicate Spiritualist effects or unmask them also had a misogynistic aspect, in keeping with fears of the "feminization" of daily life in the Progressive Era. Mediums tended to be women, and their workplaces often were their home parlors, the only place of power that society then accorded them. The press depicted the typical Spiritualist society member as female, past her prime and laughable. An 1893 cartoon featuring Maskelyne shows him in one corner strangling a serpent labeled "humbug" with the subtitle "*He is Rough on Spiritualists.*" Farther down some matronly women surround the conjurer above the subtitle "*The Ladies of the Spiritualistic Societies Will Persist in Claiming Him as One of their Own.*" One of the matrons says, "Why should you not own that you are a medium?" As in this cartoon, journalists tended to treat stage magicians as virile, top-hatted gentlemen while depicting Spiritualists as matronly, superstitious women—or effeminate men—prone to "intuitions" and to romantic but wrong-headed views of the world.

Houdini went to great lengths to strip mediums of respectability. His most-publicized nemesis was the Boston medium Mina Stinson Crandon. In the 1920s, few popular mediums were willing to brave a *Scientific American* panel that included Houdini, magazine editors, and several Harvard scientists. The magazine was offering $2,500 to any medium who could prove genuine psychic powers. Those who tried were "busted" by Houdini and subjected to public humiliation by pamphlet. On Conan Doyle's recommendation, the committee agreed to look at the work of Mina Crandon, called "Margery," to protect her anonymity.

Mina Crandon was the wife of a well-to-do Boston surgeon, Le Roi Crandon, and twenty-seven years his junior. By most accounts she was quite attractive. She was blonde, had blue eyes and a good figure, and was amusing and playful. She wore a silk gown during séances. According to her husband's records, during one séance, her breast began to glow with some mysterious substance, and afterward she insisted that one of the male séance attendees study her breast for his séance notes. Sexual energy, important to the charismatic appeal of preachers like Aimee Semple MacPherson in this same era, undoubtedly added to Mina Crandon's allure.

The spirit helper she relied on, her deceased brother Walter, was rude, foul-mouthed, and temperamental. Conan Doyle noted that such lower-class license was common in channeled spirits. Such personas delighted the middle-class séance attendees and provided a way for mediums to release frustrations. Le Roi Crandon described his wife's helper as follows, "As Walter says he (W) is no 'little sunbeam' or 'gladiola' but a full grown man who 'wears a 11$^1/_2$ shoe on a supernormal foot.'"[38]

Neither was Le Roi Crandon a "little sunbeam." This wealthy Boston surgeon was arrogant and dismissive. Writing to Conan Doyle, he commented, "The minute the materialistic and coldly scientific paper such as the Scientific American opens its more or less respectable doors to admit the validity of psychic phenomena the whole matter at once assumes a kind of respectability for many of the morons who inhabit the Main Street of America."[39] Constantly seeking Conan Doyle's approval, Crandon lambasted Houdini's *A Magician among the Spirits* and wrote to the British author, "My deep regret is that this low-minded Jew has any claim on the word American."[40] In another letter Crandon fawned over Conan Doyle, insisting that "all the faithful over here look on you as the great leader of this present world movement."[41]

Crandon also was, to say the least, protective of his wife, whom he called "Psyche." After explaining to Conan Doyle how he had required all the *Scientific American* panelists to submit their notes to him after each séance, he added, "If they ever make any announcement not consistent with these notes you can readily see I have the material to crucify them. We are not wasting any time in compliments or politeness. It is war to the finish and they know I shall not hesitate to treat them surgically if necessary."[42]

Le Roi Crandon refused to take time off from his medical practice or to let his wife travel alone to New York City for the *Scientific American* tests. Instead he urged the New York members of the panel to stay in Boston at his expense. While Houdini was off touring, the other *Scientific American* panel members became appreciative spectators at Mina's séances. In addition to Houdini, the panel included Boston engineer David Comstock, the somewhat skeptically minded psychic researcher Walter Franklin Prince, the less skeptically minded Hereward Carrington, and William McDougall, a Harvard psychologist with a taste for Spiritualism. Soon they were under the sway of their charming hosts. J. Malcolm Bird, the panel's secretary and an associate editor of the *Sci-*

entific American, became a fervent believer in Mina's abilities and wished to award the prize to her. Bird's interest in Mina Crandon may have involved more than simple admiration of her psychic powers. During séances, attendees often joined hands and sat in a circle at a table with the medium. Psychic investigators, when attending, usually flanked the medium to "control" his or her hands and feet (by holding or touching them) and so spot trickery. During the *Scientific American* trial séances, Le Roi Crandon sat to the medium's right and held her right hand, while one of the other panel members held her left. Throughout most of the test séances, Bird arranged to stand and control the right link (placing his hand simultaneously on Le Roi Crandon's and Mina's) while his left hand was at liberty, as one Houdini partisan remarked, "to roam." Le Roi Crandon and, most likely, Bird were colluding with Margery by the time Houdini joined the circle. Soon after Houdini's arrival, the panel agreed to dismiss Bird for collusion.[43]

No love was lost between Houdini and the Crandons. Crandon wrote to Conan Doyle, "Houdini is apparently all that you and other gentlemen have ever said of him, to which I shall be pleased to add a choice collection of adjectives."[44] Séances with Houdini were held in late July and in late August 1924. During the tests the spirit of Walter, speaking in the dark, swore at Houdini, accusing him of sabotaging an electric bell-ringing apparatus and of placing a folding ruler in a cabinet Margery was locked in. Walter thundered, "What did you do that for, Houdini? You God damned son of a bitch. You cad you. There's a ruler in this b[c]abinet, you unspeakable cad. You won't live forever Houdini, you've got to die. I put a curse on you now that will follow you every day until you die."[45] This was hardly the comic relief that séance attendees might have desired from the colorful Walter.

Houdini also came into conflict with another of the *Scientific American* panelists, Harvard psychology professor William McDougall, a psychic researcher and an ardent champion of a romantic brand of psychology much like that of William James, his predecessor at Harvard. McDougall viewed the rising mechanistic school of psychology, eventually to be embodied in behaviorism, as a threat to his worldview, which included the notions of the reality of both the human soul and free will. Though not prepared to endorse Margery, McDougall questioned the integrity of Houdini's eventual exposé of Margery. Houdini responded to McDougall with ridicule. The *Boston Herald* ran a photograph of Houdini holding bonds worth $10,000. Five thousand dollars would be given to the Crandons if he could not duplicate all their séance effects, and another $5,000 "to a Harvard professor if he will consent to be thrown into the river nailed in a packing case."[46]

Houdini undoubtedly was in Boston to discredit Margery and not to crown a genuine medium. One of Houdini's colleagues, magician Joseph Dunninger, later noted that when Houdini was certain someone was a fraud or a threat to his authority, he would find a way to destroy the person's credibility, with or without proof of fraudulence.[47] The Crandons' malice made the job easier. Houdini approached his work with zeal. His pamphlet, *Houdini Exposes the tricks Used by Boston Medium "Margery" to win the $2500 prize offered by the Scientific American,* describes some of the rigors he underwent in order to reveal her frauds. One of "Walter's" tricks was to depress a button on a box that then completed an electric circuit to ring a bell. The other panelists believed that Margery kept her feet far from the box when the bell rang in the dark under the séance table. Houdini thought otherwise and prepared with a fetishist's taste for pain and detail: "Anticipating the sort of work I would have to do in detecting the movements of her foot I had rolled my right trouser leg up above just below my knee. All that day I had worn a silk rubber bandage around that leg just below the knee. By night the part of the leg below the bandage had become swollen and painfully tender, thus giving me a much keener sense of feeling and making it easier to notice the slightest sliding of Mrs. Crandon's ankle or flexing of her muscles."[48]

Houdini remarked that for the séance she "wore silk stockings and during the séance had her skirts pulled well up above her knees." And when he did feel her foot moving in the darkness, the moment of recognition had a conceivably erotic charge. "I could distinctly feel her ankle slowly and spasmodically sliding as it pressed against mine." Houdini's thrill during this game of footsie was at the very least that of a hunter who had finally caught his prey. The pain that he had submitted himself to had helped to guarantee this pleasure. Although Crandon later referred to Houdini's pamphlet and articles as "sewage," the magician succeeded in discrediting Margery. The prize, almost hers, was denied. "Walter" got in one last dig at a séance held on October 4 that year, when he remarked, to the amusement of the Crandon circle, "Say, write a letter to H— [Houdini] as follows: We have read your fiction with interest. W— [Walter] says to give you his love and that he will see you BEFORE LONG. He will have tea nice and hot for you and also a long fork."[49]

Houdini Lives!

In the two remaining years of his life, Houdini fulfilled his dream of traveling with a large-scale magic show. His tour revealed his fascination both with magicians of the past and with Spiritualism. A Houdini night of magic at the Shubert Princess Theatre in Chicago for 1925 included large stage illusions in the first act, escapes in the second act, and a third act dedicated to the exposure of the tricks of fraudulent mediums. The stage illusions relied on historic apparatus he had purchased. The second act included his meal tickets: Metamorphosis, his Needle Swallowing trick, and his latest escape, the Chinese Water Torture trick, in which he was bound in wooden hasps, manacled, and suspended upside down in a water tank. The final act was based on his anti-Spiritualist lecture tour. Under the subheading of "Do The Dead Come Back?" his program noted, "He is not a skeptic and respects genuine believers. He does not say that there is no such thing, but that he has never met a genuine medium." The program also included Houdini's $10,000 Challenge, "open to any medium in the world (male or female). He will wage the abovementioned sum, the money to go to charity, if the spiritualists will produce a medium presenting any physical phenomena that he cannot reproduce or explain by natural means."[50] Perhaps reflecting frustrations with hecklers he'd faced during his lecture tours, the playbill included the following notice: "At no time, however, will he discuss the Bible, or Biblical quotations, before the audience."

While no medium ever collected on Houdini's $10,000 Challenge, a challenger of a different sort proved his undoing. In 1926, when Houdini brought his show to Canada, several McGill University students visited him backstage in Montreal. One of the students, Wallace Whitehead, subjected Houdini to a grilling. He first tested Houdini's publicized skill of being able to predict the plot of an entire mystery novel, if only given a summary of events from the first few pages. He next asked Houdini "his opinion of the miracles expressed in the Bible, and looked taken aback when Houdini declined to comment on 'matters of this nature.'"[51] And while Houdini lay on his side on a couch, nursing an ankle he had recently broken performing the Chinese Water Torture, Whitehead asked if it was true that the performer could withstand hard punches to the abdomen. Before Houdini could fully rise to his feet, Whitehead began to viciously punch him. The magician's appendix was ruptured. He did not seek immediate treatment for this fatal malady. The myth of Houdini's own physical invincibility, heralded in dozens of dime novels and publicity posters, finally led to his death on October 31, 1926, Halloween.

During his life, Houdini made pacts with his wife and with several friends to attempt to communicate with one another from beyond if there indeed was an afterlife. His wife, Bess, offered ten thousand dollars to any medium who could tune in to a message from Houdini that would be in the agreed-upon code—which he and his wife had used together in a mind-reading act. In January 1928 Bess Houdini wrote to Conan Doyle describing her efforts. One night at midnight when she called out to Houdini, there was a loud report like a shot in the bathroom and she discovered the mirror had split open. She informed Conan Doyle, "It is the first time anything has occurred that has the slightest bearing on our compact. I called and pleaded again, and again, but that was all I heard."[52] The message Houdini had promised to send, in code form, was "Rosabelle Believe." This was derived from the lyrics of a love song popular during their early courtship, the lyrics of which he had had inscribed in her wedding ring.

In January 1929, one year after the mirror-cracking incident, minister Arthur Ford of the First Spiritualist Church in New York City gained the coded sequence for "Rosabelle" during a séance and reported that it came from Houdini. He delivered the full message several weeks later, and Bess Houdini temporarily was convinced. Preempting a later obsession with Elvis's immortality, newspaper headlines on January 8, 1929, read "Houdini Lives!" Two days later the *New York Sun* announced that the message was fraudulent. Apparently, Ford had admitted to a female reporter that he had not received the coded words from the spirit world. Ford, however, insisted that the reporter had been out to get him and that he had been maligned. Houdini's friends claimed that the codes had been revealed in biographies of Houdini before the séances. Bess Houdini also ultimately declared that she never received a satisfactory message and in so doing underlined her and her husband's doubts that there was "anything in this belief in Spiritism."

Though Houdini had been plagued with imitators during his career, after the magician's death, imitating Houdini gained a new dignity. The re-creation of Houdini's tricks and the busting of phony mediums became a rite of passage for young stage magicians. Such imitators ensured that the nineteenth-century anti-Spiritualist stance of magicians would be maintained. Yet few of Houdini's imitators had a relationship with the occult world as ambiguous as that of Houdini. Houdini did not simply follow the tradition of policing the lines between "honest" and dishonest trickery. His own fascination with the otherworldly fueled his zeal for exposing frauds and searching out the genuine article.

Enchantments in the Age of Disenchantment

For convenience, historians often isolate the late nineteenth century in America as an age of realism, when notions of progress rooted in simple facts and rational laws prevailed.[53] According to this description, the American Civil War and its horrors put to rest religious-based hopes in the perfectibility of humanity and ushered in a desire for scientific progress based on documentation and objectivity. This description, however, ignores the fact that from the beginning of the scientific project, the technological vision of progress has overlapped with a religious vision of the progress of the soul and millennial hopes for humanity. Likewise, the push toward realism and mass production in the nineteenth century engendered and provided the means for promoting a countertrend: a public taste for illusions of all sorts.[54] New technologies offered means to make commodities of such illusions and helped prompt the twentieth-century culture of the image.

Both Spiritualist performers who ordered their tricks from magic supply catalogues and anti-Spiritualists took advantage of this public thirst for illusions. Spiritualists presented their wonders as the authentic work of spirits, while magicians offered them as playful re-creations of inauthentic originals. Magicians could be thought of both as stripping away illusions and as relying on technology to re-create miracles in the same way that Bowery theaters and resort hotels at the turn of the twentieth century presented re-creations of spectacular events like the Johnstown Flood or the eruption of Mt. Vesuvius at Pompeii. The performer who asked "If Not Spirits, What Is It?" in Genoa, New York, might have appealed to an individual spectator who required neither a rational nor an irrational explanation of such stage spectacles. Seeing may not have led to "belief" but to pleasure. In such cases "wonder" no longer need serve as a prompting to intellectual inquiry.

The age of realism was also the age of deceptions and, as such, ripe for the stage magician. Much like the Wizard of Oz in L. Frank Baum's classic tale, the magician was an expert in deception and manipulation. Baum, who edited a journal for department store window designers, depicted his wizard as a Barnumesque character, a wonderful humbug, able to manipulate and influence the innocent munchkins with the apparatus of deception, using skills like those of the growing advertising industry at the turn of the century.[55] Within the context of rising commercialism, the magician presented a triumphant model—the master of deception from whom advertisers and corporations had much to learn. The stage magicians' ability to flourish while wonder showmen like hypnotists and Spiritualists sputtered likely had to do with the modernist core of their performances. The magicians' stance as "honest tricksters" who only simulated occult wonders with natural means enabled them to offer the public a glimpse of the apparatus of deception enveloping them.[56]

Magicians gained from their identification with science. The "science" that they supported was the matter-of-fact brand then promoted in magazines such as *Popular Science Monthly* that often featured explanations of magic tricks and other theatrical stagings. Like the publishers of such popular science magazines, magicians reveled in anti-Spiritualist exposés. Not only could these men duplicate the Spiritualists' spectacular effects, but they also offered another—possibly misogynistic—thrill as offstage and on, the male magician stalked, then stripped the garments of honesty from, the Spiritualist medium, exposing her as a naked fraud.

If the magicians reinforced cultural stereotypes, then they also exposed the manipulation inherent in commercial culture. The magicians were not precisely educating their audiences, but "wising" them up to the art of deception. From the magicians' point of view, the passion for wonder had become an appetite that could never lead to illumination but which might, if properly arranged, provide light entertainment. Wonder was no longer located in "objects of wonder" but in the magicians' skills and craft. Recognizing the disenchantment of the world, the stage magicians were offering a variant on the wonder show: one based in natural processes that exalted human ingenuity and potential.

CHAPTER FIVE

The Mind Reader

Psychologist Joseph B. Rhine began his rise in psychic research in 1927, when he and his wife, Louisa E. Rhine, examined "Lady Wonder—the Educated Mind Reading Horse" of Richmond, Virginia. Adults who paid one dollar and children who paid fifty cents were allowed into Lady Wonder's stable, where she spelled out answers to questions and solved mathematical problems by flipping lettered or numbered cards with her mouth. The Rhines concluded that "Lady" had no thinking ability but did show signs of telepathy. They reported that even when hidden behind a screen, the horse was able to pick out numbers shown only to her trainer.[1] William McDougall, then head of Harvard's psychology department, as well as one of the *Scientific American* panelists who had tested the medium Mina Crandon, encouraged Rhine's research on Lady Wonder. A longtime enemy of the materialist vision of psychology that had crystallized as behaviorism, in 1928 McDougall left Harvard to chair the psychology department at Duke University. Rhine soon after joined him and remained as the head of Duke's parapsychology laboratory until 1965. At Duke, Rhine's extensive experiments with ESP in the 1930s helped bring renewed legitimacy to psychic research—a field on the wane since its earlier champion William James's death in 1910.

Though his work was frequently attacked, Rhine managed to make psychic research respectable enough for academic debate. Instead of relying on anecdotal evidence about sightings of ghosts and apparitions, dream messages, or bewildering reports from dark séance rooms, Rhine set up ESP experiments with easily quantifiable results. Such research had a progressive component, as it stressed human achievements, deemphasized psychic research's seemingly morbid interests in the utterances of the dead, and questioned the paradigm of "abnormal psychology" that most psychologists relied on to explain the uncanny. His experiments also had a subversive component, as they relied on scientific methods to question the mechanistic and materialistic worldview that then still pervaded the scientific project.

Others had pioneered the route for Rhine. At the turn of the twentieth century, advances in communication technology had strengthened intellectual interest in mind reading and the paranormal. In the 1890s, the scientist and psychic researcher Oliver Lodge proposed a theoretical model for radio and an analogous model for thought projection. "Disturbances in the ether" might be projected from mind to mind or radio set to radio set. A decade later, when the public was celebrating Marconi's first success in broadcasting telegraph signals, a husband-and-wife team of mind readers, the Zancigs, became enormously popular in England and America, prompting one headline, "Is It Telepathy Or What? Probably What."[2]

Just as the advent of telegraphy in the 1840s marked the beginning of the Spiritualist movement and fascination with the "Spiritual telegraph,"[3] the promise of radio technology at the end of the nineteenth century prompted interest in miraculous human mental powers. The performances of stage mind readers fueled these speculations and dramatized the millennialist hope that humanity would unfold new mental powers.

The Rhines' willingness to examine the "mind-reading" horse Lady Wonder twenty years later illustrates how scientific "explorers" and entertainers fueled interest in both ESP and the paranormal, as the spheres of scientific research and entertainment interpenetrated. If technical innovators like Lodge helped legitimize these interests, the turn-of-the-century mind-reading show helped set the terms of the debate about the authenticity of parapsychology and ESP; although Rhine was able to capitalize on this interest and make ESP research a briefly respectable field, his work never managed to shake free of its occultist and popular culture origins.

"The Curtain of the Mind Uplifted"

In the nineteenth century, not only did entertainers contribute to public fascination with the paranormal, but often they—however genuine or phony their presentations—were the primary explorers of a "frontier" that skeptical scientists thought of, at best, as a swampland. American neurologist George M. Beard put the case in a slightly derogatory light when he remarked, "In the history of science, and notably in the history of physiology and medicine, it has often happened that the ignorant and obscure have stumbled upon facts and phenomena which, though wrongly interpreted by themselves, yet, when investigated and explained, have proven to be of the highest interest."[4] Often the "ignorant" were performers able to sway a public and influence researchers, however briefly. Who were the "ignorant and obscure" performers that managed to intrigue the public and the scientific community about parapsychology? What were their acts like? How did the marvels they presented influence popular interests and research?

Nineteenth-century mesmerists initiated the dramatic presentations of mind reading that led to a faddish interest in telepathy by the century's close. These early magnetizers, who existed in a twilight zone between showmanship and science, claimed to be in telepathic communication with their subjects and also claimed that their somnambulists showed heightened sensory awareness and intelligence, and possible clairvoyance. Soon stage magicians mimicked the mesmerists' acts, and the mesmeric and magic streams merged in the "mind-reading" specialists who emerged in the late nineteenth century.

Public fascination with mesmeric clairvoyance in the mid–nineteenth century led French magician Jean Eugène Robert-Houdin and Scottish magician Professor John Henry Anderson to add "Second Sight" acts to their performances. Robert-Houdin would wander among audiences, examining items that spectators showed him, then ask his blindfolded son to identify them. The performers used an elaborate code, with simple variants on seemingly banal statements as cues. By the late nineteenth century, some performers began to specialize in mind-reading acts, often to the bafflement of scientific experts.

Dr. S. S. Baldwin, aka "the White Mahatma," a comedic psychic performer who ran something of a medicine show in the late nineteenth century, was an innovator in the mind-reading game. A clipping from 1882 in England notes, "Mr. Baldwin is an American endowed with a perpetual flow of the native humor of his country, which serves to divert his audience as much as his skill astonishes them."[5] Like Katterfelto and countless performers before and after, Baldwin loved to coin and use impossible words like "somnambumancist," "somatic indigitation," and "asomatous sejunction." Unlike his contemporary, Bodie, Baldwin hinted that his performances and utterances were deceptive. One of his programs stated, "Professor Baldwin is an escomateur and 'deceptionist' who desires to produce an exciting, interesting, and bewildering entertainment, and *all* his talk must be understood as being to that effect, and must be taken 'cum grano salis.'"[6] Baldwin revealed the tricks of Spiritualists as one part of his act. Yet his wife, Miss Kittie Baldwin, who also performed in the show as a clairvoyant, was subject to "Somnambumistic Visions." An admitted trickster, Baldwin developed the "pad system" later passed off as genuine by many other mind readers, such as Anna Eva Fay. This act required that audience members scribble out questions on pads of paper. The spectators would keep the questions, sealed in envelopes, but assistants would collect all the pads. The bottom of the pad was treated with wax, so the question could be read and fed to the mind reader. Anna Eva Fay's daughter, Anna Fay, added the innovation of hiding a telephone receiver in her costume to get more details.

One of Anna Eva Fay's assistants, Washington Irving Bishop, went solo in the 1880s.[7] His "muscle reading" act stirred public fascination in America and in England, where he evoked the ire of Spiritualist hunter John Nevil Maskelyne. Bishop, who at times used the term "Bishopism" to describe his ability to read thoughts, also patiently submitted to the tests of scientific panels in England. If Baldwin styled himself a "native" wit whose speech was "redolent of the humor characteristic of Mark Twain and Artemus Ward," Bishop preferred a presentation keyed to high sentimentality. One program includes the lyrics to one of Bishop's ballads, each of the four stanzas starting with "Good night! my baby; sleep, for love is here / To guard thy slumber, tiny soul."[8] Bishop was a polished performer who denounced Second Sight acts and even wrote a pamphlet, titled "Second Sight Explained," which revealed the codes a magician might rely on.[9] This feat is a testament to Bishop's popularity, as rarely would a "mystic vaudevillian" dare to take on the magicians, who preferred to do the "unmasking."

Bishop's specialty was "muscle reading." Blindfolded, he could find hidden objects or spell out words that subjects touching him were concentrating on. The vogue for muscle reading began in the 1870s with the performer J. Randall Brown, many of whose assistants, like Bishop and Stuart Cumberland, went solo. Bishop, however, demonstrated muscle reading at its finest. In the early 1880s, Bishop, blindfolded, with two subjects touching his wrists, drove a carriage at a rapid clip through the streets of Manhattan to find a hidden gem. A combination of muscle reading and the ability to put on a blindfold that one could see through helped others perform such publicity stunts, which became a traditional opening act for muscle readers. In England, Bishop gave performances before enthusiastic audiences, including the queen.

Unlike Baldwin, who insisted that his performances be taken "cum grano salis," Bishop insisted he was the genuine article, a mind reader. He was eager to let scientists examine him in England, and a writer for *Nature* indicated that Bishop's performances often resembled "in miniature a soirée of the Royal Society."[10] Bishop told a Royal Society panel that included Francis Galton, George J. Romanes, and Ray Lankester that he was not averse to the hypothesis of muscle reading, though he had no idea himself how he managed his feats. As such, he was like an artist who responded to queries of "How do you do it?" with the answer, "I prefer it remain a mystery, even to myself."

Neurologist George Beard, for one, was annoyed by all the scientific attention Bishop was receiving in England. As early as 1874, Beard had deduced that "unconscious muscular action" could explain the mind-reading act put on by "the celebrated Brown" who toured America in the 1870s and for weeks "held the American people by the nape of the neck, controlling the press as absolutely as a Napoleon or a Czar."[11] Beard had examined mind reader J. Randall Brown before a New Haven Music Hall audience of one thousand in 1874, and over the protests of Yale faculty and the audience, insisted Brown's abilities were not examples of "thought transference" but genuine examples, instead, of unconscious muscle reading. When Bishop, according to Beard an inferior performer to Brown, began to grow in popularity, Beard was perturbed that muscle reading was suddenly "exciting so much inquiry in the neurological world" while his articles about Brown had gone ignored.[12]

Brown, Bishop, and others, according to Beard, worked strictly by sensing which direction to move in and when to grab or point to an object, this all based on noting tension or relaxation in the muscles of his or her guide. Yet their acts could elicit thrills from audiences. As Beard mentioned, often the blindfolded mind reader, after carefully making contact with a subject, would then tear across the room or through the aisles of a crowded hall at top speed and stop at precisely the right spot to find a watch hidden in someone's hand. In other cases the mind reader might lift a hat off one audience member, move across the room, and leave it on someone else's head, as was "willed" to him.

The renewed interest in mind reading in the late nineteenth century helped these performers flourish. Back in New York, Bishop continued to stir up controversy and to use an elegant tone in his staging and promotion. An eight-page program for Bishop, full of noble sentiments and diction, for a New York performance of 1887 suggests how an audience member's excitement may have built while seated in the comfortable theater surrounded by the murmur of other spectators' voices and the swishing of program pages as they waited for the controversial Bishop to appear on stage in his tails and medal-daubed coat.

The program begins: THE ENIGMA OF THE CENTURY. Other boldface headings included "The Curtain of the Mind Uplifted!" and "Mr. Washington Irving Bishop, the Original and World Eminent Demonstrator of the Phenomenal Power of Mind Reading has consented to give a public demonstration." The program lists a page and a half of important personages before whom Bishop had performed, including the czar of Russia and his family, the queen of England, dozens of other members of European royalty, along with such American luminaries as the preacher Henry Ward Beecher, Oliver Wendell Holmes, and, to appeal to New York's uptown German Jewish population, Rabbi Gustav Gottheil of Manhattan.

In formal diction, the program describes many of Bishop's muscle-reading feats. The illustration on the back shows the handsome, blindfolded Bishop, in formal attire, with a hand over his shoulder so his fingertips just meet those of the elegant Princess Alexandra of England, "Her Royal Highness the Princess of Wales," while he spells out with his other hand the secret "endearing name" by which she called her sister. Such an illustration rhetorically suggested that Bishop's talent, personality, and courtliness placed him on a social plane equivalent to that of aristocracy. Another of his popular feats was to re-create a murder scene that the audience had previously established, determining who had been killed by whom, using what weapon.

Bishop's stagings encouraged a romantic more than a scientific appreciation of his abilities. Some accounts describe Bishop as being keyed up during performances and drained by the level of concentration required. He told the English scientists he was in a state of "dreamy abstraction" while performing. Professor Leonidas, the turn-of-the-century hypnotist and pamphleteer who occasionally put on performances of "contact mind reading" like those of Bishop, insisted that he would receive a "visual hallucination" when he neared a place in a hall where an object had been secreted. Leonidas also said of the driving stunt, or "street test," that "it is rather weakening, but this is a weakening business. I am not giving my hearty endorsement to this class of work. It is not productive of long life and happiness."[13]

Leonidas might have had Bishop's life story in mind when he penned these lines. Bishop's life was short and his end macabre. After a controversial tour during which critics charged Bishop with fraud, he collapsed on stage toward the conclusion of a performance at the Lamb's Club in New York on May 12, 1889. Attempts were made to revive him by "electric shock" and brandy, but he was soon pronounced dead. Those intimate with him, however, knew that he, like a hero in a Poe tale, often fell into cataleptic spells.[14] His distraught mother, also subject to such spells, accused the doctors who performed Bishop's autopsy of murdering him, likely because they wished to study Bishop's phenomenal brain. Adding a particularly macabre touch, Bishop's mother declared, "I have a witness who heard my boy cry out 'Mother, help,' as the surgeon's saw entered

Contact mind reader Washington Irving Bishop lightly touches the fingertips of the Princess of Wales while spelling out the secret nickname for her sister. Bishop referred to his many performances before aristocracy as "researches into the little known Science of the Human Mind." Performing Arts Collection, Harry Ransom Humanities Research Center, University of Texas at Austin.

his brain [during the autopsy]. . . . In my son's clothing was at the time a paper directing that no autopsy should be held, as he feared that just such a mistake should be made."[15] The following year, Bishop's mother, a devout Spiritualist, began distributing photographs captioned "The Murdered Mind Reader" showing him at rest in his coffin.[16]

Just as hypnotism, largely on the basis of Charcot's theories, led to public concerns about mental derangement, Bishop's death touched off a small panic about the connection between mind reading and catalepsy.[17] If anything, these new worries increased public interest in telepathy. Dozens of similar acts were launched. Hypnotists also turned to muscle reading as a secondary act. Hypnotist P. H. McEwen mixed mind reading with his demonstrations, as did Professor Leonidas. Leonidas cautioned hypnotists, when appearing as mind readers, to drop the "Prof." before their name and to take the title of "Mr." more appropriate to this intuitive profession. Leonidas's book codified the muscle-reading act and explained how the "street test," whether performed on foot or in a carriage, was crucial to drumming up an audience. He also offered the up-and-coming mentalist this copy for eventual posters: "There journeys a stranger from the far east, a man of mystery, a student of Oriental Sorcery, an adept in the fields of Mental Power, a reader of unuttered thought. A Seer. From the east he comes, and unto the east he shall return."[18]

Contact mind reading led to yet another fad for "no contact" mind readers who performed the same stunts without actually touching their guides. In the early twentieth century, George Newmann, aka "the Great Newmann," a hypnotist, magician, and mind reader, launched his own traveling show in the West, the Mystic Vaudeville Company, which included stage illusions, anti-Spiritualism, hypnotism, mind reading, and motion pictures. He stressed that he was the "original no contact mind reader." He performed the blindfold driving test early in his career with a carriage and horses, and later with an automobile, but only "under favorable weather conditions," and his posters warned citizens to "Stay on the Sidewalks and guard the kiddies."[19] Dropping the gothic trappings of Bishop's act, Newmann, as a twentieth-century man, preferred to emphasize the psychological entertainment he provided.

Whether or not the proliferation of mind readers is entirely to be credited, telepathy was "in the air" in the 1890s. Performers piqued interest, while the technically minded relied on new technology as a metaphor for the likelihood of the phenomenon. Several years after Bishop's death, Mark Twain wrote an essay indicative of this change in public attitude. "I have never seen any mesmeric or clairvoyant performances or spiritual manifestations which were in the least degree convincing," Twain wrote, ". . . but I am forced to believe that one human mind (still inhabiting the flesh) can communicate with another."[20] In the essay he thanked the Society for Psychical Research in England for having "convinced the world that mental telepathy is not a jest, but a fact, and that it is not a thing rare, but exceedingly common."[21] Twain remarked that he'd hoped to include some notes on telepathy in *A Tramp Abroad* (1878), but his editors had persuaded him not to, and he admitted that he had also feared the public's attitude. However, by 1891 notions concerning such matters had changed mightily.

Twain's essay included several anecdotes conceivably psychic in origin, generally on the theme of the "crossed letter," with one party responding to a request before the letter has arrived. Twain, an aficionado of inventors and an occasional visitor at Tesla's workshop, concluded his essay with a suggestion for a mind-reading device that may well have been serious: "This age does seem to have exhausted invention nearly, still it has one important contract on its hands yet—the invention of the *phrenophone;* that is to say a method whereby the communicating of mind to mind may be brought under command and reduced to certainty and system."[22] Twain argued that mind reading was likely activated by a "finer and subtler form of electricity," which inventors needed to capture and work with. He concluded his essay with the note that "while I am writing this, doubtless somebody on the other side of the globe is writing it too."

Twain, to some extent, was following the lead of the technical press. Turn-of-the-century interest in telepathy was not limited to the occult fringe. Articles in technical journals outlining the new possibilities of technologically enhanced senses appeared as early as 1891. In that year, the trade magazine *Electrical Review* ran a column from an anonymous contributor known as "the Prophet" who eloquently addressed the theme. The writer outlined how the human ear could hear only a narrow range of pitches and suggested that a recording device might be developed to capture sounds beyond this range. As the Prophet put it, "For aught we know, the air may be at all times filled with most beautiful music. . . . Shall we ever be able to listen to the . . . mysterious music of nature who has for centuries wasted its sweetness upon the dull ear of mankind? Science answers with the phonograph, and says that by its aid we may annex, perhaps, another world."[23]

In 1892, the *Electrical Review* published two articles by leading engineers that speculated on the possible mechanism behind telepathy. In the first, the telephone engineer J. J. Carty, later to become the chief engineer for research and development at AT&T, developed a model for telepathy using as his basis the workings of telephone exchanges. He noted that in phone exchanges, messages sometimes could get crossed when the current in one wire induced a corresponding current in another wire it was not touching.[24] This could happen even when both wires were "perfectly insulated." The author speculated that nerves in the brain were also sheathed to prevent induction, but mishaps likely could occur. "If it is conceivable," he went on to inquire, "that one nerve might act upon another without contact, why not one mind upon another?" Carty apparently anticipated the similar theory set forth by the physicist and psychic researcher William Crookes in 1896. Just as telegraphy had stimulated interest in Spiritualism, telephone and hopes for radio were stimulating interest in telepathy.

The second technical article on the feasibility of telepathy was based on a talk at the Franklin Institute in Philadelphia, given by Edwin Houston, a cofounder of one of the early electrical companies that eventually joined General Electric. Although Houston noted that he was indulging in speculation, he argued that when a subject is deep in thought, cerebral energy "is dissipated by imparting wave motions to the surrounding ether."[25] These waves could be imparted to another brain by way of "sympathetic vibrations," as when one tuning fork causes another to vibrate, or as in the "electric resonance" of Hertzian waves. Houston was arguing from the model of radio, not yet invented but frequently discussed. He went on to propose the possibility of a thought-recording machine. A lens would somehow impress thought waves on a "suitably sensitized plate," like one of Edison's wax cylinders, and a means could conceivably be found to "project" waves from the sensitized record so that others could receive the stream of thought.

Love, Marriage, and Telepathy

In December 1901, Marconi succeeded in broadcasting the brief telegram "S" across the Atlantic Ocean and so initiated the radio age. Radio was still in its infancy five years later when the Zancigs, the greatest of the turn-of-the-century mind readers, played upon the public's fascination with mental telegraphy and preference for it over the spiritual telegraph. Leaving behind muscle reading, the Zancigs revived the magicians' Second Sight act, in which one performer would covertly send coded messages to the other, through the use of either banal phrases or nonverbal signals, in order to convince an audience that telepathy was taking place. The Zancigs' act played upon notions of domesticity and

In this Victorian parlor setting, Julius Zancig stares at his pocket watch, while Agnes Zancig strains to determine telepathically what time it has been set to, perhaps by an audience volunteer. Performing Arts Collection, Harry Ransom Humanities Research Center, University of Texas at Austin.

love to explain their abilities. Successful appearances before the king and queen of England made their London run in 1906 and 1907 hugely popular.

The Zancigs, Julius and Agnes, were Danish immigrants to America at the turn of the twentieth century who met and married in Portchester, New York. She was working as a governess and he at a variety of menial jobs when they began to develop the elaborate visual and spoken codes on which they based their Second Sight act. During performances they both wore glasses, he rather spiffy in a white suit reminiscent of the tropics, she wearing simple flowered hats and high-necked, lacy Victorian dresses that draped on the floor. Their cowritten stage biography, *Two Minds with but a Single Thought* (1907), includes many photographs of the duo suggestive of the art of telepathy. In one such portrait, they face opposite sides of the frame in profile, he in his white suit with hand to the side of his brow, concentrating furiously, "telegraphing," while she looks out with a simple, open expression, hand cocked gently to ear, "receiving." Other photographs show them being recorded via phonograph, to capture the brilliance of their predictions, to suggest their modernity, and to allow linguistic experts to study their speech for codes.

They reached the height of fame in 1906, when they began a long, controversial run at the Alhambra Music Hall in London. As in one of Barnum's museum displays, the issue of "authenticity" keyed public fascination. In London, the Zancigs were the object of ceaseless attention and speculation and received daily newspaper coverage. Not everyone was swept away. According to one dour report titled "The Zancig Fever," "M. and Madame Zancig . . . only a few months ago . . . were giving an open air performance in the Isle of Wight, thankful for a few stray coppers that came grudgingly from a seaside audience, and in the twinkling of an eye they are exciting the interest of the most enterprising newspapers of the day, setting the most distinguished scientists by the ear, filling the coffers of a music-hall, and receiving the hall-mark of a 'command' performance before Royalty. It is the most triumphant thing accomplished by humbug in recent years."[26]

Their popularity in London began in 1906 after the Zancigs gave a demonstration at the *Daily Mail* that led to an article titled "The Cleverest Music-Hall 'Turn' on Record. Thought Transference Through Closed Doors. Tests at the 'Daily Mail.' What is the Mystery." According to reporters, when Julius and Agnes were facing opposite corners of the room, he was able to telepathically project thoughts to her. She would identify objects and read passages from books he held. The report also stated that when placed in another room, Agnes succeeded in divining a line from a book that he had chosen. The reporters made much of the Zancigs' insistence that their ability was the product of love. "Long before they really ascertained by demonstration the possibility of this mysterious occult power, they knew that between them was that beautiful sense of harmony of which Plato discourses—a perfect affinity of soul; indeed, 'two hearts with but a single thought, two hearts that beat as one.'"[27]

Julius Zancig knew how to both flatter and slyly poke fun at his English spectators' credulity. In this article he remarked, "What a difference there is between American and English audiences. The Americans seem to care only for the amusement and wonder of it; the English take a deep and intelligent interest in it as a scientific phenomenon; they want to study it and think about it. English audiences seem to be more philosophical and appreciative of the scientific side of the exhibitions."[28] The *Daily Mail* ran daily coverage of the Zancigs. And on December 29, 1906, the newspaper began a five-part series written by Zancig titled "The Story of My Life." The London papers also ran articles that offered various theories of how the Zancigs managed their feats, and articles that asked why the Zancigs' manager was forbidding them to run test demonstrations before a scientific panel.

Media speculation continued. An article titled "Increased Mystification Last Night" described the audience's excitement, chatter, and attempts to control Zancig—for example, by demanding that he hold his hands behind his back—to prevent the use of a visual code. "Men rose out of the stalls and got near to Mr. Zancig as he examined articles, and noted his every movement with intense concentration. That one and all 'gave it up' was conclusive proof that nothing whatever of the mystery had leaked out."

The paper also ran hypothetical explanations of how the pair succeeded. One of the most bizarre was the "ventriloquism" theory that William Kennedy, an admiral—who one hopes was a better navigator than philosopher—promoted in his letter to the *Daily Mail*. Kennedy explained, "Mr. Zancig throws his voice to his wife, whose lips move and apparently utter the words, which are his." He also explained how she was able to write answers silently on a slate. In these cases, he "throws a whisper to her."[29]

Another zany explanation, which stressed the technical, came from an article that Zancig quoted in his book. This reporter insisted that the couple, rather like dogs, could emit and "hear vibrations" outside the ordinary range of human hearing.[30] A minister from Fulham became a true believer and scoffed at doubters who proposed as solutions "ventriloquism, codes, Morse signals by means of the eyelids and other equally absurd theories." This correspondent wondered why audiences ruled out "the only obvious solution to the problem—namely, thought communication in a highly-developed condition." He feared the reason was the same that kept people away from churches: that they had no appreciation for spiritual matters.[31] Like this minister, Sir Arthur Conan Doyle, never one to play it safe, later championed the Zancigs as authentic mind readers, despite the fact that his friend Houdini, who had performed with the Zancigs and once ran his own Second Sight act, assured him it was pure hokum.

More discerning observers explained that the couple used oral and visual codes. Julius Zancig's white suit made it easier for his wife to see his hand positions. Her thick Danish accent allowed her to fudge the last syllables of names, making coded transmission even easier. Reporters and audiences often recalled as "silence" Julius Zancig's seemingly harmless reiterations of phrases like "Tell me what I see" or "Now concentrate on what I see," which did in fact impart codes. Glasses helped Agnes Zancig appear to be facing forward when glancing to the side for cues. Julius Zancig's glasses were an excuse for him to "affect shortsightedness, and to move from left to right and up and down in focusing on the coin."[32]

The *Throne* continued its attack on the Zancigs and challenged them to prove their authenticity before a select panel, making note of Agnes

Agnes Zancig, left, and Julius Zancig, background right, perform their mind-reading feats while being recorded on phonograph so that experts might study their utterances. The Zancigs' dual autobiography was published in 1907. Performing Arts Collection, Harry Ransom Humanities Research Center, University of Texas at Austin.

Zancig's "frequent" hesitations and the many "contortions of her partner in the curry-cook's costume."[33] The Zancigs' manager intervened and insisted there would be no demonstrations.

Zancig played it both ways. At times he denied the team possessed occult powers, at times he affected a simple innocence and refusal to give the power a specific name, or again attributed it to that more familiar and unassailable thing called love. "Really though, we don't understand what the power is . . . we can only account for our power by explaining it on the ground of most happy union, perfect harmony and the development of a latent gift, which we firmly believe is shared by everyone."[34] In his book he said that when people attributed the couple's ability to telepathy or to codes, he would respond humbly, "Yes, your guess may be right, but we leave it to you."[35] During another interview, Zancig shaded his answer to imply genuine psychic power: "All the guesses as to how it is done are wrong; all save those that attribute it to something unmaterial."[36]

The manager of the Alhambra, Alfred Moul, forbade the Zancigs to give demonstrations before skeptical panels. Moul remarked to the press, "I have been loth [*sic*] to entertain the suggestion that they should submit to the investigating dissecting knife . . . some of the conditions which it has been sought to impose are absolutely unreasonable." Moul also said he did not want an ugly dispute to arise, and so turned down the *Throne*'s five-hundred-pound wager that it could prove the Zancigs to be frauds. Moul added, "Probably more people [in the audience] think it is a trick. . . . Of course, there are others who think it is a genuine case of thought-transference."[37] Moul also commented that it was not his business to unlock their "secret box—if they have one." And, referring to the muscle-reading acts of Washington Irving Bishop and others, he added, "It suffices for me and, apparently, the London public that they are in possession of entertainment faculties on lines which place them as far beyond ancient methods of thought reading and 'together hand in hand we will roam to find the blessed pin' old business as the first specimen of the omnibus stand in comparison to a thousand guinea motor car."[38]

Zancig, usually unflappable, also took potshots at his critics. He remarked in one article, "We are perfectly aware that there are hundreds of people who can do the work that we do. We never denied it. Yet they do not seem to have struck the public as very wonderful—are we to blame for that?"[39] Their successful run at the Alhambra, the press frenzy, and two congenial demonstrations they gave before King Edward and Queen Alexandra led a publisher to ask Zancig to prepare a book. Zancig recalled the queen asking if his book would "tell us the secret of how it is done?" He told her, "Of course, I could not give away the secret."[40] And he answered the same question coming from a reporter, with, "That would be letting the cat out of the bag."[41]

The book that the English public eagerly awaited, *Two Minds with but a Single Thought,* appeared in 1907. Zancig called it an "honest and open attempt" to teach its readers to be "mental operators." Anticipating psychologist J. B. Rhine's conclusions by about thirty years, he insisted "everyone in the world, under certain conditions, can impress upon the mind of some other person his own mental images." Telepathy was a "strange, subtle inherent faculty, latent in every normal individual."[42] The book again placed love at the foundation of the Zancigs' ability. The author reiterated the advice Zancig once offered in an interview, and which undoubtedly could be of use to anyone: "I should say to those who would develop the power: Find your other half, the alter ego, the one person who is needed to bring complete harmony into your life. Then, the rest is practice."[43]

Borrowing heavily from Zancig's "The Story of My Life" articles in the *Daily Mail,* the book describes how, after marriage, Zancig and his wife discovered, developed, and reaped the benefits of their mental link. Long before their days of glory, their ability surfaced in homely ways. He describes how he and his wife were continually surprising each other by purchasing tickets to shows the other one had also just purchased. One day when Zancig was in the midst of such a purchase, the box office clerk finally jumped in and said, "The joke's against you, Zancig . . . your wife bought the tickets ten minutes ago."[44] Likening telepathy to physical culture, Zancig claims he began to practice "strengthening" exercises. In public places like theaters, he would stare at the back of strangers' heads and get them to turn. Zancig found that "highly strung, nervous, brainy people [were] more responsive," women especially. Zancig also describes some of the taxing aspects of the business. His readers learn that it is difficult to transmit numbers and letters because of the danger that they can be scrambled or reversed in transmission, making, say, a *W* out of an *M*. Practice is in order. He also encourages any endeavors that require concentration or memory. Hence, chess, music, and games of chance are all of value to the would-be mental operator. Ultimately, as in most stage magicians' biographies, the entertainers make a trip to India, where they are feted, and swap secrets with Brahmans and sages. He concluded with visions of the duo's pleasant days in India and their opportunities to entertain sultans, kings, and queens.

Domestic bliss did seem crucial to the act's success. After Agnes's death, Julius tried to break in a series of partners but could not duplicate his former success. Eventually he became a psychic who gave consultations in Asbury Park, New Jersey. He was arrested in 1923 and found guilty of assaulting an elderly man also in the "theatrical profession." Zancig persevered, and in 1926 wrote a pamphlet called *Crystal Gazing*. There was no hedging on his psychic abilities now.[45] He pitched the book to an audience fascinated by the occult and astrology. He began the preface by claiming, "The World is at the Threshold of a New Spiritual Era and this is agreed upon by the most religious and [by] scientific clubs and societies." In this pamphlet he also attempted a scientific explanation of the sixth sense. "The photographic plate," he wrote, "can register impressions which are beyond the perception of our highest sense of sight. The X-rays have put us into relation with a new order of impression-records quite beyond the range of our normal vision. . . . Nature does not cease to exist where we cease to perceive her."[46] After explaining the principles and uses of the fortune-teller's crystal, he put in a pitch for the superior line of crystals that he had available.[47]

An Otherworldly Meteorological Bureau

In the late 1920s and early 1930s, at the same time that Julius Zancig, in his decline, was offering crystal ball consultations in New Jersey, scientist Joseph Rhine was renewing interest in telepathy, clairvoyance, and even Spiritualism with his experiments in the psychology department at Duke University. Rather than attend séances in darkened rooms like the psychic researchers before him, Rhine set up experiments in card guessing that could be analyzed statistically. In deportment, Rhine, the handsome, sober-minded founder of the American field of parapsychology, was the exact opposite of raffish showmen like S. S. Baldwin and Julius Zancig. While the Zancigs never failed to transmit a thought, Rhine's best laboratory subjects were able to guess cards at only slightly above chance. Rhine's colleagues agreed that he was a levelheaded man of integrity. But, inevitably, Rhine's work evoked professional hostility. And as he slowly became professionally ostracized, he directed his appeals instead to the public, further undermining his professional standing.

Why was the hostility he provoked inevitable? Philosophers of science have pointed out that a new hypothesis will inevitably be controversial if it questions a tenet of other scientists' worldview, that is, their ontological assumptions.[48] Rhine's romantic premises, which he shared with other psychic researchers, ultimately had more in common with the occultists they often deplored than with the majority of their scientific colleagues from whom they wished to win acceptance.[49] Rhine confessed that his conclusions ran counter to the known laws of nature. Such a stance was philosophically tenable, but as science either intolerably arrogant or heretical. Denunciations of Rhine in psychology journals stepped up when his second book became a Book of the Month Club selection and the Zenith Radio Corporation began a radio series about psychic phenomena that intentionally coincided with its publication. The radio shows included tests of the public's telepathic abilities and anecdotes about psychic occurrences, and encouraged audiences to purchase special ESP cards that Rhine's lab was marketing. Though Rhine attempted to distance himself from the show, its Barnumesque flavor tarnished his enterprise. And yet, Rhine's effort at bridging science and the paranormal can also be viewed as a heroic undertaking.

Before looking at Rhine's work and its impact, it would be of value to briefly outline the history of the field of psychic research. The first members of the Society for Psychical Research (SPR), founded in 1882, were intellectuals connected to Cambridge University in England. Henry Sidgwick, its founding president, was a professor of philosophy with a skeptical cast of mind. Another of its founders, Sidgwick's friend Frederic Myers, was a poet and a musician with a more romantic bent, tormented by his own agnosticism. Among the scientists and intellectuals who joined, there was a split between "doubters" and "believers," guaranteeing a controversy whenever any member declared he or she had witnessed a bona fide paranormal event.

William James, who helped found the American branch of the society and later served as the combined societies' second president, reflected the split between skeptic and believer within his own ample personality. He had been raised in the Swedenborgian Church, which followed the teachings of Swedish mystic and astral traveler Emanuel Swedenborg, so seemed a likely candidate for accepting spiritual phenomena as genuine. Despite this background, James professed agnosticism. He had a rigorous training in medicine, psychology, and philosophy, and was no dupe like Sir Arthur Conan Doyle or other "true believers." The governing skepticism of leaders like Sidgwick and James ended up alienating convinced Spiritualist members such as biologist Alfred Russel Wallace, and he and other believers often left the ranks. Other early members included physicist Oliver Lodge, Lord Rayleigh, J. J. Thomson, Lewis Carroll, William Gladstone, Alfred, Lord Tennyson, John Ruskin, and Sigmund Freud.[50] Freud's notion of the subconscious owed a great deal to the hypnotic studies of Pierre Janet and others who often published in the *Proceedings of the Society for Psychical Research.* Depth psychology was born of the studies of hypnotism and psychic research.

The Society for Psychical Research took as its founding statement in 1882 the following: "to investigate that large body of debatable phenomena designated by such terms as mesmeric, psychical and spiritualistic," and to do so "without prejudice or prepossession of any kind, and in the same spirit of exact and unimpassioned enquiry which has enabled Science to solve so many problems, once not less obscure nor less hotly debated."[51] They planned to return to the scientific study of wonders that Francis Bacon had once urged. Committees were set up to investigate thought transference, hypnotism, apparitions, physical mediums, and Reichenbach's experiments in animal magnetism. Articles published in the society's *Proceedings* from the 1880s into the 1890s show research interests in such topics as "veridical hallucinations"—the warning messages or images that a person suffering grave injury or death might project to loved ones; spirit communication after death to the living; clairvoyance—the ability to see hidden or distant things; and telepathy —the ability to transmit thought from mind to mind. William James later likened their work to that of a "meteorological bureau" collecting data on sightings of ghosts, apparitions, and so on.[52]

But the early psychic researchers' true passion was to seek evidence for "survival" of the soul after death. Ghosts were somewhat uninteresting because these manifestations seemed "stuck" in fixed patterns that did not necessarily indicate survival of personality or intelligence. Telepathy, on its own or enhanced by hypnosis, was a curiosity but did not ultimately shed light on the question of "survival." Inevitably, these researchers were drawn to the work of psychic mediums in the hope of witnessing "evidential" phenomena of the survival of the personality after death.

By the 1890s, psychic researchers had categorized Spiritualist mediums into two categories: physical and mental. Physical mediums presided at séances in dark rooms in which strange physical phenomena occurred: whether the sound of "rappings," table tipping, the appearance of strange lights, noises, voices, the apparently fleshy outgrowths from the Spiritualist's body, or the emergence of visible spirit bodies. The medium "Margery," or Mina Crandon, discussed in the previous chapter, is a superb example of a physical medium. Her helping spirit, "Walter," theoretically was not just providing her information, but operating on the physical plane, ringing bells, tapping people's knees, leaving thumbprints in wax, and so on. Several decades of testing such mediums led to flurries of excitement, charges of fraud, and no conclusive results. Discussing such work, Freud remarked that in all likelihood, "in occultism there is a core of facts . . . round which fraud and fantasy have woven a veil which is hard to penetrate."[53]

The psychical researchers' efforts to find genuine mediums made of them a professional "audience" of the séance performance. No doubt the researchers' thrill from attending their first séance dulled after repeated exposures. Recognizing that little was to be gained by spending time in dark séance rooms, trying to keep track of a medium's hands, feet, neck, and movements, the SPR largely gave up chronicling physical mediums' doings and turned to "mental" mediums whose work consisted in providing messages from the spirit realm. They sought "evidential" information, which, if not available through fraud and trickery, would indicate a genuine paranormal power in the medium. The abilities of one remarkable mental medium, Leonora E. Piper of Boston, eventually converted James and several other psychical researchers into true believers.

William James first discovered Piper in 1885, and to supplement his own sittings sent strangers to her who used pseudonyms to try and determine if she turned up genuine, "evidential" material. James was convinced she did. In 1887, Richard Hodgson arrived, fresh from busting the enormously popular Italian physical medium Eusapia Palladino. Mrs. Piper gave the highly skeptical Australian native detailed information about his friends and relatives in Australia and eventually made of Hodgson a convert. When Piper was brought to England, James, Hodgson, Lodge, Sidgwick, and Myers all agreed that she was the genuine article. When she related to Lodge an event he had no knowledge of concerning the childhood of one of his uncles, Lodge later verified the information and hired detectives to try to turn up the same information by asking questions. They failed. Throughout the 1890s, Hodgson conducted thousands of sessions with Piper. Private detectives frequently followed her to try to determine if she had a network of "spies" feeding her information. She eventually went on salary to the SPR so they could even more carefully chart her work.[54]

While proof of "survival" was the main goal of James and other psychic researchers, telepathy had not been left out. Numerous experiments, apparently successful, at hypnosis at a distance were carried out and reported in the journals. In the early 1880s, Janet experimented with the famous hypnotic subject "Leonie" in Havre, France, and found he could often hypnotize her at distances up to five hundred meters. At times, Leonie would carry out telepathically transmitted posthypnotic suggestions.[55] In this same decade Oliver Lodge, the British physicist, also wrote of his successful experiments with telepathy. The society's committee on thought transference also investigated the Creery sisters, who at first visit appeared to demonstrate telepathic abilities. A follow-up visit, however, revealed they were using elaborate signals. William James remarked that the investigations were a wash, even though "many of the earlier successes recorded of these children occurred when they were singly present. . . . Collusion under such circumstances can not well be charged."[56]

James summed up the accomplishments of the society in 1896 by suggesting that psychic researchers no longer had a greater burden of proof than their skeptical opponents. Their meteorological bureau had turned up enough evidence to make their pursuit respectable, and specific reports, like those on the well-documented Mrs. Piper, tightened their case. He noted, "If you wish to upset the law that all crows are black, you mustn't seek to show that no crows are; it is enough if you prove one single crow to be white. My own white crow is Mrs. Piper."[57] James argued that to dismiss such cases out of hand was to show a close-minded bias.

Looking back at the Society for Psychical Research's investigations, Joseph Rhine felt that much of this evidence, however sketchy, was of value. After reviewing past experiments in telepathy, he remarked in his 1934 treatise *Extra-Sensory Perception,* "Curiously enough, however, the facts seemed to require proof over and over—many, many times."[58] He was determined to be the last one to have to again prove telepathy, and, appropriate to the hero of a tale, he described this effort as his "quest."[59]

"Is Sense Necessary?"

Rhine was born in 1895 in Ohio to a religious farming family. Early on, he had considered a career as a minister. However, he and his wife, Louisa Banks, decided instead to work toward careers in forestry and earned doctorates in botany at the University of Chicago. They became interested in psychic research after reading Bergson's *Creative Evolution,* which offered a modern version of vitalism to replace a simple mechanistic worldview. The Rhines were also quite impressed by a lecture on Spiritualism that Conan Doyle gave during his tour of America in 1922.

Conflicted between his two interests, Rhine began teaching botany at the University of West Virginia but left to pursue a career in psychic research. After several years of professional frustration as a would-be psychic researcher, he received a Richard Hodgson fellowship at Harvard. After sitting in on several of the medium "Margery's" séances in Boston and concluding she was a complete fraud, he broke off contact with the American Society for Psychical Research, which had championed her. In Cambridge he befriended Walter Franklin Prince, who had also left the American Society for Psychical Research over the "Margery" controversy. After a year in Cambridge, the Rhines had become rather skeptical about psychic phenomena. He wrote in an unpublished article, "I think too, we are tiring of chasing the Psychic Rainbow or the Philosophic pot of gold."[60] Soon after, though, visits to the mind-reading horse Lady Wonder and other endeavors gave the Rhines another glimpse of the rainbow—at its end was Duke University and the patronage of William McDougall. Rhine arrived at Duke at McDougall's instigation, made use of its facilities, took seminars, helped McDougall in his unorthodox research interests, and eventually was hired on.[61]

Shortly after his arrival at Duke, Rhine began the telepathy experiments that made his name and helped found parapsychology as a field. He and his colleague, psychologist Karl E. Zener, developed a set of twenty-five cards based on five simple symbols, destined to eventually take on near occult significance. The "Zener deck," later marketed commercially by Duke's parapsychology lab, with royalties accruing to Rhine, included a circle, a rectangle, a star, a "plus" sign (which also looks like a cross), and two wavy horizontal lines. In tests of clairvoyance, subjects were required to stare at the backs of each of twenty-five such cards taken off the deck and guess at its symbol. The odds for each correct answer were one in five. Rhine further refined the clairvoyance test by placing screens between subjects and experimenters, and by requiring subjects to guess the entire sequence of the deck of twenty-five cards without any cards being removed or revealed.

In tests of telepathy, the experimenters "chose" a Zener symbol to think of and subjects had to guess what was in their mind. Again the odds were presumably one in five. Telepathy tests, begun with subject and experimenter facing off across a table, later were refined with subject and experimenter in separate rooms, communicating only a "ready" symbol with a telegraph key, and recording their results separately. Later, telepathy experiments were run with subject and experimenter in separate buildings, and at distances up to hundreds of miles from each other.

After three years of testing numerous subjects, Rhine published his findings in 1934. Relying on mathematical analysis that established that his subjects' results were astronomically beyond normal chance limits, Rhine announced that he had proven without a doubt the existence of ESP. He proposed that it was a common faculty, latent in most people, to be found in one in five students tested at Duke and, presumably, similarly distributed in the general populace.[62]

Rhine's prize subject was a young Methodist ministry student at Duke named Hubert E. Pearce Jr., whose mother was subject to occasional psychic experiences. Rhine described Pearce as "sociable and approachable" and somewhat "artistic" with a particular interest in music. Pearce depicted himself as often having "hunches" but was not otherwise aware of any psychic power. A photograph of Rhine working with Pearce shows the student bent over, looking somewhat rumpled, while the taller Rhine sits calmly with pen poised, carriage erect. This photograph does not suggest that the laboratory was straining for good results. But Pearce offered them. After 10,300 calls for clairvoyance (412 times through the Zener deck), Pearce scored correctly on 3,746. One in five odds would have made his correct calls total only 2,060. In clairvoyance, Pearce averaged 9.1 correct calls per 25.

These results were debatable as sometimes the experimenters indicated to Pearce every 5 cards how well he had been doing—perhaps hoping to help the subject "focus in" on his power, but inevitably helping him (consciously or subconsciously) make calculations regarding what cards might reasonably be expected to remain in the 25-card run. At other times, the entire deck of 25 was gone through, one by one, before results were checked. In the most carefully guarded results, in which Pearce was required to guess the entire sequence of the deck without it ever being moved, after 1,625 cards (65 times through the Zener deck), he had guessed 482 correct, making his average 7.4 out of every 25 cards, well above chance. In 1932 while "performing" before witnesses, including magician Wallace Lee, out of 1,800 guesses Pearce had 578 correct, a rate of 8.0 per 25.[63] Rhine placed astronomical odds against such data being a result of chance.

Twenty years after the Zancigs were at the height of their fame, Joseph B. Rhine began his experiments on ESP at Duke University. Here Rhine sits calmly, poised, on the right, while one of his star subjects, student Hubert Pearce, slightly disheveled, attempts to determine the order of a stack of Zener cards. From J. B. Rhine, *Extra-Sensory Perception*. Courtesy Harry Ransom Humanities Research Center, University of Texas, Austin.

In a sense, Rhine had brought a performance piece, that of the mind-reading exhibition, into the laboratory. Perhaps giving credence to the Zancigs' claim that their strong affinity for one another aided telepathy, Rhine's "star" telepathy results came from a student couple, George Zirkle and Sara Ownbey, who were engaged to be married. Flanking photographs of the couple in action in Rhine's first book create a scene reminiscent of promotions for the Zancigs. Zirkle (the "percipient") sits quietly in an armchair, eyes closed, face relaxed, while Miss Ownbey (the "agent") sits alertly at a table, a scoring sheet before her, her hand on a telegraph key prepared to signal that a new trial would begin. This team reversed the usual gender categories and offered a dreamy, abstracted man receiving while a brainy woman sent. With Ownbey as agent projecting Zener symbols and Zirkle receiving, Rhine announced that in 3,400 trials, Zirkle averaged 11.0 hits per 25. Rhine also had a loud electric fan running in the room to guard against the often-cited danger that "unconscious whispering" might account for such good results for telepathy. With a wall between them, separated by a distance of about ten feet, Rhine reported Zirkle averaged 14.6 "hits" per 25 for a run of 750 trials. Separated by thirty feet with two walls between them, Zirkle averaged

16.0 hits per 25 for a run of 250 trials. Again, Rhine insisted that the odds of such a performance being a statistical fluke was impossibly high.

Content that he had proven his case for both sorts of ESP, clairvoyance and telepathy, Rhine went on in his books to correlate the waxing and waning of this new power with a variety of factors, to offer possible explanations for the force, and to ponder its philosophical implications. Rhine's major contribution to the "mechanics" of ESP was to rule it out as a form of electromagnetic energy, whether "brain waves" or some other unknown force. He also ruled out more occult or outdated scientific notions such as that of "Odic Force" originated by Baron Reichenbach to explain mesmerism and animal magnetism. Rhine was content to prove that ESP existed and to let others argue about its mechanics, insisting, "Of first importance, perhaps, are the facts pointing to the absence of any yet known energy principle in E.S.P."[64]

To reject explanations that relied on radiation, Rhine began with the theory of physicist William Crookes, presented in 1897, that "telepathy might be due to high frequency vibrations of the ether generated by molecular action of the brain of the agent and received by the percipient's."[65] Though this theory was helped by the detection of "brain waves," Rhine insisted that even if this could make sense for telepathy, it made no sense for clairvoyance. No "known radiation" was conceivable to explain how one card in a pile might emit energy to distinguish it from others. Rhine experimented with X-rays to show that even after ten-second exposures, X-rays could only indicate the dim outlines of a card itself, but not its ink

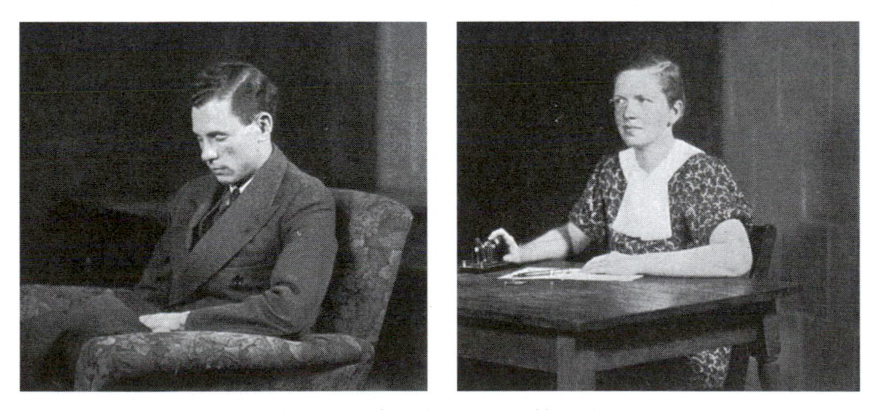

Rhine's best telepathy team, the engaged students Mr. Zirkle and Miss Ownbey. While the photograph is reminiscent of performers' publicity photographs, this team reverses the usual gender stereotypes, with the dreamy, abstracted Mr. Zirkle "receiving" and the brainy, alert Miss Ownbey "transmitting." From J. B. Rhine, *Extra-Sensory Perception*. Courtesy Harry Ransom Humanities Research Center, University of Texas, Austin.

markings. Likewise, the "angle" at which percipients sat to the card had no seeming influence, as it might with faculties such as sight or sound. Rhine further argued that his experiments showed that clairvoyance and telepathy worked at least as well and perhaps better "at a distance" than they did close up. This seemingly defied the rules of wave radiation weakening in ratio to distance according to the "inverse square" law.

Rhine challenged physicists, stating, "If anything were known that *could* change one's [mental] responses that was not of the known energies, it would promptly be declared another kind of energy, because it 'does work' and 'effects change.' This would have to be done to save the coherence, unity and comprehensibility of our basic physics. At this point, we are, then, it seems, faced with the need of another order of energy, not radiant."[66] He also reasoned that if causation without energy seemed impossible, researchers would be wise to "concede that there is still possible growth in the basic concepts of the field" of physics.[67] Perhaps wishing to avoid the appearance of being out of his depth, Rhine refused to speculate whether the new physics, based on quantum mechanics and relativity theory, could find a place for ESP. In 1937, going over these same results, Rhine suggested that his research supported McDougall's contention that "in mental processes a non-mechanical and, as he calls it, teleological but not mystical mode of causation is in operation."[68]

Unable to explain the mechanics of ESP, Rhine was happy to suggest how it was most easily evoked in subjects. He insisted that to get best results, a subject had to be interested in and not bored by the proceedings and had to believe ESP was an inherent capacity. Likewise, caffeine was shown to stimulate flagging ESP, while depressants suppressed the faculty. The subject should be "relaxed" but also in a state of "concentration." His laboratory workers had the right attitude of curiosity and excitement to encourage such bashful phenomena as ESP. Rhine insisted that in those early years, he had the cooperation of the entire psychology department. "We knew we had something by the tail that was too big for all of us, but we were having riotous fun pulling and holding on, twisting and prying, to get a better hold, a further advantage, a more complete capture."[69]

Rhine's speculations became more cosmic in his book's conclusion. ESP, he argued, helped establish the "non-physical" nature of mind. The ESP percipient's mind relied on a "peculiarly non-mechanistic procedure."[70] He noted that ESP was not "space-bound" and probably not "time bound"—making precognition or prophecy possible. He also gave some credence to the tenets of Spiritualists. "If the percipient's mind is, as hypothetically suggested in Chapter XII, a relatively free agent that can, under certain conditions, go out space free, escaping material limitations, it might well be expected to be able to find in this spaceless order of reality whatever (if any) strange forces or entities there may be. If there are incorporeal personalities, it could 'contact' them. If there are reservoirs of knowledge, it might tap them, by a more transcendent clairvoyance."[71]

Rhine repeated this notion in his next book, *New Frontiers of the Mind* (1937), tailored to a more popular audience, saying that there was no proof of survival of the soul after death, but that "what we have so far found in the ESP research would be at least favorable to the *possibility* of survival of personality after death."[72] ESP research could at least assure us that noncorporeal entities might have some method of communicating with each other and with corporeal personalities—that is, mediums.

Reaction to Rhine's first volume, published, presumably in small numbers, by the Boston Society for Psychical Research, was somewhat limited. A few academic attacks on Rhine's mathematics and procedures were launched in 1934 and 1935, while his proponents, including a leading statistician, mounted defenses. Newspaper science writers lauded his work and suggested that it jibed well with the model of the universe that the new physics had established, which was equally antimechanistic. Rhine's second book, aimed at a popular market, stirred up far greater interest and controversy. His publisher coincided the book's release with the first broadcasts of Zenith Broadcasting Corporation's weekly radio series on psychic experiences, which included on-air promotions of Rhine's work and of his Zener cards. Duke University president William Preston Few, however, persuaded Rhine not to accept any official advisory position with the broadcasts.[73]

A Book of the Month Club selection in October 1937, this second book, *New Frontiers of the Mind: The Story of the Duke Experiments,* sold 150,000 copies and provoked dozens of critiques and defenses in both academic and popular periodicals. *Time* was able to report on the "Rhine Question" in early October 1937 and follow that story with the presciently titled "Battle on Rhine" the following April. In *American Scholar,* in the spring 1938 issue, Rhine supporter Gardner Murphy, a professor of psychology at Columbia University who had also been a Hodgson Fellow at Harvard, wrote "Dr. Rhine and the Mind's Eye," praising Rhine's exploration of "non-material" notions of mind, while Joseph Jastrow responded in a later issue of *American Scholar* with "ESP, House of Cards," which belittled Rhine's procedures, results, and conclusions and labeled him "irresponsible" and his book "educationally deplorable."[74]

Not surprisingly, Rhine's book and the surrounding publicity stimulated an annoyed response from behaviorists. B. F. Skinner reviewed

New Frontiers of the Mind for *Saturday Review* under the title "Is Sense Necessary?" Skinner first gave Rhine mild credit, noting that "he has taken a disputed subject matter out of the realm of casual observation and anecdote into the experimental laboratory."[75] He went on, however, to repeat early complaints that Rhine's researchers should not have told test subjects how successful they had been doing every five guesses, to argue that confusion over the names of Zener symbols might cause researchers to mistakenly record positive results, to insist that Rhine should not have dropped subjects whose early data was not promising, and to call into question Rhine's mathematical premises.

Skinner then offered the argument that the experimenters may have suffered "hypnotic delusions," and questioned Rhine's "apparent" refusal to allow "accredited scientists" to observe his laboratory. Skinner also mentions that Rhine was clearly "biased" and let "anecdotes set the tone of the book." Skinner found greatly irritating Rhine's habit of "explaining away" bad results, as the parapsychologist theorized how sleepiness, boredom, illness, emotional upset, or the entry of strangers to the laboratory adversely affected subjects. All this showed that Rhine made the grievous error of "presupposing what he undertakes to prove." Skinner concluded, "When a supposed fact is so prodigiously at odds with established knowledge as extra-sensory perception, the proof required is proportionately greater . . . minimum requirements of proof have hardly been approached."[76]

When Rhine's use of probability was defended by several published reports from mathematicians, critics shifted tactics. Dael L. Wolfle of the University of Chicago allowed Rhine his mathematical premises, and also said he saw nothing wrong with Rhine selecting subjects on the basis of how well they scored on screenings. "If extra-sensory perception is an ability possessed by only a part of the population, Rhine has a perfect right to select that part for study."[77] Wolfle's critique then combined ridicule with skepticism about Rhine's precision. He noted that the "fruits of Rhine's work" included a "widespread revival of interest in the occult" and a "weekly radio program."[78] Shifting to ad hominem tactics, Wolfle stressed Rhine's background in "forestry" and his reported desire to live a "free and natural life," and complained that Rhine was not a member of the American Psychological Association. More seriously, Wolfle pointed out that few other researchers had confirmed Rhine's works. He also stressed the possibility that unconscious cues might be involved, noting that "scoring [was] highest when sensory cue possibilities were greatest."

Another critic, Harold Gulliksen, belittled Rhine's popular appeal, his Book of the Month Club connections, and the "What-Is-It?" motto of the

Zenith broadcasts.[79] Gulliksen admitted the case was not closed—eight laboratories had confirmed Rhine's results while six did not—and then focused his critique on the clerical methods used at Duke, making a fairly convincing case that some of the positive results might have been due to unintentional errors. When percipients "called aloud" guesses, recorders could conceivably mark their answers wrong. Even if attempting to honestly record answers, they had incentive to occasionally "hear wrong" in favor of a positive result. Rhine's descriptions of methods used were vague, sometimes requiring answers to be written by both experimenter and subject, sometimes relying on verbal calls. "It is characteristic of Rhine's reporting that one cannot always tell which method was used."

Gulliksen, like Wolfle and other critics, also delved into the literature to question Rhine's conclusions about the genuineness of the horse Lady Wonder's telepathic powers. After Rhine had returned to "Lady" and discovered her responding to cues from her trainer, he did not conclude that the act had always been a fraud, but instead insisted that this only proved that Lady had since lost her telepathic abilities and that her trainer was desperate. Gulliksen quotes Rhine stating, "It is a poor kind of cheating which grows worse with practice." Gulliksen then responds, "It is a poor kind of observation that doesn't increase in acuity as it proceeds, possibly discovering trickery not at first noticed."[80]

Gulliksen also belittled Rhine for encouraging experimenters to use the "official, Rhine-patented, Duke laboratory of parapsychology ESP cards" available in bookstores and stationery stores. Gulliksen remarked that B. F. Skinner had noticed that "the figure printed on the [official] card can, under favorable lighting conditions, be read from the back as well as the front of the card."[81] He closed his article with valid complaints about Rhine's equivocations; for example, at one point Rhine stated that ESP abilities close up or at a distance are the same, while at another he stated that ESP abilities improve with distance. Gulliksen noted that "such contradictory statements tend to diminish rather than enhance, the scientific prestige of an experiment."[82]

In 1938, the year after *New Frontiers of the Mind* was published, the eastern branch of the American Psychological Association took up the controversy at its conference. *Time* reported that at the meeting "three papers were read on the Rhine question, all of them hostile." *Time* also remarked that Steuart Henderson Britt showed how Duke's ESP cards could be read from the backs by sight or touch and "proved this point when he correctly read 24 out of 25 ESP cards whose faces he could not see. Psychological chuckles filled the hall as he did so."[83]

Finally, some detractors based their critiques less on science than on politics. In the *New Republic* Norbert Guterman wrote, with some anger, "What first strikes the reader of Dr. Rhine's book is the great disproportion between these monotonous, not to say trivial, 'readings' of cards, and the extraordinary hopes they have aroused. The innumerable groups in this country who practise occult sciences responded to them with enthusiasm."[84] At least temperamentally an activist, Guterman went on to express his hope that ESP research would not become a "national pastime" and to insist that Rhine was reasserting a "passive" model of the mind that encouraged helplessness before social forces. He concluded, "The 'scientific' language used by the latest variety of psychic research must not hide from us the fact that its social thinking is on the same backward level." Along the same lines, Joseph Jastrow wrote in *American Scholar* that "the social responsibility for misleading the public into the belief that telepathy has been established is serious."[85]

Yet responses to Rhine's work were not all hostile. One of Rhine's great defenders was the *Scientific American* and its publisher Orson D. Munn, a longtime champion of psychic research. The *Scientific American,* which had offered prize money to successful Spiritualists in the 1920s, began running articles about telepathy in the early 1930s that took the matter quite seriously. For example, Walter Franklin Prince had published a favorable review of novelist Upton Sinclair's *Mental Radio* in the magazine in 1932. Sinclair's book described experiments in telepathy he had devised for his wife. The book included paired sets of drawings, both originals that Sinclair had drawn in a separate room and the duplicates that his wife, Mary Craig Kimbrough, rendered through clairvoyant efforts. And in 1933, foreshadowing the Zenith Broadcasting Corporation's semi-farcical telepathy tests, *Scientific American* printed a series of public "tests" of its readers' telepathic abilities. Readers were asked to fill out forms and return them for tabulation. Experts on *Scientific American*'s panel, which included psychic researcher Prince, Columbia psychologist Gardner Murphy, and professional mind reader Joseph Dunninger, concluded that the first test results "show something that cannot be ascribed to pure chance."[86] The second test suggested that *Scientific American* readers lacked ESP. The magazine bowed out of the testing business, concluding their readers did not "have it." In a slightly peevish side note, the editors complained that though they had asked each "agent" to project "the name of some simple and familiar object which he can readily visualize," many readers chose to project such phrases as "bootlegger," "your life's ambition," "pain,"[87] "free love," and "wind."[88] *Scientific American* then brought to a close its "active participation in research for telepathy" but still presented the topic even-handedly.

Though their own tests were a flop, Munn gave Rhine's Duke experiments glowing coverage. In 1934, Prince wrote an article for Munn describing Rhine's remarkable results at Duke.[89] The following year, Munn ran an article by Rhine describing his ESP tests on the British Spiritualist medium Eileen Garrett. Rhine's findings were that she was endowed with ESP—but not as impressively as some of his homegrown subjects at Duke. In the article, Rhine typically went out on a limb to state that his ESP findings had "a positive bearing on the spirit hypothesis."[90] And in June 1937, when Rhine's first book had already stirred up controversy, *Scientific American* ran a short editorial, "Telepathy Comes of Age," that stated that Rhine had come to the rescue of psychic science, once the "Orphan Annie in the psychological and therefore the whole scientific world." But now, Rhine's work was "winning a place in the sun for that science."[91] And in 1938, when the "Battle on Rhine" had begun in earnest, *Scientific American* let Rhine defend his methods—both his mathematical procedures and his results—which he insisted held even when whittled down to include only tests which were absolutely "cue proof."[92]

Along with *Scientific American*'s endorsement of psychic research and parapsychology was a laudatory article about Rhine in *Popular Science Monthly*, "Can We Read Each Other's Minds?" The writer framed his look at scientific research with the sort of anecdotal evidence that drove academics like B. F. Skinner wild. The author highlighted Rhine but also mentioned the successful work of investigators such as J. E. Coover of Stanford, Gardner Murphy at Columbia, and G. W. Estabrooks at Harvard. In light of the training advice given in many manuals teaching telepathy, at least one of Coover's Stanford experiments is worth mentioning. He arranged a test for ten subjects who believed they could sense when "anyone stared at them from behind." He situated the subjects with backs turned to him and told them to record whether they were being looked at at a given signal. He tested each student one hundred times. The results showed the answers were rarely correct, and, contrary to the training regimens that mind readers like Zancig and later performer Joseph Dunninger encouraged in their students, "there is little to substantiate the common belief that we can 'feel' the stares of others."[93] The author goes on to describe Rhine's work and a visit to his laboratory and concludes that his investigations "form an important milestone."[94]

While endorsements from the *Scientific American* and *Popular Science Monthly* would not impress B. F. Skinner or Rhine's other academic antagonists, they are nevertheless revealing, as other popular science magazines could promote a narrowly materialistic vision. For example, Hugo Gernsback's *Science and Invention* magazine of the 1910s and 1920s assumed a hostile stance to psychic phenomena. While Munn let Rhine discuss his work with a medium in *Scientific American*, Gernsback's earlier article, "Can the Dead Be Reunited?" ridiculed such notions. It began with such premises as, "Let us assume the souls of men are only as large as an ant," and, "With ten trillion souls abounding somewhere above the planet, it must be assumed that there would be practically no standing room."[95] From the premise that souls would then have to wander through outer space, Gernsback concluded, "Your odds of getting a royal flush, are 16,686,166 times better than your chances of communicating with a particular soul." Clearly such a crudely materialistic mode of speculating was more in line with the beliefs of B. F. Skinner than with those of J. B. Rhine or Walter Franklin Prince. The debate in the academic world had its counterpart in the realm of popular science forums.

Behaviorism versus the Other-dimensional Mind

Rhine's academic defenders tended to ask: who is really showing bias here, the psychic researchers or the mechanistic hardliners? Vernon Lemmon of Washington University opened his defense by remarking that Rhine's work stirred resentment because the stereotypical answers that psychologists could once give regarding ESP, i.e., we "are not interested," or "carefully controlled experiments have yielded negative results," would no longer hold.[96] He went on to defend Rhine's use of probability and his decision to "winnow" subjects, remarking, "If you are studying maze learning in rats, you are justified in rejecting crippled or diseased rats, or in fact, any that seem unlikely to show any learning ability."[97] He directly rebuffs Skinner's critique by noting that "the first psychologist ever to study maze learning was not deterred by the thought that to look for it was to assume its existence in advance."[98] Even Rhine's decision to stop testing when a subject was growing fatigued could be justified, for "in a learning experiment it is usual to stop work on a subject when he shows signs of undue fatigue, or when external distractions show signs of interfering with his performance."[99]

Lemmon concluded by showing his distaste for Rhine's complete rejection of a radiation theory of ESP, calling the alternatives "if not frankly mystical . . . [then] repugnant to the scientific mind."[100] Like several newspaper science writers before him, Lemmon turned to quantum mechanics to bolster a theory of ESP based in physics. Lemmon started from the premise that if a single quantum of energy is needed to induce a telepathic reaction, distance will not be a factor, since a quantum departing remains a quantum when it arrives regardless of distance. He cited one experiment that established that the energy equivalent to a single quantum of green light would be enough to arouse a physiological response.[101] He also noted that physicists had recently posited, "No energy ever starts from a source until a receiver is ready for it." With this assumption, "If A and B are mutually ready for a transfer of energy, it will occur regardless of distance."

Finally, Gardner Murphy of Columbia University, another Hodgson Fellow and former student of McDougall's, staunchly defended Rhine's work. Murphy depicted Rhine's work as an important victory in the ongoing war that open-minded researchers were waging against the mechanistic worldview that moderns had inherited from the Enlightenment. Murphy praised Rhine's work and insisted that confirming results came from numerous groups.[102] He also insisted that B. F. Skinner's discovery that some decks could be read from both sides of the card wasn't devastating, since "all the serious work emphasized by Dr. Rhine and his followers has been carried on with cards out of the reach of the senses. . . . In the critical tests *neither* [side of the card] is seen."[103] Murphy concluded that Rhine's opponents were entrapped in a worldview that blocked them from accepting evidence contradicting its tenets. Murphy likened research into extrasensory perception to Archimedes' lever that could move the world. In this case the lever's action would involve replacing "17th-century naïve mechanism by other conceptions more characteristic of 20th-century scientific adventure."[104]

Murphy's stance was in the tradition of William James, who frequently argued that scientific "absolutism" was ultimately based on faith or taste and not on rationality.[105] Such skepticism required its adepts to deny the validity of any experiments that produced facts that seemed contrary to natural laws. James admitted that "the excesses to which the romantic and personal view of Nature may lead . . . are direful," like the "Mumbo-jumboism" of Central Africa. But James added that "the oftener one is forced to reject an alleged sort of fact by the method of falling back on the mere presumption that it can't be true because so far as we know Nature, Nature runs altogether the other way, the weaker does the presumption itself get to be."[106] To James, a priori mechanistic convictions can bias judgments as powerfully as religious convictions.

As James admitted, though, the problem of correlating the possibility of a seemingly "magic" effect with a standard scientific worldview was troubling. This was an issue that Sigmund Freud debated when responding to the plausibility of ESP. Freud and other psychoanalysts believed they had experienced telepathic moments in their work. Aware that psychoanalysis was still suspected of occultism, Freud cautioned his followers away from the acceptance of a phenomenon like telepathy until it could be explained mechanically, lest such an admission open the door to other occult beliefs. To Freud, the occult explanation put one's critical faculties to sleep, helped falsify perception, and forced one to agree regardless of confirmation.[107] Later, Freud inched cautiously toward a more open-minded stance, noting, "It seems to me that one is displaying no great trust in science if one cannot rely on it to accept and deal with an occult hypothesis that may turn out to be correct."[108]

Most critics, however, conformed to James's negative description and argued that such phenomena were impossible, hence useless to debate. During the earlier debate over muscle reading, George M. Beard had explained that since the whole notion of one mind transferring a thought to another without using sensory channels was contrary to nature, he had been led to the correct theory in the case of entertainers like Brown and Bishop. They were "muscle readers" not "mind readers." Beard argued that if we insist on the broader principle that human beings' capacities differ only in "degree" and not in "kind," we can rule out telepathy. "Mind-reading, in the usual meaning of the term, is a faculty that in any degree does not belong—indeed, it is never claimed that it belongs—to the human race, it cannot, therefore, belong to an individual."[109] He went on to argue that because of this, "for one person to read the thoughts of another would be as much a violation or apparent violation of the laws of Nature as the demonstration of perpetual motion, the turning of iron into gold, or the rising of the sun in the west."

In 1877, Beard had anticipated the position that Skinner and other critics would tend to sixty years later. And it may precisely be due to Beard's assumption—that humans can only differ in a manner of "degree" but not "kind"—that Rhine was led to assert that ESP was in fact a faculty latent in all people. He and science fiction writers seemed to see this ability emerging in a new evolutionary stage, presaging a millennial shift. Freud, whose worldview lacked such cosmic optimism, worked from the opposite assumption when he later argued that telepathy, if it existed, was most likely an evolutionary throwback, likening it to the manner in which insects, antennae and all, might communicate.[110]

If the skeptics' assumption that the hypothesis of ESP was in itself fishy seems close-minded, their questions about its seeming subjectivity provided stronger grounds for pointing out hypothetical flaws in all of Rhine's methods. Rhine, like earlier adherents of Spiritualism, admitted that psychic phenomena, in general, were "bashful" and had to be encouraged to emerge. Subjects had to be relaxed, confident, and encouraged. Unsympathetic researchers, like Willoughby, were unlikely to uncover ESP. Rhine's results would most likely be reproduced by other individuals with the same instinct for a religious or romantic worldview. The pro-Rhine camp would be just as likely to reject as definitive an experiment that indicated subjects did not have ESP as their critics rejected work that pointed to its reality.[111]

Skinner and other diehard materialists undoubtedly were biased, yet their conservatism seems reasonable. Reversing Gardner Murphy's metaphor, skeptics could argue that ESP research was really a lever for reintroducing superstition and the irrational into the world. What appeared to be facts to Rhine and his supporters, to detractors, by definition, had to stem from flawed experiments. To them, Rhine's efforts to create scientific proof of mystic phenomena were similar to the efforts of Renaissance anatomists to find the physiological seat of the "soul" in the human body. Charles Fort, a critic of science whose ideas became faddish during the 1920s, pinpointed the aversion rationalists might have toward the psychic with his remark, "If science shall eventually give in to the psychic, it would be no more legitimate to explain the immaterial in terms of the material than to explain the material in terms of the immaterial."[112] The impulse to establish a romantic science contained the seeds of its own failure, as the impulses towards an orderly, rational universe and a "live" or spontaneous universe did not appear easily compatible.

Others, like Freud and Lemmon, took a middle position. They reasoned that a physical explanation for telepathy must eventually emerge and so establish grounds for verification. In such a case, Rhine's strategy could be likened, however humbly, to Newton's decision to add the unproved assumption of "force at a distance" for his theory of gravitation. While many of Newton's contemporaries called the assumption of force at a distance a throwback to the occult—that is, to Renaissance notions of universal "sympathies and antipathies" at work—Newton insisted that if such an assumption led to simple, universal mathematical laws, it was better to add force at a distance without even trying to explain why it might emerge.

Rhine, who wished to add the "occult"—or in his preferred term, the "teleological" force of ESP to the modern worldview, like Newton before him, attempted to establish quantitative rules that telepathy followed. Rhine hypothesized this faculty was strongest at the beginning of the session, determined that it could be depressed with sodium amatyl and revived with caffeine, and found that ESP could be affected by illness, by the appearance of strangers, and so on. His critics found such "rules" to be merely ad hoc explanations of failures. And, in fact, Rhine was not really positing a new force in a mechanistic world. He proposed instead that mind operated in another "dimension" unbounded by time and space. This was a revolutionary idea and far easier to reject than Newton's theory. Newton's conception of "force at a distance" helped explain the movement of the planets, comets, tides, and cannonballs; all that Rhine specifically could explain was some seemingly good guessing results at cards. For mechanists, whose entire worldview was threatened by Rhine's data, it was far easier to assume that the good guessing that occurred in Rhine's laboratory was the result of error, delusion, or luck.

Many of Rhine's backers attempted to show how the new physics might ride in to rescue the romantic science project of the nineteenth century and Rhine's "dimensional" explanation. According to such defenders, "subjective" science was no longer an oxymoron. The new physics, well established by the 1920s, insisted that indeterminacy was built into the scientific process—at least on the quantum level. Light could be thought of as a wave or as a particle, depending on one's mathematical needs. Mass could be redefined as energy. Heisenberg's principle implied that verification had very real limits and at the quantum level interfered with outcomes. Other scientists posited that every possible outcome of a chance interaction might actually occur, generating parallel universes. The rational outlines of the mechanistic universe, once firmly outlined, had become hazier.[113] Perhaps Gardner Murphy was right in demanding a new conception of mind in keeping with the continually opening "adventure" that psychology and physics were undergoing. Whether Rhine's work offered the "new" or was a throwback to the older ideals of romanticism, however, was open to debate.

A Showman Responds

However great his integrity and courage, Rhine needed private contributions to both endow and maintain his parapsychology lab, and to do so, he became something of a showman. His decision to let Zenith Broadcasting and his publisher hawk ESP cards to which he held copyrights hurt his reputation as a diligent, sober scientist. His decision to market his lab's "product" made his science share characteristics with the artistry of Barnum; and his research no doubt gave great encouragement not only to sober parapsychologists and psychic researchers but also to occultists and professional mind readers.

Rhine also penned popular articles for *Forum* and other magazines. In such articles, Rhine made no effort to rely on a sophisticated, above-the-fray tone. With titles such as "Are We 'Psychic' Beings?" "The Practical Side of Psychism," "The Gift of Prophecy," "The Evidence for Prophecy," and "Don't Fool Yourself," Rhine created his own version of psychic research "goes pop." He valorized his own work and also included much anecdotal evidence, guaranteed to please popular readers and drive critics into a frenzy. One article concluded with the editorial note: "Professor Rhine is preparing further papers on the scientific investigation of clairvoyance and survival after death."[114] He employed popular vernacular to make such pronouncements as, "To the genuine sportsman, in this jungle of the mind, I am sure the high frequency of danger only adds to the zest."[115] Such articles may have increased his celebrity status but clearly diminished his scientific standing.

Rhine helped found parapsychology in America, but the field has led a troubled existence ever since. His peers in the psychology department at Duke eventually ostracized Rhine and his laboratory, and his cause was not helped by later scandals involving falsified results. His desire to be the last to ever have to prove again the reality of such faculties as telepathy and clairvoyance was not realized. Like a "ghost," psychic research appears trapped in a repetitive pattern, its adherents continually trying to make the "final" case that the phenomena they study actually exist. To each such "final" proof, critics can respond that the "impossible" results must point to shoddy methods. Final confirmations "proving" or "disproving" ESP become subjected to an infinite regress of retesting and new "final" outcomes—that only will confirm the beliefs of those making them.

With parapsychology's scientific basis continually reexamined, the main beneficiaries of the "Battle on Rhine" ultimately were psychics and performers. On vaudeville stages and on the newly opened venue of the radio broadcast, hypnotists and mind readers—exhibiting success rates far more spectacular than those in Rhine's laboratory—continued to boost the possibility of "new frontiers" to be explored and exploited.

Rhine's doppelgänger, his pure show business shadow, was Joseph Dunninger, the leading American stage mind reader after Julius and Agnes Zancig. Dunninger, born in 1892, began as a magician and so largely avoided the wrath of magicians even when he presented his stage faking as genuine ESP. He had been an acquaintance of Houdini's, a fellow buster of Spiritualists, holder of the dubious honor of being the first to hypnotize a subject via radio, the author of a pamphlet that gave instructions on how to hypnotize a parrot, and one of the judges on the *Scientific American* telepathy panel in the early 1930s. His articles about phony Spiritualists were amusing and offered a common touch, as when he described a female medium in Georgia who, after collecting a ten-dollar fee, answered his hidden question, "How much money do I have in my wallet?" with "Ten dollars less than you came in with."

More bold than Zancig, perhaps because he feared less the attacks of magicians, Dunninger insisted that he had powers of "hypnotic clairvoyance."[116] A few years after the Zenith Broadcasting Corporation's show on psychic testing concluded, Dunninger went on the air in 1943 with a half hour NBC network mind-reading show on WJZ in New York, which began with this tag: "Who is Dunninger? The Man with the Miracle Mind." He later brought his act to television. Unlike Rhine's subjects, who could guess nine out of twenty-five cards correctly, Dunninger was almost always right. John J. O'Neill, a prominent science writer—and Tesla's first biographer—tested Dunninger with playing cards and found him to be accurate in 95 percent of his guesses.[117]

Dunninger's reminiscences about the "Battle on Rhine" suggest that he regarded Rhine as a show business rival. The mind reader observed of Houdini, "Always Houdini was a challenger; to prove that he was right, he had to prove that someone else was wrong. He became great when he made a rival look small. . . . By getting rid of the tough adversaries, he cleared the field for a new crop of softies that he could mow down as fast as they sprouted up."[118] Dunninger's observations on how Houdini dealt with competitors also apply to Dunninger's handling of the Duke work. Dunninger insisted that the Duke laboratory had often requested his presence, but Dunninger had never complied, noting harshly, "I declined, rather than be identified with the residue of crackpots and publicity seekers who put in an appearance there. I say 'residue' with emphasis, because nobody—and I mean nobody!—who had real telepathic ability would have reduced themselves to the kindergarten stage of experimentation then underway at Duke." In a critique reminiscent of Willoughby's, who described confusion in his psychology lab regarding the Zener card symbols, Dunninger blamed the Duke ESP cards primarily for inducing boredom. He insisted he would have submitted to tests if normal decks of cards had been used. Familiarity guaranteed better results in transmitting information, telepathic or otherwise. Speaking like a true magician he remarked, "Playing cards, with people who know and like them, are a language in themselves."[119]

The fact that Dunninger regarded Rhine as a rival also suggests a popular culture that could prefer temporary marvels and exotics to worldview-bending new realities. In his work, Rhine sought to scientifically investigate telepathy and the paranormal; in the public view, he succeeded more in bringing the performance pieces of telepathy and clairvoyance into another exotic setting: the scientific laboratory. To characterize Rhine's research as a mirroring of activities that had earlier taken place on entertainment stages is revealing. The mind-reading acts of Bishop and the Zancigs promoted a form of populist science as drama: would the performers' mysterious mental "faculties" succeed or fail? Would they continue to maintain the illusion they promised? What if their "power" really was genuine? How would scientists or medical experts explain? Rhine answered by moving the drama of telepathy into a laboratory and subjecting it to scientific method. His "players," such as the mind-reading team of Zirkle and Ownbey, offered a similar drama of "trial by ordeal" as that of the Zancigs and an alluring record of success scientifically tracked. Yet, however clear and honest his methods, Rhine as a "scientific showman" attracted the same challenges and charges of fraudulence that performers such as the Zancigs and Bishop had once gladly weathered as the price for filling auditoriums.

Rhine, in his work, attempted to mount an "honest" wonder show—one that would establish that the soul should not be overlooked as a genuine category for scientific concern—that the "psyche" should not be removed from psychology, and that a quasi-religious or teleological worldview could ground science; however, the public preferred the gee-whiz "thrill" of his ideas. Consumer culture encouraged that products be exotic and alluring, and telepathy served better as an entertainment product than as fact or even as the subject of scientific debate. For a brief period, Rhine's intercession into the world of stage wonders prompted public thought and intellectual challenges, but ultimately the "threat" became contained—not only because of Rhine's scientific critics but also because his research was too plain to remain exotic.

Likewise, as radio became normalized as a technology, public hopes weakened for the human expansion of mental powers based on "wireless" technology as a model.[120] The excitement that prompted writers like Mark Twain to hype telepathy in the 1890s had died down; radio as a medium had lost its aura, and the commercial airwaves became yet another vaudeville stage requiring entertainers to bring it to life. Telepathy and ESP were best marketed not as science but as the "thrill" performances with which Dunninger and other entertainers could enliven the air.

PART III

MILLENNIAL WONDERS

CHAPTER SIX

The Missionaries

The grim economic landscape of 1930s America encouraged a public taste for exotic escapes and romances. Movie patrons feasted on fantasies that included musicals with Ginger Rogers and Fred Astaire, the aquatic epics of Esther Williams, jungle or space adventures with Tarzan and Flash Gordon, and at least one journey to the spiritual utopia of Shangri-La. This desire for the exotic was explored in Charles G. Finney's 1935 underground classic, *The Circus of Dr. Lao,* a novel that depicted the combination of boredom, hopelessness, and sexual longing with which small-town America could greet the appearance of a mysterious circus in the 1930s. Many historians have argued that such a longing for the exotic was a natural outgrowth of consumerism: a consumer society relied on the advertising industry's ability to impart a taste for the exotic and to create free-floating desire to encourage public spending.[1] Such conditions made the 1930s ideal for performances that could instill hope that America would ultimately escape its economic horrors. Accordingly, numerous doctors, generally in laboratory coats, appeared with the sugar-coated pill that combined science with progress and a touch of mystery.

One of the great incubators for such a cure was the Century of Progress Exposition in Chicago from 1933 through 1934. Crowds flocked to this World's Fair with its theme of progress as the product of science and industry. The promotion of a positive vision of science was one of the fair's goals. After one season a newspaper pronounced the entertaining displays designed by scientists in the Hall of Science "a whoopee success —a slangy expression which would have shocked the men of science a few years ago, but which goes today."[2] Corporate exhibitors at the fair courted the public at least as strenuously as the scientists; General Electric (GE) dramatized its research laboratory with thirty-minute performances running all day in its "House of Magic." Westinghouse and General Motors (GM) presented similar technological marvels in their exposition areas. After the fair closed, GE decided to tour its entertaining science shows, and in 1936 GM launched its "science circus" to bring science wonders to small towns throughout the United States. Historian Roland Marchand has argued that the lecturers in these "magic-science" shows were missionaries for industry, promising consumers that corporate America was synonymous with progress.[3]

At the same time that the GM Parade of Progress caravan was touring America, a West Coast evangelical preacher named Irwin Moon offered his own scientific wonder show. His debut at the Golden Gate Exposition of 1939–1940 in San Francisco gained him the sponsorship of the Moody Bible Institute. This preacher relied on technological demonstrations as a source for new parables to convert the youthful and the jaded to fundamentalist Christianity. The wonders of nature and science, in Moon's demonstrations, only highlighted the greater wonders of God's creation. Moon's show not only had similarities to those of the large corporations of the 1930s, but also had historical similarities to such early wonder shows as that of the nineteenth-century electrical healer Charles Came. Both, for example, led audiences from views of the cosmic—the starry heavens—to the microcosmic—with magic lantern slides and microscopic views of paramecia and other single-celled animals. Both promoted technology and modernization, whether the early telegraph or a sophisticated eavesdropping microphone. In ninety years, the basic recipe for the wonder show had undergone little change, suggesting a naïve public long separated from the technical elite.

In their wonder shows, both the industrial corporations and Moon were engaging in image making and sales. The corporate-financed shows assured the Depression public of an inevitable return to prosperity to encourage spending; Moon, however, was fishing not for the public's pocketbook but for its soul. He documented his success with the number of pledges to Christianity that audience members signed; even more ambitious, perhaps, were Moon's efforts to remake the image of fundamentalist Christianity, to harmonize it with notions of progress, and to return the relationship between Christianity and science to the contented marriage of the antebellum era.

Professor Frost's Remarkable Astraphone

The corporate science shows offered at the Century of Progress were a collaborative production, developed by corporate public relations departments, corporate research scientists, and academic scientists eager to promote science's value to the public. To counter public perceptions that corporations were ungainly, inhuman bureaucracies interested only in profits, in the 1920s corporate public relations departments began to promote the corporate research laboratory as a romantic terrain in which white-coated technicians were new pioneers, conquering new lands.[4] Since the turn of the century the use of the term "wizard" for a scientist or inventor had been a commonplace, but it took the advent of mass advertising to make this one of the chief metaphors for explaining the work of corporate research laboratories.

When John Dos Passos insisted in his novel *USA* that GE had made its colorful head engineer Steinmetz into a "parlor magician," the novelist was condemning the public relations imagery that successfully sugarcoated ruthless business practices. The ongoing antitrust actions brought against GE encouraged the company's public relations department to find other "human interest" stories in the ranks of their scientists after Steinmetz's death in 1926. This strategy would become more crucial in the 1930s, when the goal of corporations was to assure consumers that with its close links to science, big business could perform miracles.

As part of this effort, in 1929, GE hired war correspondent Floyd Gibbons to provide colorful ten-minute talks about developments at GE's Schenectady research laboratory for the *General Electric Hour*'s symphonic broadcasts. The dashing Gibbons, who wore an eye patch as a result of a wound suffered while covering World War I, easily shifted from newspaper to radio work. Gibbons coined the term "House of Magic" to describe GE's Schenectady laboratory and commented that the GE engineers' efforts represented "the weirdest hocus-pocus I ever heard. The trouble with all these unsung wizards of the research laboratory is that a cloak of modesty screens their great accomplishments from the public they serve."[5] Though originally opposed by GE's research engineers, Gibbons's phrase, "House of Magic," stuck. For the 1933 World's Fair, GE's exhibit contained an art deco theater called the House of Magic; inside the theater, every thirty minutes, a lecturer demonstrated research products and magic.

The Century of Progress Exposition was shaped around the theme of science's contribution to industry. This theme also gave big business the opportunity to argue that hand in hand with science, it could lead America from the Depression. Early in the fair's planning stages, its management had reached out to the scientific community for aid in shaping exhibits. In 1928, the fair's organizer, Rufus C. Dawes, recruited Columbia University physicist Michael I. Pupin, who suggested that the National

Academy of Sciences, through its National Research Council, could set up a Science Advisory Committee for the exposition. Formed during World War I to bring scientific aid to the war effort, the National Research Council (NRC) had continued on as an organization that brought together academic scientists, engineers, and industrial researchers.[6] Its members were eager for public awareness of scientific contributions to the good life.

Contrary to arguments that science popularization in this era was strictly the work of corporate public relations departments, the planning of the Century of Progress reveals how the Science Advisory Committee (SAC) dedicated itself to scientific showmanship to intrigue the public.[7] In 1929, the SAC's director, Maurice Holland—also the director of the Division of Engineering and Industrial Research at the NRC—wrote to Frank Jewett, the group's chairman, describing his encounter with a Cornell physics professor, F. K. Richtmyer, who specialized in optics. Asked to join the SAC, Richtmyer had responded by asking Holland why scientists should get involved, pointing out that World's Fair crowds were more interested in the distractions of the midway than science, and arguing that the temporary nature of an exposition made the investment of time dubious. Holland responded that involvement in the exposition could increase levels of corporate funding for scientists. He predicted that "following the exposition there would be a considerable increase in scientific activity in industry," and added, "It might influence a number of potential 'angels' to direct their endowments to scientific enterprises."[8] Richtmyer was convinced. Soon numerous other scientists eagerly joined the Science Advisory Committee.

Early on, the SAC formed a strategy that would involve popularizing science without making it "cheap and sensational." They wished to balance instruction with entertainment and to maintain the dignity of science "without being dull."[9] Yet Holland was certain that to interest the public, scientists would have to leave their elitist notions behind and become showmen. His unpublished article "Science Takes off the High Hat" stressed the notion that scientists needed to become flashier to advance their cause and insisted that the World's Fair would be an ideal way to introduce "Tom, Dick and Harry and his brother and sister" to "Science."[10] Holland's article laid out the central strategy that all successful exhibitors at the fair eventually were to employ. Scientists did not deserve the common man's attention if they did not learn his language and needs. The answer, similar to the strategy Holland had used for encouraging Richtmyer, was to pierce the public's "armor plate of self interest" by finding the vulnerable "pocket nerve," which encouraged the spectator to think of personal benefit. Scientists needed to establish that science was well worth its cost, as it improved the life and comfort of the average citizen. The World's Fair scientists could do so with showmanship. Movement and color were the keys to gaining the public's attention. It was movement and color that brought crowds to automobile showroom windows on Broadway, and would do so as well at the exposition in Chicago. The point was to sell. If necessary, they would use "living actors, moving models, talking pictures, and spectacular displays of every sort."[11]

The National Research Council surprised the fair management when it hired its own public relations company several years prior to the fair's opening to begin publicizing its work. Numerous press releases were distributed nationwide to announce lectures, new discoveries, and developments in the industrial and agricultural applications of science. Many of these press releases included what historian John Burnham has called a "gee-whiz" quotient. For example, one described the wonders of ultraviolet light in which "false teeth appear black, natural ones a brilliant blue white"; another described a "gate of ice" to appear at a refrigeration exhibit.[12] The council also arranged radio talks for its scientists that were broadcast on NBC affiliates in 1930 and 1931. These broadcasts dealt with topics such as railroad technology, anthropology, aviation, agriculture, microbiology, physics, math, and paleontology. Richtmyer, the originally reluctant professor of physics from Columbia University, gave a radio talk on "How Light Puts Electrons to Work," which explained how the photoelectric effect allowed light to dislodge electrons and complete electrical circuits. His clear explanation of the technical issues also included gee-whiz touches. For example, he mentioned that when visitors came to the photoelectric cell exhibit at Chicago, they "will very likely be greeted by a mechanical parrot."[13] The "electric eye," which set machinery into motion, then a novelty, did indeed became one of the marvels of the fair.

In the radio broadcasts, the scientists often tied their discussions to the upcoming exposition. When the Century of Progress management complained that these talks could encroach on their own publicity efforts and promise marvels that would not eventually appear, Holland did not demur. He reminded the officials that the broadcasts had been approved, seldom mentioned specifics about the fair, and helped "to build the national reputation of the chairmen of the Science Advisory Council who are responsible for the philosophy of the science exhibits."[14] Not one to easily back down, Holland added that the broadcasts, which required no outlay of expenses except for the travel costs for the scientists, would reach ten million people and stimulate interest in the upcoming exposition.

If the World's Columbian Exposition in Chicago in 1893 had established the power and beauty of electricity properly managed, the Century of Progress in Chicago looked beyond the earth to cosmic ratification of its displays of industrial might. President Grover Cleveland launched the 1893 fair by tapping a gold telegraph key. The 1933 fair echoed and expanded that opening. The star Arcturus is forty light-years from Earth; hence, light that Arcturus had emitted in 1893, during the World's Columbian Exposition, would just be arriving on Earth in 1933 for the Century of Progress. Chicago organizers decided to open the 1933 fair by capturing a beam of light from Arcturus at an observatory, transforming it into electricity with a photoelectric cell, amplifying it, then relaying it to the Chicago fairgrounds along Lake Michigan.

The inaugural lighting ceremony, held May 28, 1933, proved quite popular. Dignitaries telephoned four separate observatories, and at a signal, each observatory relayed the "Arcturan electricity" toward the fairgrounds. This was demonstrated on a large illuminated map of the country; bright red streaks of light streamed from the observatories' locales and converged on Chicago. This signal then triggered a spotlight from the roof of the Hall of Science that shone on each of the fairground's dark buildings, making each burst into light. Reflecting on this elaborate procedure, Will Rogers pointed to the great distances that existed between scientists and laymen, with his comment, "'Course it may all be just a gag, but it's a good one, anyhow. These scientists I expect have more fun out of us than we do out of them. Neither really knows when the other is kidding."[15] The gag was so good that the fair management, responding to public favor, continued the Arcturus lighting ceremony every night for the two-year run of the fair.

Scientists involved in the fair sought valiantly to dispel Will Rogers's sense that scientists and laymen belonged to two separate cultures. Edwin B. Frost, director of the Yerkes Observatory and one of the heroes of the opening ceremony, became overcome with ideas to dramatize the fair's relationship to the stars. He proposed to the fair management that he might create an "astraphone"—an organ that would play the music of the stars. The instrument would require that telescopes be trained on major stars, each of which would be keyed to a musical note and so play a "well-known hymn" for the crowds. Frost suggested using stars prominent in the summer sky such as "Vega for high soprano, Arcturus for baritone, Antares for Basso Profondo; then by various arrangements, we could let the stars sing together."[16]

The symbolic value of this opening was not only to remind Chicagoans and the world of the past glory of the World's Columbian Exposition, which might as well have been forty light-years distant, but also to show the new and surer reach of humanity in the 1930s. As one guidebook argued, "Science, patient and painstaking, digs into the ground, reaches up to the stars, takes from the water and the air, and industry accepts its findings."[17] A less-flattering formulation of this process was found in the fair's official motto: "Science Finds—Industry Applies—Man Conforms."[18] This foreboding motto encapsulated the fair theme of science serving humanity, particularly through the industries. Key to the development of this theme was the need to present not only the displays of large corporations or "applied science" but also separate displays of "pure science."

The Science Advisory Committee originally urged the management to create at the fair's center a "Temple of Science" that would reflect "scientific idealism" and offer thoughtful visitors a quiet schooling in both the principles of science and their importance to "modern life."[19] The SAC also recommended that the Temple of Science include a central rotunda, "the approaches to which might contain an allegorical representation of the transition from ignorance, superstition and tradition . . . to present day conditions when the spirit of science permeates every phase of life."[20] The Temple of Science was eventually scaled back and named the Hall of Science. But it still retained a central position in the fairgrounds. Large allegorical friezes of mythic figures depicting such concepts as "Energy" and "Light" adorned the building, and a large sculpture of a heroic figure slaying a serpent—*Man Combating Ignorance*—was stationed outside the hall's main approach. If Edison's float "Elektra" for the Columbus Day parade of 1892 showed angels harnessing the power of dragons, the exhibitors of the 1930s wished to show a complete triumph over—no longer a partnership with—the earth's forces and the older sources of wisdom.

Following Holland's dictums, the Hall of Science displays sought to be lively. Static arrangements were avoided; instead, a variety of media depicted processes and movement. Likewise, fundamental scientific principles, whenever possible, were explained and then linked to a "result well known to the public." For example, an exhibit outlining the chemical process of catalysis might include Crisco as one of its products.[21] All was motion. In the physics section was a massive cup-shaped black billiard table with a central rotor that set in motion two hundred white billiard balls; depending on the rotor's speed, the balls gathered in the center or bombarded off each other and the edges, suggesting the process of condensation and evaporation of water molecules. The physics exhibit also included a revolving platform, kicked into motion like a potter's wheel. On it stood an athlete with dumbbells; as he extended his arms the rotation slowed, and as he brought them toward his body, he sped up, revealing the laws of momentum. Watching the athlete repeat these motions and the shifts in speeds created a "weird" spectacle, according to one reporter.

Many exhibits also featured robots. A "transparent man" made of plastic and nicknamed Oscar offered an anatomy lesson in the medical section. Dr. William Mayo pronounced Oscar, the transparent man, to be "a perfect specimen of manhood." Oscar became a popular focus for spectators and writers, prompting such witticisms as "Aren't all men easy to see through?" or empathetic pronouncements about his lack of a "private life."[22] Another medical exhibit featured an eight-foot talking tooth that gave lectures on tooth decay. In the chemistry section, a fifteen-hundred-pound robot "with the serious, intellectual face of a scientist" and moving lips lectured about food chemistry while a projection screen in its torso offered images of the digestive process to complement its speech.[23]

Just as Charles Came's magic lantern displays of microscopic life had helped establish a mood of wonder in the 1840s, the microscopic level of existence also was featured in the 1933 fair. The biology exhibit included the crowd favorite of the "micro-vivarians": its six screens revealed magnified microscopic animals. Spectators saw "these monsters of a minute world dart about, forage for food, fight, reproduce."[24] This Darwinian display of realism, only one step removed from the human struggle for survival during the Depression, appealed to the crowds. Georg Rommert of Munich tended to the microscopes and the hydras, amoebas, paramecia, and other microscopic creatures that he carefully cultivated from specimens of pond water, ditch water, and moss scum. Journalists assured readers that the display did not involve the use of mere motion pictures, but "the best projections of living micro-organisms ever displayed."[25] Rommert, whom journalists alternately called a magician and a showman, explained how he lectured all afternoon, would then rush home to get more specimens, and return to lecture in the evenings. "Each of these creatures has a different diet. Each prefers a different temperature. I must keep them hungry, so they will perform for the crowd. But they must not get too hungry, or they will be too weak to make a good demonstration."[26] Reflecting the Darwinian world of exhibiting, Rommert was a success, called back for the 1934 season and given a larger exhibit area with better seating for the crowds.

In these and other Hall of Science exhibits, the SAC's strategies paid off. Articles that evaluated the first year of the exhibit noted that the Hall of Science had been a popular destination. Furthermore, despite predictions to the contrary, "most visitors took the science exhibits seriously last year, and more pencils and notebooks were seen in the great hall than in any other part of the fair."[27] If such reactions are to be believed, the SAC succeeded in reaching both fairgoers with short attention spans and those with genuine interests in learning. Such a conclusion supported the views of Northwestern University physics professor Henry Crew when he wrote for the inaugural issue of the *World's Fair Weekly* that young people of the 1930s were less likely to be in awe of science or to regard it in terms of "miracles" or "mysteries." The professor argued that in the process of building their own radio sets, "boys" learned a great deal of physics, and they would bring this hands-on knowledge to the exhibition. Of the exhibits, Crew promised, "You look, and you listen, and right away the whole thing is as apparent to you as a card trick that has just been exposed by the performer."[28] But Crew's optimism and year-end celebrations of the science exhibits contrasted with another article that described changes in the Hall of Science for 1934. The author remarked that exhibits had been revamped to fit what appear to be Holland's basic rules: "1. Is it simple? 2. Can it be tied up with some common experience of the average visitor? 3. Does it move?"[29]

A minority of the end-of-year-one evaluations looked less favorably on the science emphasis of the fair. A letter to the editor of a Chicago newspaper from early January 1934 argued that the celebration of science at the Century of Progress had gone too far. Titled "Science Bankrupt" and signed "An American Citizen," the letter began with a description of Oscar, the transparent man. The writer noted it took three years to build Oscar and that he was a "wonderful piece of workmanship," but then added, "And yet he is not perfect . . . he cannot think, he cannot speak, he cannot walk, has no life and is not able to judge right from wrong. Now, if God had created him, he could do all these things."[30] The writer praised University of Chicago president Robert Hutchins for his advocacy of the humanities over the soullessness of science. He quoted Hutchins as saying that "science is bankrupt" and expressed approval toward Hutchins and the other new humanists of the 1930s who questioned whether "progress" and science could always be linked without further question. This letter writer raised the seemingly timeless issue of whether humanity's ethics could keep up with the responsibility of overseeing new technologies and their social impact. Perhaps with fairgoers such as "An American Citizen" in mind, fair organizers had wisely named the SAC's proposed "Temple of Science" the "Hall of Science." Undoubtedly, the fair management would have steered this malcontent away from Oscar to the Hall of Religion, which also had done remarkable trade, with ten thousand visitors daily. The management had originally questioned that anyone would bother with the Hall of Religion except "the remnant of the late Victorian era."[31] But the hall became a popular destination because of its inclusive policy, lessons of tolerance, and staging of events such as a pageant that celebrated Chicago's large Jewish population.

While Franklin Roosevelt urged the fair management to continue the exposition for an unplanned second season and gained congressional appropriations for federally sponsored exhibits, big business used the 1934 season's inauguration as a forum to convince the public that business and science allied could solve the country's woes without any federal help. GM's president Alfred P. Sloan sponsored a conference of business leaders and academics at the fair's second year inaugural. Its purpose was to argue that prosperity would return without any need for heroic experiments in government. Sloan insisted that government intervention would only lead to a stagnant system, while unbridled capitalism provided dynamism and new solutions on its own. According to the GM executive, "New Deal attempts to regiment the nation and reduce its affairs to a static condition are the one sure way of preventing the scientists . . . from accomplishing advance."[32]

At the conference, Glenn Frank, president of the University of Wisconsin, critiqued the new humanists' theories and insisted that "the machine has not betrayed us, we have betrayed the machine. Science and technology have given us the means by which we may emancipate the race from poverty, drudgery, and insecurity." Frank added, "[Let us not] be a people strangled by our own success."[33] Rather than call a halt to scientific research and technological advances "until they no longer put so many strains on the traditional structure and functions of our social order," Frank urged a partnership between the hard sciences and social sciences so that "adjustments" could be made in advance to avoid the "social and economic havoc"—such as job layoffs and increased unemployment—that innovations could otherwise bring.[34]

During the winter break and beginning of the 1934 season, many corporate exhibitors had redesigned their displays to bring them closer into accord with the appropriate philosophy of showmanship. Business writers approached the fair as a grand experiment in social science and salesmanship and wrote with approval of these adjustments. One article showed a crowded midway and asked, "What exhibits stop this moving throng and why?" The writers announced that at the fair Mr. Average Citizen was "King" because he voted with his feet and avoided exhibits that lacked movement or human interest.[35] The article revealed that Mr. Average Citizen responded to gimmicks like electric eyes, showmanship, movement, and pretty women. Robots could draw a crowd, but thrilling spectacles did better. Marionette shows that featured new products in a dramatic framework also were successful.

Exhibitors dramatically changed their display strategies to accommodate the masses. Chrysler originally installed an outdoor track for demonstration rides, but in 1934 it added hourly stock car races by Barney Oldfield and other drivers, who would "race around hair pin turns while tires smoke and brakes squeal."[36] In 1933, Standard Oil offered a movie that discussed the contributions of oil to industry, but in 1934 added an outdoor wild animal act that included "Allen King and his den of ferocious tigers and lions."[37] The authors argued that the wild animals did an excellent job of conveying Standard's motto of "Live Power." The Safety Glass trade group invited passersby to throw baseballs at windows.

Hopes pinned on automatic exhibits and robots seemed misplaced. The authors of "King Customer" noted that "Sunbeam's potato peeling Mixmaster exhibit has three pretty girls on individual stages that are worth a dozen robots."[38] Overt salesmanship also vanished. Whereas the GM exhibit included salesmen who sold about three thousand automobiles, the Ford exhibit that opened in 1934 became the fair's most popular and lavish. Car dealers were conspicuously absent. The article praised Ford's strategy, noting that at the pavilion, "There is not even 'low-pressure' selling . . . [but] let none mistake. Ford's gigantic gesture is magnificent merchandising."[39]

Exhibitors were in a Darwinian struggle for the crowd's attention vying with the midway's bawdier attractions. The Streets of Paris provided sidewalk cafés and smutty postcards; peep shows could reveal naked or barely clad Venuses on clamshells; Sally Rand—and her imitators—wore only white body powder for fan dances, and at a freak show the "largest collection of strange and curious people ever assembled" could be seen for twenty-five cents.[40]

The fair's moral, for salesmanship, was clear. The writers of "King Customer" concluded that "no one ever lost money under-estimating the intelligence of the American public."[41] Marketers needed to conform to "King Customer's" low tastes and push drama over logic; this diminution made of the public a cruel tyrant, yet one curiously worthy of respect. Savvy marketers were having to "recognize the Customer as King and are learning to serve him in the ways he wants to be served."[42]

Conjuring in the House of Magic

Corporate research laboratories, in particular, assumed that the customer was king and offered miracles and wonders rather than dull scientific explanations. Miraculous science was on display at Westinghouse, General Electric, Bell Telephone, and American Telephone and Telegraph. Many of these research laboratory shows were given dramatic names. Westinghouse presented its research developments in a "Hall of Miracles," GE stuck with its "House of Magic," and GM eventually created a "Room of Mystery."

An advertisement for Westinghouse urged the public not to leave without seeing the "latest developments in electrical science direct from the famous Westinghouse Research Laboratory on 'Miracle Hill' in East Pittsburgh." The advertisement promised the public air-conditioning, a look at "black light," and an automatic steel rolling mill. It also promised to reveal developments in one of Tesla's pet fields, often derided by Steinmetz, "the transmission of power by radio."[43] The Hall of Miracles also featured a cobalt magnet that defied gravity as it hovered above an electromagnet, and demonstrations of radio beams deflected with mirrors. The exhibit also featured a bank teller window with a twenty-dollar bill resting on it. Visitors were invited to grab it, only to find that a photoelectric eye caused a gate to descend on their hand before they could reach the bill. An X-ray apparatus also allowed visitors to see the bones in their hands. And a magnetic strain gauge sensitive to one-millionth of an inch showed how a railroad track bent under the weight of a visitor.

GE, which had been hyping its House of Magic since 1929, was in the vanguard of such exhibitors. Its streamlined theater, topped with a snack bar, had an art deco electric sign, "House of Magic," that shone forth beneath the interior columns of the electricity building. Inside, the GE theater was modest, seating an audience of about three hundred. Every thirty minutes, a staff of young engineers gave performances; chief among them was a genuine stage magician, William A. Gluesing. Each thirty-minute show at the House of Magic would include six or more acts that featured high-frequency coils, oscilloscopes, and stroboscopes.[44] GE also gave special hour-long performances to more technically astute audiences—for example, to a group of visiting transportation engineers in town for a conference or a group of visiting engineering students.[45]

Gluesing, GE's resident entertainer, spoke a variant of the sophisticated Hollywood dialect of the 1930s—with elongated syllables and a vaguely English accent that was, incongruously, both lugubrious and light. His patter stuck to the grand themes and relied on various rhetorical reversals to create an aura of paradox and "magic"; for example, he would announce that "the pure science of yesterday is the applied science of today, and the pure science of today will be the applied science of tomorrow."[46] Although subtopics such as medical advances or radiation helped organize these performances, press release accounts make them appear to be a grab bag. Many of the demonstrations depended on the

Crowds mill about the entrance to General Electric's House of Magic at the 1933 Century of Progress Exposition in Chicago. Inside the art deco theater GE entertainers offered performances of gee-whiz science. A Century of Progress Records (COP 487010), Special Collections, the University Library, University of Illinois at Chicago.

high-frequency currents that Tesla had favored in his demonstrations. For one popular trick, Gluesing held a slender glass tube filled with helium glowing with lavender light. He then would tilt the tube and grasp its center, which would promptly turn black, and then continue to pass his hand down the tube, "squeezing the light out of it." He then informed the public he would "hold" the light in his palm and magically "cast" it back in the tube. Afterward he would explain that a nearby "coil of copper, heated by the electric current has been broadcasting electrons. You didn't know it until you saw the effect of the electrons lighting helium gas. I passed my hand down the tube to show you how, when the bombardment of electrons is cut off, the light goes out."[47]

A GE publicity photograph of Gluesing shows him looking somewhat subdued in his pinstriped suit as he holds a star-topped "magic wand." The electrical current of the human body, in this case the magician's, was sufficient to light this "Thyratron" electron tube. Photographs of Gluesing at work also suggest he worked in magic tricks that served as metaphoric parallels to new technologies.

This strategy is apparent in the GE film *Magic Versus Science* (1932) that featured Gluesing and an unnamed actress. She arrives at his art deco headquarters, greets him, and thanks him for letting her borrow a

William Gluesing, magician in residence at General Electric, holds a "science magic" wand, a "Thyratron" electron tube that glowed with light induced from the human body's electric currents. Journalists admired the "scientific vaudeville" that GE offered in its House of Magic. A Century of Progress Records (COP 487027), Special Collections, the University Library, University of Illinois at Chicago.

leatherbound "book"—clearly a prop—also titled *Magic Versus Science.* After helping her remove her wrap and revealing her low-cut evening gown and smart shoulder vest, he performs several opening tricks and announces the House of Magic themes. He explains why GE is willing to spend so much money on research, since "the pure science of today when applied in the future will give us more new products and still better quality." He also explains to his guest the distinction between what he does as a magician and what scientists do; the magician, he says, tries "to make the calm and ordinary seem baffling and mysterious," while the scientist makes the "baffling and mysterious seem calm and ordinary."[48] Gluesing then performs a series of magic tricks involving such props as cards, a thimble, a candle, a magic wand, and linking rings. His pretty but inept visitor can duplicate none of them, and her humiliation helps establish his authority.

Following each trick, however, he shows her new technology that will defy both her and his wishes. In one sequence he makes a lit candle vanish in a tube, using suitably obscure language such as "now that the candle is in it's really out." He then invites her to look at what science can do and brings out an apparatus that includes an electric eye and air blower that makes it impossible to light a candle. After her failure, she invites him to try and he also fails. "Even the magician is not able to fool science," he intones.

In this, as in the other short sequences, he next reveals the science behind the technology, relying on phrases such as "this is the explanation" and "get this straight." After his explanation, he shows her that by blocking the electric eye or "blindfolding it," she will be able to light the candle without triggering the blower. The film's strategy in each sequence is to build Gluesing's aura as a magician but then to introduce technology that places him back at the level of the average person; science removes magic's power and restores it to the informed citizen. Gluesing, in his good-natured presentations, reverses the usual power dynamic for stage magic: he establishes his superiority only to have it stripped away by innovative technology; in the film he regains some of that lost power when he serves as a teacher to his hapless student, who clearly never bothered to read the borrowed book.

Journalists appreciated his showmanship. In the illustrated column "Strange As It Seems," cartoonist John Hix featured Gluesing operating apparatus with which "popcorn can be popped between 2 containers of ice . . . without the application of heat."[49] And the authors of "King Customer at a Century of Progress" wrote approvingly of GE's "scientific vaudeville."

The apparent hodgepodge of marvels continued in 1934. Referring to scenes in western serials in which the heavies shoot out the lights in saloons, a press release announced that Gluesing would fire a "light gun that reverses the order of the old Wild West, [as it] is shot at a photoelectric target to turn on a signal light." This trick was a modern variant of one that magicians had perfected in the mid–nineteenth century that involved activating an electrical circuit to cause theater gas lamps to suddenly illuminate.[50] In another House of Magic feat, high-frequency currents were used to "burn steel wool, [and] light [a] lamp held in hand."

Seeking a common thread, press releases would offer themes and morals. One quotes Gluesing saying that certain "electrical novelties need only slight modifications to make them useful to the medical profession in combating illness."[51] Likewise, nuclear radiation keyed one 1934 performance in the House of Magic titled "Voice of the Atom" that included a Geiger counter and a demonstration of how lead would muffle the "voice of the atom." Gluesing and his staff also demonstrated ultraviolet lights, stroboscopic lighting effects, and devices that showed sound waves as visual patterns, allowing audiences to "see" orchestral selections on a screen.

The House of Magic's strategy of turning products and corporate research into theater was so successful that in 1934 GE opened other small "theaters" in its exhibit where performances were offered. One such performing space, for example, offered "The Romance of Lamps and Lighting," a show that featured a historical collection of lamps that ranged from "oil lamps of the stone age" to modern mercury lamps, and the tiny "grain of wheat" lamp used during surgery. In the performance, engineers explained what light was and how colored lights combined to make white. They also showed "how lighting can alter the expression of the face of a statue—even to the point of making it appear to laugh with rapid changes of light."[52] GE also opened a theater to present demonstrations of cookery and household appliances.

GM did not originally orchestrate their research marvels as carefully as GE. However, by 1934 they too arranged an entertaining show of "science magic." Since the names "House of Magic" and "Hall of Miracles" were already taken, they dubbed it the "Room of Mystery" and put it in place for 1934, proudly announcing it was an air-conditioned and "light-controlled" room. Its centerpiece was an ultraviolet "fountain" "made of many minerals used in the Research laboratories in Detroit" that "provide[d] weirdly beautiful color effects."[53] GM also exhibited a stroboscope that made a "rapidly whirling crankshaft" seem to "stand still" and explained that automobile engineers used the stroboscope to "detect moving parts that are out of alignment, causing vibration."[54]

Contrary to gloomy predictions, the Century of Progress was both a financial and an ideological success. Financially, it was the first and only World's Fair to fully pay off bondholders. The fair had employed more than forty thousand people and generated as much as $700 million in tourist revenues for local businesses, and, not surprisingly, city politicians attempted to make the fair a permanent fixture in Chicago.[55] Flourishing during the Depression, the fair suggested that demoralization need not be inevitable, nor the economic situation hopeless.

The clear value gained through such exposure emboldened corporations. After the exposition's close, GM concluded that it would befit a motor company to recycle its magic science show with a touring unit. After a year of planning, GM's "Parade of Progress" premiered in Miami,

Florida. Eighteen streamlined trucks then led the caravan of thirty-three vehicles on a nationwide tour of towns with populations between ten thousand and seventy-five thousand, where town officials were eager to cooperate and GM could dominate public attention. The show's arrival in town would be announced with a parade down main street that would include local city officials, business leaders, and GM dealers. After the parade a large tent was set up for the shows. Admission was free and attendance superb.

The caravan promoted the GM message, strongly advocated by Sloan and research laboratory director Charles Franklin "Boss" Kettering, that contentment only bred stagnation. Revealing new technology and redesigned products could make people discontented with their lot, and such agitation was a boon to an economy that depended on planned obsolescence. GM's and Kettering's beliefs that sales pressure created a more dynamic society was not particularly original—it brought to the larger economic picture ideas that were standard fare for sales manuals of the early twentieth century. For example, one manual from 1916 noted that "to sell means to awaken a desire for the hitherto undesired thing." This manual added, "What men do when they *want* is making history."[56] Amplifying these notions, Kettering argued that when "people begin to want things that they do not need . . . they begin to become more alert mentally, more willing to work, more willing to do the unusual."[57] Kettering's speech at the tour's premiere in Miami explained the purpose of the Parade of Progress as follows: "During the Depression people got the idea that the world was finished. We are trying to prove that it is not. . . . We are trying simply to sell you confidence in America, American industry, and American resourcefulness."[58]

Kettering, the head of research at GM, was a gifted speaker with a folksy manner who enjoyed giving amazing science demonstrations and who had trained many of GM's lecturers. He had offered one of the first such lectures in 1916 to a Society of Automotive Engineers meeting held on a cruise ship on the Great Lakes. The lecture involved such tricks as "freezing a flower with liquid air, freezing mercury into a hammer and driving a nail with it, and burning iron wire in liquid air." Like Tesla, Kettering used high-frequency electricity to light lightbulbs that he held; Kettering also offered the trick of frying eggs in a skillet that rested on a cake of ice.[59] In addition to displays of automobiles, the Parade of Progress featured Kettering-styled lectures, and its posters announced: "See Frozen Motion"—a reference to stroboscope effects—"Bend a Railroad Bar by Hand," and "See the Law of Gravity Defied!"[60]

Adopting Ford's strategy at the Century of Progress, the GM "science circus" did not involve any direct sales of automobiles. After opening in

February 1936 in Florida, the traditional winter resting grounds for many circuses, the GM Parade of Progress, also titled the "Caravan of Science," visited approximately one town a week through 1938, before settling in at the 1939 World's Fair in New York City.[61] A "Midget Caravan" was also developed to visit towns with populations of five thousand or less. The caravan spread goodwill and was designed to remind the public of GM's message that the "world was not finished."

The Parade of Progress, visited by as many as seven million people, was a masterful public relations campaign. With about one-third of each town's populace attending the free shows, the visiting "science circus," like the mythical circus of Dr. Lao, was a powerful distraction. GM did not offer Lao's bill of snake charmers and exotic acrobats, but instead offered exotic technologies and products. The Parade of Progress succeeded as part of a larger corporate effort to instill in an anxious public the confidence that technology and corporate innovation would bring new opportunities.

The Million-volt Man

On the midway of the 1939 Golden Gate Exposition in San Francisco, in a modest auditorium, a young evangelical preacher named Irwin Moon offered the throngs an escape from nude fan dancers, human oddities, and other carnival displays. Instead of traditional fire-and-brimstone sermons, however, Moon surrounded himself with electrical equipment; for the climax of one sermon, he jumped up on a transformer that sent a million volts of high-frequency alternating current through his bare feet, causing forked lightning to explode from the metal thimbles on his fingertips. At the fair's close, Moon's cosponsors, the Christian Business Men's Committee and the Moody Bible Institute, deemed the show a success. One of the groups' leaders reported that Irwin Moon's four Sermons from Science had helped these groups distribute nearly two million evangelical pamphlets during the fair's two-year run, prompted "hundreds" of conversions, and evoked "valued prayers" for the indifferent as well as for the "convicted but not converted."[62]

Moon indulged in the same sort of gee-whiz science perfected earlier in the 1930s by lecturers from GE, GM, and other corporations. An early testimonial letter that Moon solicited insisted that his lectures and demonstrations were "on par with those given by representatives of the GE House of Magic and others representing organizations of that type."[63] Moon's posters heralded such marvels as "SEE Steel floating in air . . .

Tiny living creatures enlarged over 2,000,000 times" and "HEAR Music made with a flashlight . . . Molecules moving in a bar of ordinary steel."[64]

In his sermons he offered demonstrations of gee-whiz science and technology but added theological commentary. To Moon, technology offered modern source material for parables. Moon was an evangelical minister who refused to shy away from science. Instead, he sought to convert his audiences by demonstrating that the hidden wonders of the world were all part of a divine pattern. His goal was to reach educated young people of his generation who were likely to have drifted away from religion and would shy away from the typical evangelical revival meeting. Moon would have been thrilled to attract the sort of bright young people, often from conservative Christian backgrounds, then surrounding ESP investigator J. B. Rhine at Duke University. As his publicity posters often put it, Moon offered the "First Century Gospel in a Twentieth Century Manner."

His career reveals how scientific wonder shows could be linked to religious themes. Science demonstrations and education were the bait, conversions the goal. But to succeed, the science demonstrations needed to instill wonder and awe. Moon's version of the "million-volt" demonstration, for example, repositioned the Tesla demonstrations of the 1890s, which showed the inventor "in the Effulgent Glory of Myriad Tongues of Electric Flame After He Has Saturated Himself with Electricity." If Tesla chose to cast himself as a demigod, Moon cast himself as a modest Christian obedient to the awesome powers of God. Even though such "Tesla coil" tricks were common to vaudeville of the early twentieth century, Moon's religious recasting of the exhibition was remarkable enough for a photograph of Moon to appear in *Life* magazine a year prior to the Golden Gate Exposition.

Moon's demonstrations offered an updated version of Charles Came's traveling show of the nineteenth century. Came had mixed lectures about the solar system and phrenology with magic lantern slides of the Holy Land, slides of microscopic life, and sparking electrical effects. Came's electrical equipment, free "shocks," and "thunder house" that collapsed when struck by a bolt of electricity gave way to Moon's million-volt demonstration. And Moon, like Came, evoked the "wonder" of science by demonstrations of the great and small—the macrocosmic and microcosmic, the universe as revealed by the telescope and microscope. Such contrasting scales of perspective could make the human perspective seem fresh and new and of great value. For Moon, if not the religiously indifferent Came, nature and its intricate patterns gave testimony to the existence of God as the Great Designer. While Came had been fascinated

This pamphlet cover for *Sermons from Science* circa 1940 shows the athletic evangelist Irwin A. Moon performing the "million-volt man" demonstration, his fingertips spraying electric sparks. Courtesy Moody Bible Institute Archives.

with the early psychological theories of phrenology and its self-help agenda, Moon offered musings such as, "The soul of man needs to be satisfied by union with God just as chemical elements seek union with others to preserve their stability."[65]

Moon was born in 1907 in Grand Junction, Colorado. His father was an ostrich rancher. As a youth, Irwin Moon had been a football player and daredevil. He also was a tinkerer who put together radio transmitters, read radio magazines, received a ham radio license at age twelve, and played pranks that included giving other family members electrical shocks. During his youth, after "wandering into a revival meeting," he became a convert to fundamentalism. He gave up his pursuit of a scientific career and began studies at the Moody Bible Institute in Chicago and several Bible institutes in Los Angeles.

In 1929 Moon became the pastor at the Montecito Park Union Church in Los Angeles. According to his own lore, it was while trying to reach out to young members of this church that he began to revive his own interest in electronics and science. In 1931 he offered the church's youth group a lecture on "The Microscope, the Telescope, and the Bible." This lecture involved colored slides with images of the microcosmic and macrocosmic, and he soon expanded it to include stroboscopes, photographic equipment, and sonic instruments.[66] He was invited to other churches and schools. In 1937 he resigned his pastorate to lecture. In 1938 his spectacular million-volt demonstration was featured in *Life* magazine, and for the next two years he performed daily on the midway at the Golden Gate Exposition.

Moon's determination to bring science to the pulpit went against the grain of fundamentalism's anti-intellectual and antiscience stances of the 1920s. While science and fundamentalism had coexisted until the early twentieth century, the Scopes trial polarized these two affiliations during the 1920s. Though a strong-willed person, Moon's decision to preach with science-based lectures was not without its personal cost. Looking back, his wife commented that after his conversion he had "given up on it"—that is, his interest in radio and other scientific gadgets—as it was "not the right thing."[67] One press release also remarked that after his conversion, "the prized radio transmitter was given away. The electrical and mechanical playthings were packed up."[68] Moon reported agonizing over his later decision to use scientific equipment for religious purposes. According to an account in the *Sunday School Times*—based on one of Moon's press releases—he earnestly explored whether science ran counter to his faith. He at first "wondered often whether the Lord had wanted him to give up his science. Perhaps the Lord only wanted him to be willing to give up those laboratory toys." The fact that his technical know-how endeared him to young people emboldened him. As the article notes, "Prayerfully and cautiously he began to use his God-given scientific ability." He continually searched his Bible to determine if "science had a place in his ministry."[69] His wife insisted he "began to see how to use it [his scientific knowledge] to show the accuracy of the Bible."[70]

The turning point, according to one account, came when he looked into a secondhand store window and saw an empty mahogany carrying case that would be ideal for one of his bulkier transformers. He had only a dollar and ten cents in his pocket. According to the story he prayed to

God that this be a test. If the money in his pocket was enough and if the case indeed fit his ungainly transformer, he would let this be a sign that he was permitted to bring science into his ministry. The case cost him ten cents; he rushed home with it and the transformer was a perfect fit. The sign Moon had needed had been provided. "Irwin Moon would be a preacher-scientist."[71] Whether apocryphal or not, this story implies how gravely Moon weighed his ambitions against the fundamentalist culture that surrounded him. The bitter battle over the teaching of Darwinism in the 1920s had galvanized most fundamentalists; in the 1930s, members of this community regarded science as synonymous with atheism. Moon's "parable of the transformer" points to how difficult this decision had been for him. It shows him attempting to fit science into the larger carrying case of evangelicalism. He had crossed a border and could justify it best with the image of a transformer that shifts voltages up or down to allow electronic equipment to connect and run efficiently. The carrying case and his prayer, however heartfelt, permitted him to "transform" and so validate his desires and ambitions. Ultimately with such instruments he hoped to transform his public by initiating conversion experiences.

If this parable were not enough, in its profile of Moon the *Sunday School Times* offered another story that indicated providence at work in Moon's life. He decided on his own that Sermons from Science would be a great exhibit for the Golden Gate Exposition. However, he had no sponsorship. On the spur of the moment, Moon drove to consult with Tom Olson, a leader of the Christian Business Men's Committee of the Bay Area. Upon his arrival, Olson announced that Moon was the answer to his organization's prayers. The committee had recently been told that the building dedicated to "Business Efficiency" that they planned to base themselves in at the fair would not be built and they were casting about, in fact, praying for help with a new approach. Moon's energetic and innovative Sermons from Science would be an ideal vehicle, and a simple building on the midway an ideal, if paradoxical, forum. Providence had again emerged into human affairs. This story was a little too neat—and had been tailored slightly, to amplify its "providential" quotient. A Christian Business Men's Committee pamphlet indicates this second "miracle" was more matter-of-fact. Moon telephoned and made an appointment to see Olson in southern California. During the appointment, after hearing Moon's proposal, Olson urged Moon to come with him to San Francisco to propose the Sermons from Science project to the Christian businessmen's group at the committee's annual dinner for the following week.[72] The blind workings of providence are not as apparent in this rendering. Regardless, Moon soon was performing at the Golden Gate Exposition. As the *Sunday School Times*' author put it, "A church had been built that looked like a laboratory on a street that was a midway. It was like Christ going to the publicans and sinners."[73]

At the Golden Gate Exposition, Moon perfected three lectures that would remain at the core of the teachings he and other disciples eventually gave under the auspices of Moody Bible Institute. The first, which dealt with the wonders of light and color, was titled "Christ the Light of the World." The second, focusing on "sensation," centered on sound, perception, and voice recordings. The third lecture, his "million-volt" demonstration, was titled "The Scientific Necessity for the New Birth." Each lecture relied on equipment and techniques familiar to GE's Gluesing and other industrial science showmen. Moon, however, added a layer of moral commentary to each of these demonstrations. If the shows of GE and GM with their grab bag of tricks had offered the general message that corporate science was a force for social good that had helped displace a dark age of superstition, Moon added that to create genuine goodness, science needed religion's guiding hand.

Moon grouped his sermons around several key themes: the limitations of a materialistic worldview; the limitations of the human sensory apparatus; and the reality of unseen forces. Moon and his later Sermons from Science disciples liked to say, disparagingly, "Today it is popular to consider man as nothing more than a pile of chemistry and physics—a sort of beefsteak with a nervous system."[74] Moon would attempt to show that such materialism had dubious scientific standing. He also insisted that the gap between religion and science was unnecessary—a Christian science could be attained.

To make his case for a Christian cosmos, Moon relied on logical arguments that clashed when carefully inspected, but which served for his dramatic presentations. He promoted both an argument for separate spheres and a conflicting argument "from design" of God's existence. The "design" argument insists that God's hand can be seen in all the minutiae of nature, whereas the separate spheres argument had been inaugurated to ease religion of the burden of scientific proof for its claims about the natural world and its origins. The Creator may not always be apparent in nature's workings nor the Bible infallible. Uneasily mixing the two arguments, Moon proclaimed, "A true scientist knows that he is only reaching for truths which always have been there and have been there because of God. Religion seeks God and truth in another direction."[75] Moon's separate spheres subordinated science to religion, unlike the version that the physicist Millikan had earlier promoted. Like Silliman before him, Moon felt it necessary to seek God below the surface of all scientific truths.

Moon's logic, ultimately, was associative, poetic; while his "arguments" conflicted, when they were enacted on the stage they had a powerful impact on his audiences. More important to Moon's purposes and arguments by analogy was the theme of the limits of materialism and of the human senses. This argument had previously encouraged Spiritualists and parapsychologists—science established both the limitations of man's senses and the reality of the "invisible." If science could prove that invisible forces operated in the universe, it was then not far-fetched for the religiously minded to make a similar argument for unseen spiritual forces. By the 1950s, Moon had refined this formula to insist that "many of us have missed God simply because we haven't been in tune. That's why God said, 'unless a man be born again, he cannot see the kingdom of God.'"[76]

Two of Moon's sermons focused on the limitations of human senses. In his sermon on "light," he offered optical illusions and lighting effects to remind audiences how limited were their senses and to urge them to accept the reality of the unseen. Throughout, he would draw out the morals from such demonstrations. His discussion of the prism led him to preach of the "pure white" light of Jesus that contained all the other colors of light. To illustrate simple miracles captured in light, he showed his own efforts in time-lapse motion pictures, filmed in his house. Footage included that of a seed sending down roots, stems and leaves growing, and a plant then flowering, as well as footage of close-ups of caterpillars fighting and "a fly's foot magnified three million times."[77]

Moon and his associates also featured stroboscopes and displays of ultraviolet light. A moral could be drawn from ultraviolet light, which could make plain stones beautifully iridescent, and so suggest the hidden wonders of the world and of the soul. In his "sound" lecture, Moon also lectured on the limitation of human senses, using electronic eavesdropping and recording devices and other sonic displays. These served Moon as material for a sermon on how God records all of one's deeds—proving one's need for repentance.

Of these sermons on sensation, George Speake, a disciple of Moon who carried on the sermons after Moon began to concentrate primarily on evangelical filmmaking in the late 1940s, would explain, "My purpose is not to amuse people with parlor tricks but to show that eyes and ears are such tragically feeble instruments in some realms of nature that man is reduced to the functional stature of a jellyfish in a symphony orchestra."[78] Moon and later Moody lecturers also intended to amuse spectators with their parlor tricks. In his early lectures, Moon liked to use a powerful directional microphone to record whispered conversations of audience members prior to a performance and play them back during the show. During his "sound" lecture he would also play higher and higher frequency tones, asking the audience whether they could still hear them. Inevitably, people would continue to raise their hands after he had turned the equipment off. When he toured military bases, during and after World War II, Moon would encourage soldiers to inhale helium and transform their manly voices into "lisping falsettos," or as his publicity material also put it, "to speak in the helium dialect."

The climax was the crowd-pleasing "million-volt man" sermon, in which Moon electrified himself with a million volts so that sparks flew from his body. In 1941, John Hix in his syndicated "Strange As It Seems" column, who had once featured GE's Gluesing, also featured a sketch of Moon in his suit with hands upraised and fingers shooting out flames. "Pillar of Fire!" ran the heading.[79] During these performances, barefoot or wearing only socks, Moon would climb up on a copper transformer. When the performer shouted, "On!" an assistant would throw a switch, causing the electricity to shoot through Moon's body and thimble-tipped fingers. On other occasions Moon would hold a piece of wood in his hands that would burst into flames. This "death and resurrection" display pointed to a clear moral. Moon would explain that he survived the electricity's wrath because he was using a high frequency. At a lower frequency, electricity of the same voltage could easily kill a man. Moon's lesson was that he and his audience needed to be "in tune" with God. All of mankind was in need of a rebirth experience—or cosmic reattunement—through which to leave sins behind. The later Sermons from Science lecturer George Speake added that in order to survive this trick, he needed to obey natural laws of electricity and physics; did not he and his audience, therefore, need to follow spiritual laws as well? In such cases, ignorance was not bliss. Survival depended on awareness of law. Speake then would use the million-volt demonstration as a springboard for defining faith. He would ask the audience, following his electrification, if they now believed that standing on the transformer was safe. If they did, he would insist that faith must lead not just to belief but to action. He then would ask if they now were willing to stand on the transformer themselves.

Through the creation of such demonstrations, Moon attempted to heal several cultural rifts. Not only did he need to live up to his billing as the "Harmonizer of Science and Religion,"[80] but he also was insisting on the importance of scientific truths in a fundamentalist context hostile to science. To fulfill his agenda, he needed to convince sophisticated audiences that feats like watching "living objects die under invisible death ray" or "drab gray rocks display beautiful colors" had some educational value. But his fundamentalism and his refusal to be a Christian mod-

Keith Hargett, who toured with Sermons from Science briefly in the 1950s, offers his version of the "million-volt man" effect. Courtesy Moody Bible Institute Archives.

ernist—a believer who also accepted modern scientific theories including evolution—required him to lecture on the scientific validity of Genesis. This last task led to some of his seemingly more absurd claims, for example that Methuselah could have lived 969 years because at that time on Earth "a vast vapor canopy" shut out ultraviolet rays and slowed down the aging process. Likewise, that vapor canopy accounted for the fact that no rainbow had appeared until Noah and his crew saw one following the biblical Flood.[81] Moon's wit and often-unexpected explanations most likely threw hecklers off balance while pleasing the already converted.

In the first years of Sermons from Science, Moon was frequently on the road, visiting such towns as Portland, Seattle, Grand Rapids, Chicago, San Antonio, El Paso, Houston, Dallas, Charlotte, Atlanta, and Buffalo. He did advance work, sending press kits that included advertisements, promotional photographs, and press releases that provided newspapers with pleasing copy. A typical newspaper advertisement could read: "Hear the Man who Thrilled Thousands at the Treasure Island Fair! SEE 1,000,000 Volts Discharge From a Human Body! SEE Metal Caused to Float in Space! SEE and HEAR Your Voice Projected on a Beam of Light! See and Hear Your Voice Recorded Inside a Tiny Thread of Steel! IRWIN A. MOON presents 'SERMONS FROM SCIENCE' Unique! Startling! Convincing!"[82]

His appearances were either solo or in tandem with other evangelists. He would present each of the three or four lectures on separate days. Success, on this circuit, involved the number of converts gained, or the number of the indifferent one could succeed in prompting to think seriously about religion. At the end of a performance, the evangelist would encourage people to "accept Christ." A 1942 evaluation of a Moon appearance indicated that "about 25 hands were raised to the invitation to accept Christ on Thursday night. Many stood and voiced their acceptance on Friday night at both services."[83] Sixteen years later, Moon's disciple George Speake reported of a performance series in Cincinnati: "They started out with quite a weak crowd on the first night there with only about 400 out but it did build up to about 1000 on the last night with a wonderful response to the invitation—over 100 indicated a desire to make a decision for Christ and many were in the inquiry room with them."[84]

Moon, who trained at Chicago's Moody Bible Institute, worked within an evangelical framework developed by Dwight L. Moody in the late nineteenth century. In 1873, Moody, formerly a successful shoe salesman, leader of a popular Sunday school in Chicago, and president of Chicago's YMCA, teamed up with gospel singer Ira Sankey and began a series of urban revivals that attracted huge audiences. During an early tour in England he drew crowds of twenty thousand or more. His subsequent revivals in America also drew crowds larger than ten thousand. Moody would advertise in newspaper amusement sections, hire choirs of five hundred singers, and rely on his own considerable storytelling and preaching skills to gain converts and to build an evangelical empire that included the Moody Bible Institute, the Northfield School in Massachusetts, and the Moody Press.

Crucial to Moody's style of evangelism was the "inquiry room." He noted that "few are converted under the sermon, they are only impressed; that is sowing the seed and the personal conversation is the reaping."[85] In the inquiry room, which might be adjacent to the theater or in a separate church or building, Moody would hold a more intimate sermon for audience members eager for personal dialogues. Inquiry room assistants, whom Moody may have drawn from the local ministry, then circulated with Moody, speaking to small groups of curiosity seekers, repentant sinners, and others hopeful of undergoing a conversion experience or of renewing their religious faith. One of the main reasons that Moody set up the Moody Bible Institute was to train assistants to work in his inquiry rooms during large revivals.[86] Moon, who relied on the pattern first set by Moody, also concluded his Sermons from Science with inquiry room meetings.

In the early 1940s, with the onset of World War II, Moon began to tour military bases, seeing young soldiers as ideal candidates for his sermons. Not only did they have a "masculine affinity" for science but likewise the "seriousness of war had set many of them to thinking about spiritual things."[87] Moon's virile approach met with approval. Army chaplains were thrilled with his lectures and wrote glowing reviews, remarking that Moon was ideal for entertaining and morally uplifting troops. For example, one chaplain wrote, "His opening lectures at the Camp Theater was to a packed house, and he held his audience night after night. On several evenings the audience remained for as much as an hour and a half asking questions after the meeting was dismissed. . . . His platform manner, with good natured and humorous presentation and rebuttal, has won him many friends as well as converts."[88] Another chaplain wrote that four thousand men "of all faiths" heard his lectures, which "were a milestone in confirming the faith of thousands who had already made their decision for God, and . . . challenged the indifference of thousands of others who have been on the 'fence', as it were."[89] Moon gained special favor with the air force and performed frequently for both officers and men. Soon he was performing along with comedians like Bob Hope under the auspices of the USO.[90]

In the late 1940s, Moon trained George Speake, an air force pilot with an engineering background, to take over the Sermons from Science while he went on to create evangelical science films for the Moody Bible Institute—more specifically the Moody Institute of Science, which Moon had helped initiate.[91] The films that he and his assistants produced soon became important evangelical tools, often shown in tandem with the Sermons from Science lectures. In both the sermons and films Moon relied on the "argument from design."

Moon purchased military surplus cameras, lighting, and optical equipment and set up a film studio in a former Masonic temple in West Los Angeles. What turned out to be a film studio began with the more ambitious effort of Moon and fundamentalist scientist F. Alton Everest to create a "Christian laboratory." The Moody Bible Institute funded this experiment in bringing together fundamentalism and genuine scientific research. Moon and Everest's initial proposal to Moody president Will Houghton in 1948 insisted that the fundamentalist rejection of science was damaging the movement's ability to reach out to and convert modern-thinking people, including scientists. Society was the big loser. The fundamentalist Christian retreat from science had led to an "overwhelming skepticism regarding the activity of God in creation." Yet they believed that "many who are steeped in such materialistic teachings would give them up if there were a reasonable alternative position stated in understandable scientific terms."[92]

The planned Christian laboratory would produce and distribute scientific films and monographs and develop a library. It would encourage, subsidize, and direct scientific research and analyze the "moral and spiritual significance" of scientific developments. The "Christian laboratory," however, saw little actual research; instead it became the place where Moon improved Sermons from Science apparatus and produced a number of films that revealed the wonders of the world. Efforts to stimulate scientific research that could mesh with creationist beliefs carried over into Everest's work with the American Scientific Affiliation that he led for the Moody Bible Institute.[93] At the film laboratory, Everest served as a science advisor to the film crew.

The Moody Institute of Science films opened up Moon's Sermons from Science approach; each film documented natural wonders and scientific breakthroughs with added moral commentary. The films used innovative techniques, whether close-up photography, underwater photography, time-lapse photography, or some of the first footage of open-heart surgery. Their films of the 1940s and 1950s had titles such as *God of Creation, They Live Forever, God of the Atom, Voice of the Deep, Dust or Destiny, Red River of Life*, and *The City of Bees*. Moon worked fervently to give the films a sophisticated polish. Accompanying the scenes was a "relentlessly anthropomorphic" narration, and each film ended with a shot of Moon at a desk with his Bible discussing the religious implications of the earlier footage.[94] The Moody Institute of Science distributed these films widely. They became useful tools, which along with Sermons from Science demonstrations provided shows for the armed forces under the auspices of the "character guidance" programs developed in the late 1940s.[95] The films were technically innovative. Eastman Kodak awarded Moon a gold medal in 1980 for "his contribution to the advancement of the educational process through many unique uses of the art of the motion picture."[96]

George Speake, who inherited and refined the Sermons from Science lectures in the 1940s, in turn trained other lecturers, including Keith Hargett, Dean Ortner, and James Moon, son of Irwin Moon. However, Speake continued as the primary lecturer for Sermons from Science from 1948 until the 1970s. Over that period of time he and associates made appearances at the Seattle World's Fair (1962), the New York World's Fair (1964–1966), the Montreal World's Fair (1967), the Munich Olympics (1972), Spokane's Expo '74, the Montreal Olympics (1976), and the Atlanta Olympics (1996). At Moody pavilions they would alternate film screenings with live demonstrations.

A script for the sermon on light reveals that these demonstrations were fairly sophisticated. The lecturers covered such phenomena as the psychological concept of "persistence of vision," which was demonstrated with rotating discs and motion pictures; they also used photoelectric cells to create strobe effects and to create music from an interrupted flashlight beam; a "modu beam" they projected across the room also became converted to music. Demonstrators discussed the range of theories of light including wave theory, "puscular" theory (corpuscular), and quantum theory. They would emphasize that each of these theories was merely an educated guess. They would also discuss the production of light and the theory of color. At this point a sodium vapor lamp would be switched on that would turn the theater from a world of color to one of gray and yellow. Next the lecturers would discuss how narrow a section of the electromagnetic spectrum humans could distinguish and respond to. This led to the conclusion that "the possibility that a spiritual realm exists apart from the physical realm is not at all fantastic."[97]

Even though we cannot trust our five senses, the lectures went on to insist that we should put our faith in unvarying scientific laws. Volunteers were invited to pick up and move a suitcase that contained a twenty-four-pound revolving gyroscope. Their inability revealed "the laws of gyroscopic rigidity." Chemicals were also shown to follow unvarying laws. In one demonstration the lecturers mixed up a solution of iodine and starch that would, after a given length of time, turn black, depending "upon the quantity and concentration of the chemical compounds." This and other demonstrations led to the conclusion that "unvarying laws control all of the heavenly bodies, and the sub-microscopic world as well." Such an ordered universe led to the conclusion that "the universe is controlled by an infinite wisdom and creative genius that is concerned with us." Humanity was "his crowning work."[98] Not only could science and religion coexist, but humanity's centrality to creation was restored. In these lectures, all of the requirements for a wonder show—to demonstrate the miracles of science and highlight technology's millennial role—were fulfilled.

More Sugar for the Science Pill

Other showmen worked with the science-magic model in the 1950s. In 1951, thirteen years after Moon appeared in *Life,* the magazine published a color photograph of a young woman in vaguely Balkan garb standing on a platform while blue bolts of electricity were shooting from her fingertips. The headline was "A Young Lady Impersonates an Electrode." Beneath the photograph, the copy explained, "Wearing thimbles to avoid burns, Betty Brown smiles as a million volts of electricity spray from her upraised fingers."[99] She was the fifteen-year-old daughter of showman Bob Brown, who ran Bob Brown's Science Circus. He performed at schools and luncheons, and made at least one television appearance on ABC in Chicago. Science Circus acts included mathematical demonstrations, a static electricity generator that made a volunteer's hair stand on end, a sparking transformer that lit a young assistant's cigarette, and the million-volt trick pioneered by Tesla and featured in Moody Sermons from Science performances.[100]

An article about Brown, "Sugar for the Science Pill," indicates that even prior to the launch of *Sputnik*, Americans were concerned with finding ways to interest students in science to maintain American technological and scientific proficiency. Befitting the consensus mentality of the 1950s, Brown's show sought to cleanse science of the negative images common since the launching of nuclear bombs on Hiroshima and Nagasaki, on Pacific atolls, and the western American deserts. The atomic industry began a campaign to reassure the public about "the peaceful atom," and Brown wished to be on the vanguard.[101] A journalist wrote that Brown "hopes the time will come when North Carolina's only Circus will be the first in America to use atomic power. Made, of course, from uranium dug up right out of our own Blue Ridge mountains."[102]

Between GE's Gluesing in the 1930s and Betty Brown in the 1950s, numerous models of magic science had emerged, employing the technological sublime to different ends. Corporations in the 1930s relied on such shows to prove the vitality of the big business model for prosperity. Irwin Moon and associates employed the technological sublime to religious ends. The lecturers were mildly critical of the status quo, as they implied that material plenty and technological superiority were not enough. American society needed spiritual values as well. Nevertheless, Moon stands as one of the great emblems of the "consensus" culture of the 1950s. That culture presented a united front against the dangers of world communism, and was an attempt to deny any great ideological rifts in America. This consensus vision required that racial hatred, class consciousness, disparities of wealth and poverty, and any form of social discontent be ignored or suppressed as "deviant." Moon's efforts to be a "harmonizer," to bring together science and religion, mainstream thought, and religious fundamentalism, as well as the warriors of the military with the peaceful messengers of the Bible, make him an important symbol of the postwar consensus years.

Likewise, Bob Brown's desire to go "atomic" with his science circus points toward public debates contrasting the "friendly atom" with more apocalyptic visions of nuclear devastation. As many of the decade's psychologists indicated, beneath the calm facade of the consensus years were turmoil and anxiety. In the same decade that Bob Brown and the Moody lecturers sugarcoated science, the flying saucer subculture emerged to mount its own wonder show, mixing technology with religious yearning to challenge a dominant culture's fiction that all was well.

CHAPTER SEVEN

Flying Saucers

In 1953, George Van Tassel arranged the first Giant Rock Space Convention, a gathering of flying saucer enthusiasts that continued to meet annually at his resort in the California desert for two decades. Van Tassel, a pilot, built a landing strip and a hotel and created a council room in a cavern at the foot of Giant Rock near Yucca Valley, California. At his first convention, over five thousand people came to discuss flying saucers, hear lectures, scour the desert skies for saucer sightings, and stop at booths to purchase books and talk to recent "contactees" who, like Van Tassel, claimed to have met beings from outer space.[1] Usually such tales fit optimistic, millennialist hopes. If technology was speeding humanity toward the end of time, it was also bringing humanity closer to heavenly promises.

Narratives of contact with aliens were central to this subculture and meshed neatly with earlier Spiritualist accounts. For example, according to Van Tassel, one evening in the summer of 1952 he had gone into a trance at the base of Giant Rock, and soon after met the "Council of Seven Lights," a group of wise beings from beyond, then circling the earth on a spaceship.[2] Van Tassel's tale, which involved not a physical encounter with aliens but an encounter on a spiritual plane with these beings and their spaceship, demonstrates the blending of futuristic technology and occult notions that was a key component of the contactee subculture of the 1950s. If the movement relied heavily on the modern mythology of space travel culled from science fiction books and movies, it had roots in the older lore of Theosophists, Spiritualists, and Swedenborgians. Much as at earlier séances, at flying saucer conventions, attendees learned that the angelic "Space Brothers" were monitoring Earth and attempting to heal the planet of its evils, not the least of which then was the nuclear arms race.

Contactees gather in the desert for the Giant Rock Space Convention in 1955, offering their books for sale. Narratives of contact with Venusians were popular. Courtesy J. Allen Hynek Center for UFO Studies.

Conventions like that at Giant Rock served as early prototypes of the New Age expositions common later, bringing together health faddists, religious fundamentalists preparing for Armageddon, occultists of various sorts, and flying saucer enthusiasts. A journalist attending a saucer convention in 1959 reported that one booth featured a prototype of a flying saucer on sale and that the attendees included a handsome couple, costumed in blue shirts coated with mystic symbols, claiming to be Prince Neasom from the planet Tythan and Princess Negonna presumably of the same planet, though she had a distinguishable New York accent.[3] Other flying saucer contactee gatherings offered even more elaborate pageantry to dramatize their optimistic vision. One such group, the Unarians, based near San Diego, still mounts an annual Conclave of Light in which attendees dress up in costumes that indicate their former lives on other planets, carry banners with the names and paintings of those planets, conduct healing sessions, and encourage Space Brothers to land their saucers.[4]

From the onset, the flying saucer movement was a stepchild of the cold war and its nuclear dread. The trend sparked in 1947 when pilot Kenneth Arnold reported seeing "saucer-like" objects in the sky while flying near Mt. Rainier. Within weeks, UFO sightings were reported in thirty-five states and Canada.[5] The rage was on, encouraged by Hollywood productions and proliferating global reports. In the first six months of 1952, sixteen thousand newspaper articles were dedicated to the sightings. The U.S. military responded with numerous efforts to investigate the sightings, eventually leading to charges they were involved in a "cover up."[6]

While serious amateur enthusiasts attempted to establish the authenticity of such sightings, another popular movement rallied around contactees like Van Tassel who had stories to tell about meetings with angelic beings. The enthusiasm for such tales illustrates the anxiety undermining the contentment of the postwar age. The possibilities of suburban plenty and the widespread use of tranquilizers could not calm all discontent. A distaste for the conformist mind-set led to rebellious attitudes not only in the bohemian literary culture of the Beats or the highbrow world of the mass culture critics, but also in the more outlandish setting of the flying saucer movement. This subculture's critique was based in metaphysical beliefs and religious longings.

In her book *Aliens in America,* Jodi Dean has argued that the NASA launches of the 1960s and 1970s and the UFO contactee movement of the 1980s offered contrasting "theatrics of space," promoting opposed narratives of power and truth.[7] Dean likened the NASA launches, with their new frontier rhetoric, to a glossy Broadway production designed to reassure the public of American superiority, while the 1980s UFO contactee movement was offering a community theater rendition of a play like *The Fantasticks.* To take this formulation back one step, the 1950s flying saucer enthusiasts were offering an even barer-bones production, more like that of a living room reading group. Their production, a pure product of the imagination that annexed the galaxy as its backdrop, gained in sweep and ambition what it lacked in stage props, sympathetic media coverage, and credibility.

For its devotees, the 1950s flying saucer movement scripted and produced a cosmic drama. While evangelists such as Billy Graham were warning of the approaching nuclear Armageddon, contactees like Van Tassel were reporting that messengers from beyond had arrived, equipped with otherworldly technology and a demand for peace on

Earth. The contactees' script dropped any pretense at science education, placing its millennial vision center stage—like earlier itinerant showmen, the UFO inhabitants were traveling wonder workers, wandering the galaxies with their utopian messages. In this drama, science melted away into desire and magic.

The cold war primed America to momentarily listen to the message from its fringe. Nationally broadcast television and radio appearances by contactees like Van Tassel, as well as their appearances at conventions, and the popularity of science fiction movies and books helped promote the occultists' unorthodox worldview. Mainstream scientists could no longer fully be trusted, opening the window to such populist science visions. Under the guise of the flying saucer movement, fringe salesmen of the 1950s critiqued cold war society and science while re-envisioning the universe in terms of a cosmic wonder show.

Mars Revealed: Or, Seven Days in the Spirit World

Carl Jung was one of the first commentators to connect the flying saucer craze to the religious impulse. In his 1959 monograph on flying saucers, Jung insisted that the movement, whether founded on pure fantasy or on actual observations, had importance for its creation of a modern legend based on "visionary rumor." Jung argued that the flying saucer offered a divided world a symbol of wholeness, completion, and perfection, even divinity. Jung saw the flying saucer as a visionary product arising from the schizophrenic political divide of the cold war era; like Irwin Moon, Jung recognized that technology was an ideal vehicle for mythology in modern times, and he added that "the possibility of space travel has made the unpopular idea of a metaphysical intervention much more acceptable."[8]

Within the context of a religious movement, contactee tales need not be interpreted as fraudulent sales pitches or delusional concoctions, but as defining narratives of a new creed. These narratives had their counterparts in shamanism, in the Spiritualist movement of the nineteenth century, and in the religious yearnings expressed in twentieth-century science fiction with its apocalyptic themes of the death of civilizations and rebirth of the new.[9] One scholar also has pointed out that many of the contactee tales followed the pattern of a shaman's initiation voyage to the upper realm.[10]

The beliefs of 1950s contactees ultimately owe a debt to the work of the eighteenth-century natural philosopher and mystic Emanuel Swedenborg, who elaborated on that basic tale of initiation. Swedenborg, one of his age's leading natural philosophers, abandoned his studies of the natural world after a period of crisis in the 1740s when he underwent mystic experiences in which he met and journeyed with spirits and angels to other planets.[11] The sweep of Swedenborg's vision and its influence is clear even from the title of one of his books: *The Earths in Our Solar System Which Are Called Planets: and the Earths in the Starry Heavens; with an Account of their Inhabitants, and also of the Spirits and Angels there: From What Has Been Seen and Heard* (1787). In this tract Swedenborg explained how each of the planets and others "innumerable" have spirits and angels inhabiting them. Swedenborg depicted a universe imbued with meaning: "He who believes, as everyone ought to believe, that the Deity created the universe for no other end, than that mankind, and thereby heaven, might have existence, (for mankind is the seminary of heaven,) must needs believe also, that wheresoever there is any earth, there are likewise men-inhabitants."[12]

Occultist writers of the turn of the twentieth century followed Swedenborg's tradition of depicting spiritual voyages to other planets. In 1880, Henry A. Gaston published his *Mars Revealed: Or, Seven Days in the Spirit World*.[13] C. W. Leadbetter, a prominent Theosophist, offered two cosmic travel books describing the astral plane and the "devachanic plane" in the 1890s. One had the title *The Astral Plane: Its Scenery, Inhabitants and Phenomena* (1895).[14] Such books continued to appear in the twentieth century, such as one in 1922 titled *The Planet Mars and Its Inhabitants: A Psychic Revelation by Iros Urides (A Martian)*.[15]

Such literature inevitably converged with the quasi-mystical conception of spaceflight to be found among early rocketry enthusiasts and in science fiction. Sociologist William Bainbridge has shown that early rocketry pioneers such as Robert Goddard and Konstantin Tsiolkovsky, both dedicated readers of early science fiction, had mystical conceptions of spaceflight and of the liberating consequences of overcoming gravity. Such ideas became common in the rocketry movement. The German rocket scientists Hermann Oberth and his follower Wernher von Braun clearly articulated this vision. Oberth's occultist studies underpinned his scientific work. He developed rockets with the fervent hope that he would be shaping a world that would enable him one day to reincarnate as a spaceship captain.[16] Von Braun, whom the military brought to the United States after World War II, was a leading rocketry expert and a spokesman for the position that spaceflight would prepare the human species for its next step in evolution—to speed off into the heavens to populate the stars. In America von Braun became a convert to evangelical Christianity and began to announce spaceflight's millennial mission: "It is profoundly important for religious reasons that he [man] travel to other worlds, other galaxies; for it may be Man's destiny to assure immortality, not only of his race but even of the life spark itself."[17]

UFO contactee tales of the 1950s fused the occult tradition with the evolutionary "spaceflight" vision. Both of these communities also turned for inspiration to science fiction, a medium that had long promoted a millennialist vision for science. Robert Wise's 1951 film *The Day the Earth Stood Still* was particularly influential to contactee narratives. In this work, a visiting spacecraft brought the enlightened messenger Klaatu to warn earthlings to end their nuclear standoff. When Klaatu was ignored in diplomatic channels, he fashioned a miniature apocalypse to gain earthlings' attention, by halting all mechanical processes on the earth for an hour. The filmmakers fashioned the character of Klaatu, the visitor that came to warn Earth, as a stern savior. When mingling with the earth's population, he took on the pseudonym of "Carpenter"—a name associated with legends of the Christian savior. Likewise Klaatu, though betrayed and later killed by soldiers, underwent a technological resurrection on his flying saucer. Klaatu, the savior, was also depicted as a scientist who found more natural kinship with Earth's scientists than with its political leader.

The plot and design schemes of *The Day the Earth Stood Still* made the point that Klaatu was a representative of beings with technological and scientific superiority. Doors emerged and vanished seamlessly from Klaatu's spacecraft, pioneering the art direction concept of "fluid metal" that became a standard for film depictions of alien technology.[18] The craft's interior was portrayed as a maze of curved womb-like walls pulsing with light and shadow, creating a high-tech gothic design scheme. Feminine and masculine traits were intermingled, and passages through the ship took on ritual significance, as when Klaatu's giant, smooth-metal-skinned robot, Gort, carried the terrified heroine Helen, played by Patricia Neal, into its interior, or when Gort later carried the lifeless Klaatu to the laboratory to be resurrected with pulsing lights. Klaatu looked human in appearance, but after his resurrection and the shedding of borrowed Earth clothes, Klaatu donned a tight, zipperless body suit of miracle fabric of the sort that aliens and aeronauts have worn in science fiction adventure tales from the beginning.[19]

Two years after the movie's release, George Adamski, a teacher of mysticism who worked at a hamburger stand near Mount Palomar in California, published the first "nonfiction" narrative of contact with a man from a spaceship. Adamski followed the script and production of *The Day the Earth Stood Still* closely. Adamski's description of the scout ship—or bell-shaped flying saucer—that left the large cigar-shaped mother ship to descend to the desert near him was reverential. "It was translucent and of exquisite color."[20] He suspected it was a metal brought "from the opaque stage to a translucent stage" like carbon made into a diamond. "The splendour as it flashed its prismatic colours in the sunlight surpassed every idea I had ever had about space craft."[21] Adamski also offered photographs of the scout vessel to bolster the authenticity of his account.[22] The Venusian spaceman who met Adamski had long sandy hair, was clothed in a seamless, otherworldly fabric, and made Adamski feel "like a little child in the presence of one with great wisdom and much love."[23] Although Klaatu, in *The Day the Earth Stood Still,* preferred to mingle with Earth scientists, Adamski rejected the idea that science and scientists were now Earth's only hope. Instead, Adamski carefully rendered the visitation as one that was thoroughly religious in nature and presented himself as the vanguard for a new chosen people.

Another contact narrative, Orfeo Angelucci's *The Secret of the Saucers* (1955), even more carefully paralleled the hero's meeting with the Space Brothers to older tales of religious initiation. Angelucci, who came from a working-class Italian American family in Trenton, New Jersey, reported a childhood full of illnesses and sensitivities—electrical storms, for example, pained him greatly. As a youth he fancied himself an amateur scientist and attempted at least one experiment that involved launching samples of the fungus *Aspergillis clavatus* into the atmosphere in a weather balloon. Later he moved to the West Coast for his health, worked at Lockheed, and unsuccessfully peddled a screenplay that detailed a trip to the moon. According to his account, while driving home from work one night in 1952, he felt ill and saw on the road ahead of him a saucer, glowing with reddish light, which ascended and disappeared. He pulled his car to a stop and saw two green circles ahead of him. When he drank from a goblet that appeared on the fender of his car, he heard voices announcing they were friends from another world.[24]

Angelucci reported having numerous contacts in the months and years to come. One night after leaving a café and walking down a lonely street, he felt a tingling in his arms and saw a fuzzy dome taking shape. He stepped in and was whisked off to outer space. The interior of this dome, he reported, "was made of an exquisite mother-of-pearl stuff, iridescent with exquisite colors that gave off lights."[25] The room was empty but for a reclining chair also made of the same "translucent, shimmering substance." His unseen escorts soon informed him that despite its beauty, Earth was a "purgatorial world" with hate, selfishness, and cruelty rising from many parts "like a dark mist." The small scout vehicle then approached a "crystal-metal-alloy" ship. Inside, vortices of green flame appeared and voices further instructed Orfeo about earthlings' need to follow a creed of love. Orfeo was told that Christ had originated as "an infinite entity of the sun" and "out of compassion for mankind's suffering he became flesh and blood and entered the hell of ignorance and woe and evil" that was the earth.[26]

Angelucci then underwent a death and resurrection experience. He heard loud music and saw bright light and concluded, "I am dying . . . I have been through this death before in other earthly lives. This is death! Only now I am in ETERNITY, WITHOUT BEGINNING and WITHOUT END." He heard a voice announce, "Beloved friend of earth, we baptize you now in the true light of the worlds eternal." Then he was bathed in peace and beauty.[27] He was taught that the flying saucers were symbols of mankind's "coming resurrection from the living death."[28] Angelucci learned that communism, too, was a symbol of the earth's fallen state and of the evils that must be overcome. A Great Armageddon was approaching, possibly in the form of an atomic war, possibly in the form of a destructive comet, if humanity did not reform.

Following the aptly named pioneers Adamski, our American Adam, and Angelucci, who met with angels, countless other chosen people began to announce that they had made contact with Space Brothers, leading one UFO writer to dub the 1950s the "golden age of UFO religion."[29]

Some contactees more directly addressed the political context of the era's anxieties. For example, in 1960, Gabriel Green, president of the Amalgamated Flying Saucer Clubs of America, a large contactee organization, decided to engage directly in politics. He mounted a brief "Space Candidate" write-in campaign for the U.S. presidency in 1960, then ran for California state senator in 1962 on a platform that insisted on ending nuclear testing, and received approximately 170,000 votes in the Democratic primary.[30] Green's political advertisements included slogans such as "Solutions instead of stalemates," "Survival instead of annihilation," and "The true Stairway to the Stars instead of missile-fizzles and launching-pad blues." The smaller copy went on to insist that his goal was "to eliminate vested interest in inefficiency so that machines and automatonic industry can be permitted to do the laborious work of man, and still distribute the abundance produced by those machines to those who need them." He called for many dizzying reform measures including improved dental care, shorter workdays, unlimited education for all, an end to traffic jams, free energy, human rights, and "The World of Tomorrow today, and UTOPIA now."[31]

Such interests firmly establish the flying saucer craze as an instance in which a segment of the American public sought salvation in technology. In contrast to the dystopian weapons building they were witnessing, Green, Angelucci, and others joined turn-of-the-century thinkers like Edward Bellamy in imagining a world in which technology would usher

Gabriel Green, president of the Amalgamated Flying Saucers Clubs of America, a contactee organization, was a 1960 presidential write-in candidate. When Green ran for California state senator two years later, he received approximately 170,000 votes in the Democratic primary.

in a new millennium. In this case, the technological sublime undermined established power. The flying saucer's inhabitants evaded efforts to sight them, confounded government authorities, and chose humble working-class people such as Adamski and Angelucci to bear their otherworldly message.

Nikola Tesla Unbound

To lend the cause greater credibility, the contactees often added the mysterious inventor Nikola Tesla to the list of otherworldly players shaping millennial technology. Like the fans of Houdini before him, Tesla followers developed numerous resurrection scenarios for the deceased hero whose public demonstrations had once made him appear indestructible. One of the more remarkable literary products of the flying saucer movement to invoke Tesla's spirit was Margaret Storm's *Return of the Dove* (1959). Though very much a part of the "Movement," as flying saucer aficionados called it, Storm's volume was less concerned with tales of contact than with the promotion of utopian technology and a vision of the unfolding millennium, or new age. In her book, Storm, a fashion journalist and occultist, drew upon Theosophical theories to deify the inventor Nikola Tesla and promote her friend Otis Carr's free energy schemes.

At the heart of this book's eccentric claims is Storm's announcement that Tesla had actually been from Venus, born on a spaceship in 1856 and soon after presented to his earth mother in Croatia. Storm explained that Tesla's mission was to bring utopian technology to mankind as part of the Aquarian Age program for the earth's redemption. Storm related that throughout Tesla's career, industrialists, politicians, and military leaders did their best to stifle the unearthly inventor's creations. But he never was vanquished. Even after death, as an ascended master, Tesla was available to help other inventors devise new technology—whether free energy devices, antimissile defense shields, or flying saucers that could enter the etheric realm. Storm's insistence that Tesla was from Venus gave him kinship with the space people that contactees then were describing. The godlike Tesla represented science that was spiritually aligned. In this way he embodied one of the goals of the more formidable of the occultist groups, the Theosophical Society, to smoothly fuse the religious and scientific worldviews.

Tesla was an ideal choice to serve as a figurehead for utopian science. Newspaper stories of his era noted how his visionary World Wireless System would soon smash the power monopolies, as virtually free electricity would be beamed throughout the earth. His hero status was also ensured when the liberator later offered up visions of machinery capable of destroying the earth. While he faded into obscurity with the general public, tales of sabotage circulated among his devotees. Industrialists reportedly had suppressed Tesla's revolutionary inventions, and the U.S. military had carried off many of Tesla's secrets after his death during World War II.

After his death in 1943, Tesla became a cult figure, often featured in popular science magazines as a forgotten genius. Pro-Tesla forces lobbied to keep his memory alive and to prevent him from being written out of historical accounts as a simple madman. In 1956, thirteen years after his death, the International Tesla Society helped arrange a celebration of the one hundredth anniversary of the idiosyncratic inventor's birth. The American Institute of Electrical Engineers dedicated its annual meeting to his memory. In 1956 the International Electrotechnical Commission adopted the name "Tesla" for the unit of magnetic flux density. *Popular Science* even offered a cartoon version of Tesla's life story in its July 1956 issue. Such publicity helped revive interest in Tesla, whom science fiction impresario Hugo Gernsback had long hailed in his publications.

Against this backdrop, Theosophist Margaret Storm picked up the Tesla story in 1959. She transmuted Tesla from a tragic or pathetic figure, a man of grand but broken dreams who ended his days living in shabby hotels, feeding pigeons, into a cosmic hero. Stories of the 1890s such as Garret P. Serviss's "Edison's Conquest of Mars" had glorified inventor Thomas Edison and made him into an adventure hero whose disintegrating ray weapon saved Earth from a Martian attack. In Storm's narrative, Tesla figured in a drama of even greater importance.

Storm placed Tesla and the cold war era into a twenty-five-million-year cosmic history. Following the teachings of Theosophy's founder Madame Helena Blavatsky, Storm insisted that Tesla and humanity's story was one of a fall from spirit into matter, followed by redemption as spirit slowly prevailed. This grand narrative of progress was to be inscribed in the evolution of five root races, corresponding to historical eras and degrees of spiritual advancement. Storm depended on this master narrative while adding her own notions about Tesla, flying saucers, and wondrous technology.

Storm depicted her own era as one that spiritually blighted individuals ruled. She insisted the planet had been spoiled by the "laggards . . . the spoilsports, the screwballs, the odd balls, the sad sacks . . . a whole assortment of wet blankets in a wide variety of sizes, shapes, and shades. They are the ones with the souped-up egos; they do not buy the idea of spaceships, music of the spheres or the singing of angels."[32] Storm related that the rise of the laggards, among politicians and scientists, "in brief, is the story of the Fall of Man."[33]

But this was not the end. Storm believed that as of 1957, the tides had shifted. "The interior of the globe has been cleansed."[34] For some esoteric reason, the axis of the earth was also successfully shifted to a new angle. Likewise, flying saucers had been bombarding the atmosphere with cleansing energies. Hosts of angels were returning from exile, and the "dove" of peace was returning with "its joyous message." She announced that the Aquarian Age, or the New Age, the time of the seventh ray, was now under way, when the earth could be transformed from a "Dark Star" to a place of joy.

Storm aligned the dove of peace not only with Noah's dove flying with its olive branch over the flood-cleansed earth, but also to Tesla and his love of pigeons. In so doing, she managed to transform what many considered a pathetic pastime of the inventor's declining years into a heroic effort. Investigators opening Tesla's stored trunks upon his death in 1943, for example, complained that several trunks, rather than being filled with notes for brilliant inventions, contained newspaper clippings and birdseed.

This love of pigeons had long troubled Tesla fans. Tesla's first biographer, John J. O'Neill, reported that Tesla, a celibate, confessed that the great love of his life was a pigeon. He tenderly cared for this pigeon when it was ill, using electrotherapy devices. Prior to the pigeon's death, she flew from the darkness into his dark hotel room and beamed intense, loving light at him. O'Neill believed that the celibate Tesla's suppressed sexuality emerged in his "abnormal" love for the pigeon. Yet at the close of the book he offered a mystical reading, commenting that the incident of the pigeon flying into Tesla's dark hotel room and lighting it up with a brilliant light was the sort of phenomenon on which "the mysteries of religion were built."[35]

Furthering this interpretation, in her book Storm explained that the dove symbolized peace and was Tesla's "Twin Ray," an enlightened partner in his redemptive work. When the pigeon died, Tesla knew he hadn't much longer to live either. But like Tesla, his Twin Ray, having ascended, was now doing scientific work in the mystic realm of Shambala.

Throughout *Return of the Dove,* Storm relied on Tesla as a symbol of the enlightened scientist to criticize the male-controlled military and scientific establishment. For example, she blamed the strontium poisoning of the soil from nuclear tests and nuclear weapons production on scientists who were laggard souls, men "dedicated to deeds of violence."[36] Her rhetoric included a subdued feminism, as she argued that the male-dominated sciences came from an unbalanced and violent perspective. The atomic scientist, a rapacious breed in Storm's estimation, "seeks to enslave the atom, just as he has enslaved his own atoms that he lives with each day. . . . He wants to split the atom; to tear it apart by brute force; to strip it bare as one would strip the skin from an orange. His way is the way of fear."[37] Once America's leaders turned from utopian scientists like Tesla to laggard scientists like this one, they had no one but themselves to blame for the current weakness in scientific education.

In her book, Storm mirrored the concerns of the mass culture critics and Beat writers in their attacks on the age's homogenized culture. Members of the neo-Marxist Frankfurt School, such as Theodor Adorno and Herbert Marcuse, argued that advanced capitalism engendered a culture industry that shaped mass tastes, adding to the conformist and virtually totalitarian dimensions of consumer society. Employing a different vocabulary, Storm described the symptoms of the fall of man in the 1950s as including an America full of bored juveniles "going delinquent," asylums full of the overstressed, and a populace subjected to a war machine draining wealth, controlling minds, and sowing fear.[38] She explained that the "average citizen . . . will just sit and sit and sit and watch television, or given a chance he will talk and talk and talk. He may appear to be talking about the state of the world. . . . Actually he is expressing the confused state of his own consciousness . . . his own troubled heart . . . his vibratory note is always the same doleful moan—the note of crucifixion."[39] Agents of the cosmic hierarchy were hard at work trying to awaken the average citizen from this slumber. But it was difficult to lift oneself from the false consciousness perpetuated by the education system, religion, the media, prisons, and mental institutions. Storm primarily blamed the military powers that "refused to reveal to the public the truth about flying saucers" and the help the saucers represented.

To a Theosophical point of view, science was an attempt to intellectually strip matter entirely from earlier notions of being or spirit; hence, restoring science and technology to spirit would be a redemptive act. Storm relied on Tesla to fill the gap. Tesla, we learn from Storm, though a man of flesh and blood, did his creative work in the "fourth ether." Indeed, Tesla had often insisted that he had an uncanny imagination that allowed him to design and calibrate all his inventions without resorting to a drafting table.[40] Storm explained that Tesla could even allow versions of his machines to run in the ether for days or years, then "test the etheric machinery and make any necessary adjustments"[41] or "examine them for signs of wear."[42]

For Storm, the flying saucer was a prime example of the confluence of technology and spirituality. If Tesla, on Earth, could design in the ether, UFOs came directly from the ether. Offering her own take on the "fluid metal" depiction of saucers in Hollywood films, Storm argued, "That is why spaceships are described as being constructed without rivets, welding, seams, or cracks around doors. They are not constructed but precipitated direct from the ether."[43]

Storm's obsession with utopian technology led her to declare that Tesla's redemptive efforts would prevail against all countertrends. In refusing to further finance Tesla's energy broadcasting stations, J. P. Morgan, a representative of dark forces, preferred what Gabriel Green had called the "vested interest in inefficiency." Tesla, however, had two major disciples who would succeed where the master had failed. Spiritual technology was soon to free the world from the errors of financiers and the "men of violence" who tended atomic energy and nurtured the arms race. Tesla's disciples would fare better.

Free Energy, Perpetual Motion, and Free Enterprise

The most important Tesla disciple was Storm's friend, Otis T. Carr, who promoted himself in the 1950s, often at flying saucer conventions, as Tesla's scientific heir. Convention-goers could marvel at a prototype of the OTC-XI, a flying saucer that Carr planned to make available to the public. Carr claimed to have received messages from Tesla and other Space Brothers to help him design his flying saucer, which ran on atmospheric energy as well as other free energy or "utronic" generators. In *Return of the Dove,* Storm hinted that Carr's innovations might even make Tesla's never-realized World Wireless System seem obsolete.

The "free energy" that Carr packaged as "utronics" was an old dream deferred, formerly filed under the category of "perpetual motion."[44] Beginning with the development of waterwheels, inventors long had explored the possibility of perpetual motion to fulfill energy needs and to create the mechanical equivalent of immortality. Such devices fit two basic variants. Perpetual motion machines "of the first kind" created a closed circuit that, once started, would overcome forces such as gravity and friction and allow a wheel to turn perpetually and perform work—for example, grinding wheat or generating electricity. "Overbalanced wheels," for example, would, through various ingenious schemes, keep a wheel weighted more heavily on its down-turning side than on its up-turning side. More promising have been perpetual motion machines "of the second kind," that is, open-circuit devices that connect the basic clockwork to outside sources such as tides, changes in temperature, or changes in barometric pressure.[45]

In the perpetual motion milieu, frauds who have appealed to occultist thinking have abounded. For example, from 1873 until his death in 1898, John E. W. Keely of Philadelphia promoted a mysterious motor that ran on "etheric force" derived from the "disintegration of water." He raised millions from financiers and the public for his company on the strength of his demonstrations of such phenomenon as musical notes causing weights to rise and fall. Of these performances, which had kinship to séances, he remarked, "I am always a good deal disturbed when I begin one of these exhibitions, for sometimes, if an unsympathetic person is present, the machines will not work."[46] Theosophists of the age admired him for combining "the intuitions of the seer with the practical knowledge of mechanics."[47] It was not until after Keely's death that it was determined that his devices relied on hidden tubes conveying pressurized air.[48]

Carr's inventing and marketing abilities were closer to the Keely model than the Tesla brand. Nevertheless, Storm regarded Carr as one of Tesla's two principal disciples, well deserving of a place in her outline of cosmic history. The other disciple was Arthur H. Matthews, a Canadian electrical engineer who had designed an interplanetary communication device à la Tesla and worked on developing Tesla's "antiwar" machine.[49] Although Storm placed her primary hope in Carr, the details of Carr's "discipleship" appear concocted from very slight material. During the 1920s, while an art student, Carr had worked as a package clerk at the Hotel Pennsylvania in Manhattan. One day Tesla, a resident, "came straight to his young disciple" and requested that Carr purchase four pounds of unsalted peanuts.[50] Tesla fed these peanuts to pigeons with Carr's help. To anyone but Margaret Storm this work with pigeons would seem no great apprenticeship for a budding inventor, but Storm elevated this homely tale to the cosmic by assuring us the two were collaborating to help assure the return of the dove of peace. During his brief apprenticeship, Carr ran errands, asked questions, and soaked up the powerful vibrations of the master inventor.

Three decades after these meetings, Carr announced his amazing Tesla-influenced technologies: the "Carrotto gravity motor" and the "Utron Electrical Accumulator." Showing a flair for paradox as well as showmanship, Carr explained that the Utron Electrical Accumulator "is completely round and completely square and generates and regenerates electrical energy."[51] Storm wrote enthusiastically, in high occult rhetoric, about Carr's devices. The world was not ready for Tesla's world system in 1900 but surely would be ready for Carr's system with the advent of the New Age. Carr's flying saucers that relied on free energy could be ideal for transport within the earth's atmosphere, but "shattering" for those who tried to leave the earth if they weren't spiritually prepared for "transmutation" to the etheric realm. Carr's free energy devices could also create a utopian revolution. Storm wrote with approval of the coming transition to clean, free energy. "Very soon now will come the big planetary housecleaning. Then down will come all the cables, conduits, wires and posts, which the public is now paying to have installed. What fools these mortals be."[52]

Although Tesla had once happily announced that his new energy distribution system would abolish power monopolies, Carr's own publicity, in the age of un-American activities, downplayed the economic repercussions of his utopian inventions. He insisted his devices would stimulate productivity and sales and enhance the American business system. Furthermore, OTC Enterprises would only design prototypes of antigravity flying saucers and free energy devices. He would encourage manufacturers all over the globe to put these inventions into production. Carr insisted that his "primary interest . . . is the opportunity of all industry the world over, to have an ownership with us in our business."[53] The flier betrayed its crank roots with its conclusion: "The best way to get the total concept of what his [Carr's] free-energy devices mean to the world, is to suppose that the wheel were just now being discovered—then consider that OTC Enterprises is putting the wheel into the air . . . in an entirely new dimension. This should be pretty good for industry."[54] He placed a hefty price tag on his flying saucers: $20 million for the first prototypes and $4 million for subsequent models.[55]

Carr and his associates found backers for their projects from a public excited by *Sputnik,* American space shots, and numerous UFO sightings. He was able to plug his enterprises on air when he was interviewed on Long John Nebel's popular all-night talk show on New York City radio station WOR; Nebel also had interviewed contactees George Van Tassel and Howard Menger. Carr and his associates helped produce at least one flying saucer convention in California and at it prominently displayed models of their products, such as the OTC-X1.[56] Such tactics, like those of Keely before him, gained Carr investors. A brochure advertising Carr's grandiose "Free Energy" research station in Maryland included a sketch of the buildings and grounds, and text that explained the station's goals of developing spaceships, interplanetary communication technology, and solar energy devices. The promotional brochure indicated that the interior of its domed building would be decorated by Salvador Dali to represent Ezekiel's vision of the fiery chariot. Likewise, a white dove and four cherubim would surround this central grouping.

Storm's relationship with Carr was clearly close. In fact, she shared a business address with him. The same Baltimore street address that appeared on her self-published book also appeared as the address for Carr's dubious flying saucer and free energy device business. One writer indicated that these headquarters, though modest, "featured numerous rooms and apparently housed various departments and in general suggested a successful operation."[57] One possible, though doubtful, explanation for Storm's book is that it was little more than a publicity stunt

Schematic drawings for Otis T. Carr's planned "Free Energy Research Institute" in Baltimore County, Maryland, 1958. Carr convinced investors to back him in his schemes to create Tesla-inspired flying saucers, antigravity devices, and free energy machines. Courtesy of the National Museum of American History, Smithsonian Institution.

for Carr, who was bilking investors in his business schemes. Margaret Storm, however, seems an unintentional partner in crime. *Return of the Dove* is far too elaborate a text to have simply been prompted by an impulse to swindle the public. It seems more likely that she believed in Carr's inventions. This would be a natural extension of her belief that a new millennium was dawning, and her acceptance of the Theosophical notion that the past civilizations of Lemuria and Atlantis had been scientifically advanced.[58]

Though ambitious, Carr's enterprises did not fare well. In 1959, he put out a press release announcing a cancellation of the demonstration flight of his prototype flying saucer, the OTC-X1, arranged to take place at Frontier City amusement park near Oklahoma City. While Long John Nebel, his assistants, and various investors gathered to watch the launch, Carr took to his sickbed. The need for "further testing and refinement" delayed the launch, which never materialized.[59] In the spring of 1959, one of Carr's business backers approached the Securities and Exchange Commission, and Carr was fined and issued an injunction to end OTC's business enterprises.[60] A year later the state attorney in Maryland considered pressing charges against him for defrauding investors.[61] He also spent time in prison in Oklahoma in 1961 after being unable to pay a five-thousand-dollar fine for "selling securities [for his UFO scheme] without registering the same with the Oklahoma Securities Commission."[62]

However bogus his enterprises, Carr's vision prefigured that of the alternative energy movement of the 1970s and the free energy movement of the 1990s. As Carr put it, in verse form: "When you fight Nature, Nature *always* fights back. If we try to share Nature's energy by injuring Nature, we will only injure ourselves accordingly. . . . Crack or split the Atom and you get frightful devastation. . . . There is only one right way. The right way is the peaceful way."[63]

A New Sisterhood of Reform

Storm's elaborate interweaving of Tesla, Carr, and UFO lore helped pattern her protest against the 1950s status quo. For Storm, Tesla, with his Venusian origins, was a symbol of a new science that balanced rationality and idealism. Contactees amplified this vision with their narratives in which Space People urged humanity to reject nuclear arms development and aggression. Storm's and the contactees' allegiance was not to nationalist governments but to Platonic realms ruled by universal parliaments and councils of higher beings. Their ultimate goal was spiritual progress. To this end they highlighted the flying saucer as a new form of the technological sublime. However sincere or insincere in intent, they offered a religious vision unique to the 1950s—one that blended occult lore with popular visions of the future now.

An article in *Harper's* from 1960 that profiled a flying saucer convention indicated the grander ambitions of the "Movement," hinted at in candidate Gabriel Green's political campaigns. With some amusement, the writer argued that the invocation of Space Brothers gave this community authority on all matters metaphysical and made them naturals to unite the era's reform cults, which included "Health Food Discoverers, the more imaginative Fundamentalist sects, Yoga . . . [and] spiritualism." Hal Draper, the author, added, "There would be room too for devotees of Dianetics, Astral Bodies, Hieronymus machines, and 'Shaverism.'"[64]

Despite the inevitable condescension, this commentator recognized that America's reformist fringe of the 1950s had grand ambitions. Members of the 1950s movement shared the sort of cosmic optimism found in America prior to the Civil War, when a "sisterhood of reform" causes such as abolition, temperance, feminism, dietary reform, clothing reform, and free love became forces for historical change—even though mainstream society then often dismissed vocalizations of such goals as rumblings from the lunatic fringe. While Christian revivalism and New England transcendentalism inspired the earlier reform movements, the 1950s reform movement relied on a folkloric wonder show script rooted in science fiction, rocketry, and occult literature.

The contactee movement relied on pamphlets and conventions to construct its unorthodox "theatrics of space." In their drama, handsome visitors from the stars in miracle-fabric outfits were stepping from aerodynamic vehicles to warn a chosen few of the infantilism of Western culture. When governments like the Soviet Union and the United States were making the most of their first fumbling steps into space, sending dogs and monkeys into orbit, the chosen few were speeding around the cosmos, guests of benevolent aliens light-years ahead of earthlings in technology and spiritual evolution. This script allowed a fringe group, America's occultists, to present themselves as a vanguard in their lectures, pamphlets, and radio and convention appearances.

This weighty task makes efforts like Storm's book seem both absurd and poignant. Clearly a fringe publication, down to the decision to print it with green ink, *Return of the Dove* nevertheless offered readers a critique of 1950s America with its docile yet terrorized public. The book directly criticized the cold war arms race and establishment science. And at times Storm's rhetoric made plausible her case that her fringe point of view was saner than that of the sober-minded "laggards" who then ruled public opinion and the nation.

The folk challenge that the "Movement" made to the cold war status quo soon after entered the mainstream. After incorporating the 1960s counterculture's fascination with Eastern mysticism, and the environmentalist awakenings of the 1960s and 1970s, the saucer and occultist "Movement" of the 1950s developed into the "New Age" movement of the turn of the twenty-first century. At the heart of the New Age movement remained a critique of materialism and technology, coupled with an interest in ecology, spirituality, health, diet, utopian technology, and, as in the widespread interest in "angels" during the AIDS crisis of the 1980s, millennial hopes of redemption through otherworldy agency. Partially because of its consumerist appeal to the good life, the New Age also built up a solidly middle-class base and no longer lurked on the lunatic fringe as it did in the 1950s. Just as earlier wonder showmen offered evidence of miraculous technology and miracles of human capabilities, this enlarged "Movement," comprising both idealists and opportunists, continued to embody hopes for the reconciliation of technology and spirituality.

CHAPTER EIGHT

The Many Gospels

In 1999, performer Austin Richards, aka "Dr. Mega-Volt," delighted the crowd at the "Burning Man" arts celebration in the Nevada desert, when he danced around in a metal suit and helmet on a flatbed truck between two large, humming Tesla coils that discharged ozone, bolts of electricity, and thunder. In that same year, evangelist Dean Ortner, heir to Irwin Moon, appeared on *Ripley's Believe It or Not!* and performed his "million-volt man" demonstration, jumping on a transformer and letting the crowd see a stick of wood in his hands burst into flames. Also in that year, inventor Dennis Lee conducted a nationwide tour in which he lectured and touted his free energy machine claims, while another part-time inventor, David Olszewski, appeared at New Age and dousers' conventions to sell his light therapy devices.

The performances of these showmen and salespeople make it clear that the grassroots wonder show had survived to the dawn of the twenty-first century.[1] On a higher economic tier, the advertising industry also has relied on special effects to generate delight and astonishment in viewers while establishing the novelty of a product and the authority of a corporate sponsor. On this mass level, wonder is choreographed to assert that all is well with the status quo. But on the grassroots circuits of the live demonstration or low-budget infomercial, the use of "wonder" more often serves as a marker for utopian dreams and hopes for social change.

This chapter brings the wonder show formula up to the dawn of the twenty-first century and the settings of small-scale promotions at conventions, on websites, and in infomercials. Similar promotions are offered in all these media, and all are open to small-scale entrepreneurs. A glance at cable television in early November 2002, for example, provided a lengthy testimonial for "Q-Ray" bracelets. According to the infomercial, these devices promoted healing by guiding the mysterious Q-Rays into affected regions of the body. Dozens of testimonials were offered, with satisfied clients describing relief from various aches and ailments.[2]

Such small-scale wonder promotions generally offer a utopian "alternative" to mainstream culture, premised on a vague critique of mainstream values. A New Age exposition's backdrop of progressive hopes and apparent miracles make an ideal setting for wonder demonstrations; at such expositions, numerous devices designed to "channel" healing energy such as the "Q-Ray" are promoted. Likewise, with increasing popularity, fundamentalist Christianity continues to offer its challenge to mainstream American values. To examine Christian millennialism and wonder in the late twentieth century, this chapter describes a meeting with evangelist Dean Ortner, who runs the "Wonders of Science" show, a descendent of Irwin Moon's "Sermons from Science." The chapter then looks at one of inventor Dennis Lee's conspiracy-minded "free energy" shows that promise utopian technology and the resuscitation of a betrayed American Dream. Each of these forums will add to the understanding of the wonder show and its long run in American public life.

Atlantis Rising

A banner that hung above the Dallas Convention Center hall in September 2001 announced, somewhat blandly:

> Whole Life Welcome
> to the Nation's Premier Event for
> Natural Health
> Personal Growth
> Sustainable Living

The Dallas 2001 show opened only ten days after the plane crashes that destroyed the World Trade Center towers in New York City. Many of the key speakers had canceled their appearances, and the exposition's organizers decided to donate proceeds to recovery efforts in Manhattan. If the canceled lectures by high-profile New Age figures such as Marianne Williamson, Jean Houston, and John Bradshaw spoiled the revivalist atmosphere, the core of the Expo, a marketplace made up of exhibitors' booths, still reflected the diverse nature and cosmic optimism and opportunism of this reformist gathering.

The first Whole Life Expo was held in 1982 in San Francisco. Each year since, the exposition has traveled to different American cities, and roughly twelve thousand people attend each conference. In 2001, for example, about one hundred thousand customers attended eight separate expositions, and the Whole Life Expo website suggests a total of two million customers have attended its expositions since 1982.[3] While these are not huge numbers, the New Age mentality is no longer a fringe affair. One study indicated that consumers interested in the "New Age," or in "Lifestyles of Health and Sustainability," make up "30% of all U.S. households, which translates into 63 million adults."[4] Such consumers purchase "New Age" publications and health products, spending, according to one report, as much as $46 billion yearly.[5]

Two recent religious studies experts have argued that the New Age is a "smorgasbord" of belief systems and practices, largely inherited from occultist systems of the past. New Age thinking reflects belief in "transcendence," or shifts from mundane reality to the sacred; in "unity," that is, in biological and spiritual notions of "wholeness" and "interconnectivity"; and in the dawning of a "New Age" or epoch grounded in new values.[6] The New Age has prospered from connecting older religious concepts of "unity" to that reflected more recently in environmentalism and ecology. With such connections, the New Age can make claims to a scientific as well as a spiritual grounding for its principles. One of the philosophical architects of the New Age movement, for example, David Spangler, downplays the occultism reflected in channeling or crystal healing and instead highlights an ecological worldview. He defines the "New Age" in terms of the emergence of a "planetary consciousness," as more people respond to the "myth of a sacred planet" and begin to confront global challenges.[7] Yet all believers in a "New Age"—as with the 1950s saucer movement—sense a millennialist shift under way.

Booths at the Dallas convention featured both environmentalism and occultism. The exhibitors included the Green Mountain Energy Company, offering subscribers a wind-power alternative to the local power company, a local Honda dealer's display of the new fuel-efficient hybrid electric-gasoline cars, a local holistic bookstore's wares, and vendors of crystals, health foods, vitamins, clothing, paintings, and beauty products. Exhibits also promoted occult teachings and alternative-healing systems ranging from high to low tech, while varied healers and psychics offered sessions to clients. The only element lacking to replicate what in the 1950s UFO contactee enthusiasts would have called "the Movement" were authors of contact narratives offering to autograph copies of their books detailing encounters with advanced Space Brothers[8]—yet the occultist formulations of that movement were amply reflected in the bookstore's offerings and at several of the booths.

Vendors of vegetable juicers wearing headset microphones offered their spiels to groups waiting for free samples of tortilla soup and other concoctions. Sellers of vitamins and antiaging potions also abounded, urging leaflets on passersby. Throughout the hall could be heard the solemn tolling of "crystal singing bowls," which emitted single tones when rubbed with a wand around their edges. The Crystal Tones company, based in Salt Lake City, marketed each of these bowls as an "advanced biosonic repatterning tool" that could be "used on chakra and meridian points" to help practitioners allow clients to "access alpha/delta states."[9] Using an earlier, more refined variant, the eighteenth-century glass harmonica, Benjamin Franklin once had entertained elites while Anton Mesmer created a soothing atmosphere for his magnetic healing sessions. The bowls, costing $179 and up, were selling briskly. No attendees were puzzled by such advertising copy as "Back by popular demand THIRD EYE BOWL SPECIAL $150."

The 1950s occultist vision was evident at a table where a middle-aged man passed out pamphlets and asked browsers if they were familiar with *The Urantia Book*. Many of the pamphlets he handed out included quotes from this work, first published in 1955, which outlines a complex cosmology that blends Christianity with the occult. The Urantians also have attempted to make their ideas palatable to the mainstream in the free magazine they offer, the *Jesusonian*. The publication's occultist worldview is softened with sophisticated typography, cosmological borrowings from Christianity, and ample quotes from authors such as William Faulkner, Johann Wolfgang Goethe, and George MacDonald. The magazine also displayed classic artwork such as Edward Munch's *The Scream* to illustrate an article titled "Is there a Hell?" and Dorothea Lange's photograph *Migrant Mother* to illustrate the article "Is there Evil and Suffering on the Heavenly Worlds?"[10]

The Urantians offer an otherworldly explanation of the origins of *The Urantia Book*. Reportedly, it was sponsored "by a commission of twenty-four spiritual administrators acting in accordance with a mandate issued by high deity authorities (the Ancients of Days) directing that they do this on Urantia [Earth] in the year A.D. 1934."[11] The Urantia teachings have much in common with the Spiritualist teachings of the nineteenth century and the earlier Swedenborgian cosmology. They stress foremost that there is an afterlife. After death, most people will undergo a period of sleep, from which the soul will be resurrected, given a new body, and, as the person's soul and intellect develop, he or she will make way through the seven "universes." The process is likened to that of education: the universe is best thought of as a university. After resurrection, for example, comes a period of "registration and orientation" when advancing souls become "mansion world students." When appropriate, the student advances to Stage 4, also named Constellation Headquarters, Morontia, the Local Universe, or, simply, "the Home of Jesus."[12] The student then moves onward through the system of seven universes until he or she finally arrives at Paradise. The group's teachings place the Urantians in the occult mainstream that dates back to nineteenth-century Spiritualism and the founding of groups like the Theosophical Society in 1875, yet their cold war origins place them firmly in the worldview of the flying saucer movement.

A more direct entry to the occult worldview that pervaded the 1950s movement was offered at the Transmission Meditation booth at the Whole Life Expo. There a sign welcomed visitors to join an ongoing meditation circle. According to their literature, the goal of Transmission Meditation is to "step down the great spiritual energies that continually stream into our planet."[13] Those meditating then send these cosmic energies of love and brotherhood out into the world. This spreading out of "positive energy"—akin to the use of animal magnetism in the nineteenth century—appeared to be the thematic basis for many of the healing systems at the exposition—both those that relied on technology and those that shied away from technology to instead rely on folk healing, channeled forces, or metaphors of "attunement." To give one example, Burnell Lee Sesker was a young man who offered a "Harmonic Body Tune Up" that involved the use of both tuning forks and the Australian aboriginal instrument, the didjeridoo. He explained that when a part of a client's body is "low energy," it will "deaden" vibrations from the tuning forks. Sesker will then treat those energy centers by striking tuning forks on crystals and then placing them near the body's energy centers. Finally he will summon deep rumbling tones from the didjeridoo to "ground" the client's energy. During one of these treatments, an elderly man handing out pamphlets that offered health guidelines to reduce the risk of cancer stood at his nearby table, looking vaguely discomforted at the spectacle of the healthy young man blowing the archaic horn while circling his middle-aged female client.

Newer technologies that relied on these same metaphors of attunement and energetics also abounded. Most of these technologies implied that the cosmic energy or vital fluid could indeed be produced and transmitted. For example, "Magnetic Health Mats" were available from the Cosmic Energy Corporation, based in Austin, Texas. Users could stand, sit, or lie down on these ribbed mats designed to activate acupressure points, while the magnets inlaid in the mat would "magnetize blood" and reduce pain, fatigue, arthritis, and so on. The cheerful representative for this product said he and his relatives traveled to health shows in Europe and the United States depending on "how much money we want to make." Nearby, the Natural Life company of Minnesota sold a "Natural Energy Stimulator" that provided "acupuncture without needles!" Good for arthritis, migraines, carpal tunnel, menstrual cramps, TMJ, "neck, shoulder pain & more!" When I confessed to one member of the sales force that my neck was sore, he proceeded to touch this stimulator to my neck, producing static electric shocks. After about two dozen of these shocks, I was cringing. But I did, indeed, feel some "tingling" and relief when he had finished. He invited me to lie down on a "Chi Energy" table when it was vacated. I watched a young woman lying there, with her heels on a machine that rocked her back and forth, and declined.

The Oxygen Research Institute of Mill Valley, California, possibly offered the most expensive technological products of the exposition. At the center of their table were several elegantly designed contraptions that included two sidelong silver gas tanks, resting on a black square pediment. A gold pyramid and metal spiral rose above the tanks. After I sat down, the woman attending the booth continued a hushed conversation with a well-coiffed woman; occasionally I would hear the whispered words "founder," "Ph.D.," then "incredible orgasm" or "bliss" and "so important." As this whispered, confidential exchange continued, it became clear I would have to exercise greater patience than I was capable of to have the technology explained. The saleswoman had a live one on the hook, and I was an obvious nibbler. The device with the tanks and pyramid was called a Life Energy Amplifier or a LEA Atlantis Highlife, and its starting price was $2,150. A pamphlet explained the workings as follows: "Rare earth magnets excite an alchemy of noble gasses, gemstone powders and energetic remedies that are sealed inside the formula electrodes, emitting high chi Far Infrared (FIR) subtle energy."[14] Different "formulas" could be blended to stimulate such emotions as Joy, Passion, Heart Opening, Peace, Bliss, Harmony, Balance, and Ecstasy. The prose mixed seemingly scientific principles with compounds and procedures more appropriate to an alchemist's laboratory; the audience was expected to assume that a smooth bridge had been built between the two thought systems so that the company's concoctions of ground-up gems and Far Infrared "energy" could offer a new form of twenty-first-century magic. The company also sold a more easily identified device called the LEA Innersex System. This vibrator, when plated with 24-karat gold or platinum, cost $4,200. The less expensive InnerQuest Turbo system, which was not gold- or platinum-plated, cost only $900. "Blisswear Far Infra Red Clothing" was also available. The Blisswear T-shirt, which included mineral particles that held the "FIR frequency" woven into the cotton, was a mere $125. The shirts were said to increase circulation and metabolism, detoxify the body, and increase alpha wave activity in the brain. Though I sat near all this equipment, I felt no noticeable change in mood or physical state. Likely, the LEA Atlantis Highlife on the table was not switched on.

My hope of meeting a wonder showman was realized when I moved on to the booth of the Light Energy Company, based in Seattle, Washington. While Pam Olszewski read a Dick Francis thriller, her husband, David Olszewski, held forth on the various light-healing devices that his company offered, some of which he had invented. One line included "full spectrum" lights that mimicked the sun's output, designed to combat seasonal affective disorder (SAD), or rainy-season depression common to the Northwest. These devices are now fairly mainstream, frequently advertised in book review supplements. But Olszewski was clearly even more enthusiastic about the light-emitting diodes (LEDs) he had designed.

Olszewski trained as an engineer. For some years he had been working as an information system manager for a petroleum company in the Seattle area. In the 1960s he became interested in alternative health and "metaphysics," including "remote viewing" (clairvoyance) and other paranormal phenomena. During a television broadcast of the Olympics in the 1960s, he noticed Soviet trainers treating injured athletes with lasers. Fascinated, he found ways to secure some of these devices and study them. In the 1970s, he began working with "soft lasers," then in the 1980s switched to LEDs—more powerful versions of the single-wavelength red lights that now glow on most electronic products to indicate they are plugged in. Many inventors were working on similar devices, but Olszewski claims to have secured patents in nine countries, including Japan and South Korea.

Olszewski was an enthusiastic salesman, and his gray eyes shone while he spoke with few pauses to those that gathered before his table. As he later told me, he had developed "a good spiel." While I sat at his booth, a woman joined us who admitted having a chronic neck injury. Several minutes further into his talk and description of his devices, a woman in a wheelchair came to the table, pushed by her partner. The LEDs, Olszewski told his growing audience, stimulated photoreceptor areas in damaged cells, which encouraged the production of proteins that could heal the damage.[15] As proof of the likely efficacy of the LED devices, he handed out numerous clippings, including an article that explained that NASA was studying the use of such devices to hasten healing of injuries to astronauts in outer space.[16] Another article focused on Harry Whelan, a physician at the Medical College of Wisconsin, Milwaukee, who was testing such devices for NASA. Whelan believed the

LEDs helped heal wounds such as painful mouth ulcers caused by radiation and chemotherapy. His studies also indicated that such devices helped musculoskeletal injuries. A clipping from a scientific journal argued that while many researchers are skeptical of the cure-all publicity surrounding earlier research with lasers, researchers had amassed "vast statistical material . . . proving that such a treatment has a positive effect."[17]

Although Olszewski tried to rein in his enthusiasm, he tended to slip into the "LED as cure-all" category of promoter when warmed up. He suggested, for instance, that shining an LED's light on one's navel could purify the blood pumping through veins. And, after learning that the woman in the wheelchair did not have a severed spinal cord, he cautiously suggested that his LED units could help her. He clearly believed in LEDs and mentioned giving a small LED unit to a grandchild to teethe on. His general spiel—both that offered at his table and in his lecture later in the afternoon—included repeated references to the fact that if you have scraped skin off your hand and treated the wound with LED light every two hours, "you can see the skin grow back."

His lecture was on a small stage in a corner of the exhibition hall. He relied on a slide projector and slides and a few samples of his equipment. He told his small audience that LEDs speed up the healing rate five times. During the lecture, he also insisted "there's nothing in the body you can't heal up with one of these. This stuff really works. I mean nothing." He mentioned skin conditions like psoriasis, eye problems such as cataracts and glaucoma, osteoarthritis, whiplash injuries, lower-back injuries, and wounds. He then looked up and added, "And oh, ladies, this light encourages the production of elastin and collagen. If you treat your skin you'll have no wrinkles."[18] In another sales riff, he began to expand on the notion that LED treatment would "return cells to normal." He went on to ask, rhetorically, "What is cancer? A mutated cell. Then we need to get it to de-mutate. It returns cells to their normal state."

Olszewski also had for sale a "Thermo Therapy Unit" that involved a rectal probe that he recommended for boosting immune systems. Reminiscent of the "Fever Machine" that GE performers promoted in their House of Magic in 1933, this device creates a "false fever" in the body, and so stimulated the immune system and the production of white blood cells. Olszewski thought this useful for treatment of AIDS and also to help cancer patients recover from chemotherapy and radiation therapy. At this point in his lecture, he showed slides to contrast alpha brain wave patterns obtained during treatments, when healing was going on, and afterward. These instruments, he said, shifted the waves to the healing pattern. He insisted that the rectal Thermo Therapy Unit would shrink the prostate, rid one of hemorrhoids, stimulate the immune system, and clear up the skin. "Every cancer clinic in the country, virtually, has similar units." At the conclusion of the lecture he directed people to his booth, and about half of the modest crowd, about twelve people, hurried to the booth to interrupt his wife's reading of Dick Francis.

Earlier, during the small command performance for the two women, and myself, Olszewski had insisted that the LED could heal allergies and linked this ability to the homeopathic theory of medicine. He said that he waved the light over his food in restaurants. If he also shone it on his hand, he said the light would excite the food molecules and bring their vibration "up to your own level so your body won't reject it." He added, "I sell a lot of lights in restaurants. People get curious when they see me with it." He also noted that the LED light did not have to be red, but that that wavelength seemed to have the most pronounced healing effects on humans, perhaps because the hemoglobin in the blood is also red. "If aliens had green blood we could switch the resonance frequency to green."

I asked if I could heal myself by getting close to the red light on my stereo.

"That's good," he said. "Nice try. But no. It's not nearly powerful enough. These here will penetrate the body two inches. And the other unit will penetrate eight inches."

The woman with the chronic neck injury seemed more interested in the devices than the woman in the wheelchair. She had been testing it on her neck during his long talk. When she stopped holding the back scrubber–shaped unit with its eighteen red LEDs on her neck, she said, "I do feel a tingling." Olszewski said, "That's a sign of healing." They began to discuss prices and the negotiations concluded with him promising her the exposition discount if she were to contact him in the upcoming months via e-mail.

When the other two customers had left and I explained my research, mentioning that my earliest references were to traveling electrical healers, Olszewski was happy to continue to share his knowledge. He said that although he did not have a horse and cart, he was probably a good modern analogue for some of the itinerant electrical healers of the nineteenth century. He said that for fifteen years he had been attending about twenty conventions a year (two a month) and restricted the venues to traditional health shows, alternative health shows, and New Age fairs. He estimated that 40 percent of his sales were to healing practitioners. He would never do state fairs. "You could talk yourself hoarse there." He said that the four or five national conventions of dowsers had been especially fruitful. "I look for people exploring new things." He insisted that at a dowser's convention he could sell products to 75 percent of his audience. He said conventions were far more valuable than print advertising or web pages, where there was a limit to what you could claim if you did not want the FDA causing troubles. "But you need someone speaking. A static exhibit won't interest anyone. Consider those two women I spoke to just now. Both of them are here with chronic health problems. They have probably tried numerous therapies. But they are ready to try something new. They have to. And I have something for them." This was his first trip to Dallas. "We'll drop the pebble into the Dallas pool here and eventually we will start hearing from doctors."[19]

The Other Side of the World

A year before the Whole Life Expo, I had driven to an unassuming suburban house in Whittier, California, to talk to another grassroots wonder showman, Dean Ortner. During my southern California visit I had hoped to meet Irwin Moon's elderly widow, Margaret Moon, in a nearby community. However, she was recuperating from a hospital stay, her daughter was busy with work and Christmas preparations, and, as an outsider to the Christian fundamentalist community, I remained somewhat suspect. Her son-in-law had asked me, "What's your background?" and I had answered, intuiting that the cause was already lost, "My background is that I'm a historian and I want to learn more about Irwin Moon." It had not played well.

Dean Ortner, a longtime showman, however, was happy to speak. He was a healthy-looking middle-aged man, with bright eyes reminiscent of both Dave Olszewski and photographs of Charles Came. Ortner was continuing the tradition of Benjamin Silliman and Irwin Moon, of performing science shows that pointed to the reality of Christian teachings; for more than two decades he had traveled year-round performing, but now he was teaching science full-time at a Christian academy and performing on weekends and during the summer at schools, military bases, and theaters.

Ortner began as an assistant to the Moody Institute's George Speake in the 1970s and started performing the Sermons from Science at the Moody Pavilion at the Spokane World's Fair of 1974. Before his conversion to evangelicalism, Ortner was a graduate student and teacher at North Dakota State University in the area of bio-nucleonics, a field that he explained involved using radioactive isotopes to study life systems. His research was focused on finding a biodegradable plant extract to replace DDT. At that time he was a die-hard evolutionist. His parents had been professors, and religion had been an intellectual affair while growing up.

However, he attended one of Billy Graham's crusades in Fargo in 1969, choosing a seat in the back row so "nothing strange would happen to me." That night, when actor and evangelist Lane Adams was preaching, Ortner felt his words were "directed right to me. I gave my heart to God and this changed my whole life."[20] He "accepted the Lord" and decided to become a missionary. This, he thought, would mean traveling to the "other side of the world." He received a scholarship to Moody Bible Institute, which was then eager and "praying for a scientist to come and carry on the live science program." He joined the staff at the Moody Institute of Science that spring and toured with Speake for a few months. They then performed for six months at the Spokane World's Fair in 1974. They alternated films and live performances throughout the day. After Ortner performed, as is common in theatrical rehearsals, Speake would give him notes critiquing his performance, to help him learn how to handle the crowds, deal with hecklers, and to offer better catchphrases and arguments.

Despite his deep-set Christian convictions, Ortner was aware that the current social temperament required him to show sensitivity to the fact that America is a "mixed society." He noted that his school audiences can be "peppered" with Jewish people, Muslims, Catholics, and agnostics. Sometimes people would come "to disrupt the program" or "trying to set you up," but they usually left "making positive comments." He nodded to pluralism somewhat grudgingly and offered an anecdote in which a "Jewish woman" at a school board meeting had protested his upcoming performance. As a result, he had learned to keep religion out of the daytime demonstrations. At schools he would offer an entertaining show and invite the audience to evening shows in a theater where he could evangelize. After years of performing he could sense an audience's mind-set and so was sensitive to "where they are at." He still remembered his pre-evangelical days when he hated being tricked into lectures that turned into preaching sessions, so he attempted to be intellectually honest.

Sermons from Science had changed its name to Wonders of Science. After Ortner had been touring with Sermons from Science for thirty years, the Moody Bible Institute phased out the program. Now he was touring with Wonders of Science only during summers. It was a one-man show. His truck pulled a six-by-twelve-foot trailer loaded with his equipment, which took about a day to set up. Volunteers sometimes helped him at each site to unload the equipment. He then gave a series of four two-hour performances. As with the original Moon blueprint, his lectures were on "light and color," "sense perception and laws of Nature," "recording devices," and the "million-volt demonstration."

Ortner was featured on the *Ripley's Believe It or Not!* television show on January 12, 1999, performing his million-volt demonstration. He favors the variant in which he jumps on the transformer, barefoot, holding a piece of wood that bursts into flames while sparks crackle about his body and sounds similar to those the rocket ships made on an early Flash Gordon serial episode assault the audience's ears.

He has added some new demonstrations. For example, he now had a pair of goggles that force the wearer to view the world upside down and backward. At schools, he will invite a star athlete to the stage, who then would don the goggles and be unable to catch a slowly tossed beach ball. He had another favorite with children when he would "freeze his shadow" on a light-sensitive screen of phosphor and then "walk away from his shadow." The audiences he has performed before have been as large as twenty thousand in downtown Detroit and as small as six technicians at a remote radar site in northern California. Like Moon before him, Ortner often has lectured at military academies and bases under the guise of the United States military's "Character Guidance Programs." At such forums his audiences have represented a broad religious and intellectual spectrum. But they were "out in the middle of the nowhere" and appreciated the entertainment. His job, he believed, was to remind audiences that "humans are not educated beefsteaks" but have a "spiritual dimension."

Ortner's posters and pamphlets combine community "outreach" for both evangelism and science education. One pamphlet asked "Does Science Make You Yawn?" and went on to insist that science should be exciting. It mentioned how the show featured "a cry that can shatter glass," "metal rings that defy gravity," "a tour of the universe with animated laser images," "a frozen shadow," and "liquid light."

In performances he has avoided the creation-evolution controversy, stating that it is better "to stay away from divisive issues." Instead he tries to prove that there is a "personality behind the scientific phenomena they see. . . . They can recognize the Creator from the fingerprints." His own thinking about evolution has changed. In his scientist days he was a convinced Darwinist. After his conversion, he adopted the modernist Christian perspective that evolution was a guided process. Eventually he came around to fundamentalist thinking, though, and now insisted that evolution had its theoretical problems. He pointed to the many "gaps" in the evolutionary record. He did argue, however, that DNA is one of God's creations and even accepted the process of mutation, yet he stated that God must put "complexity" into the system, and he argued that complex systems like the human eye or the mitochondria could not have gradually evolved.

Like Moon's Sermons from Science, Ortner's newer Wonders of Science shows stress that just as matter and energy follow natural law, humans are meant to follow divine law. Ortner also used optical illusions and other devices to show that the human sensory apparatus was limited in its scope. If we cannot trust our own senses, what can we trust? This line of argument was quite common to Spiritualist and occult circles at the turn of the century. Yet Ortner, as a fundamentalist, not surprisingly, had little patience for the New Age community or UFO enthusiasts. Though an avid reader of science fiction, he is also a school science teacher, and he said, "No Lone Ranger is coming to solve our problems from outer space." The failure of such efforts as the Search for Extraterrestrial Intelligence (SETI) suggests that the earth is "probably all alone." He has talked to people involved in "alien studies" and others who offered "abductee tales," but found their position morally dubious. Like Christian critics of nineteenth-century Spiritualism, Ortner argued that Satan could bring in the illusion of alien presences to lead people down the wrong path. The "channeling" of spirits common to Spiritualism and New Age psychics disturbed him. Channelers might well be reaching spirits, but the entities contacted might be fallen angels, not worthy of trust. People who followed these promptings, Ortner argued, were mixed up and had become less open to the truth.

Before I left Ortner's house, he commented that most of the other science sermonizers who came in the wake of Moon lasted about "two and a half years" on the job. He often put in fourteen- to sixteen-hour days and noted that you "have to be called" to do work like he has done. He also said he has not had a cold since he began.

After I had thanked him for the interview and prepared to leave, he slipped me an evangelical tract, *The Roman Road,* saying it would explain the message he tried to present in his demonstrations.

Welding with Water Gas, or Fog Alarms in the Night

In October 2001, inventor and entrepreneur Dennis Lee appeared in Austin, Texas, on stop forty-two of his fifty-state tour, in one of the sprawling white buildings of "Promiseland," a Pentecostal church lit up at night in shades of green and purple.[21] One truck in the parking lot had a hand-lettered sign: "Free Electricity? Ask Me." About 150 people filled the red-cushioned pews, which could seat about 250. Out in the lobby was a check-in table that included forms to fill out for those interested in dealerships, black T-shirts of Nikola Tesla for fifteen dollars, and twenty-dollar copies of Lee's 1994 book, *The Alternative—A True Story With Solutions to America's Most Pressing Problems.* Its back jacket included the copy "Fossil Fuels Are Polluting Our Planet," "We Are Overrun with Garbage and Toxic Waste," "The Media Is Manipulating We the People," "Courts No Longer Uphold Our Inalienable Rights," and the litany concluded with "But, What is the Alternative?"

The alternative was Lee's mixture of Yankee tinkering and magic. His show combined salesmanship with demonstrations of engines, generators, vacuums, and magnetism. Like a good wonder show worker he sought to astound his audience, to make them appreciate the wonders of the universe, and to persuade them that normal science, big business, and government need not have the last word. As with his book, *The Alternative,* the focus of his pitch was ecological sanity combined with right-wing libertarian notions. Lee framed his performance in the rhetoric of wonder and conspiracy and so played to his audience's fear that the fundamental Western dream of progress had been betrayed.

Dennis Lee, founder of Better World Technologies, and the International Tesla Electric Company, could be regarded as the 1990s and 2000s answer to Otis T. Carr of the 1950s. Though he did not promise flying saucer flights, like Carr, Lee has offered the miracle of free energy devices, and through this miracle, a vision of the world transformed. Like Carr, Lee has preyed financially on believers in otherworldly technology and sought credibility by associating his company with the aura of the deceased inventor Nikola Tesla.

For publicity, Dennis Lee has relied on a network that reaches into the free energy community, fundamentalist Christian church culture, and rural libertarian circles with his websites and weekly web radio broadcasts. Lee's radio show, which began broadcasting in December 2000, is titled *The John Galt Show.* The pseudonym Galt is an homage to the hero of Ayn Rand's novel *The Fountainhead,* a symbol of individualism and of the libertarian ideals that Rand helped articulate. Lee's broadcasts from "down on the farm" would generally appeal to rural, right-wing individuals, and to fundamentalist Christians. He notes that his company's president is "Jesus Christ," its treasurer is the Holy Spirit, while Dennis Lee is merely the director of research. His radio broadcasts have included such titles as "God Doesn't Need Us to Make Free Energy," "Free Electricity and 1.6 Million Witnesses," "Government Interference," "Partial Birth Abortions," and "God's Involvement in Our Beginning." His target audience is also indicated by the network he belongs to, the Truth Radio Network, which has the motto "Not Necessarily Your Mainstream Conservative & Christian Talk." In yet another bit of evidence that the far-right and far-left often meet, Truth Radio's web page includes the slogan "Truth Radio tells the Truth behind the Dominant Media Propaganda."[22] The actual broadcasts are rambling and numbing monologues that rely on vernacular humor and reports of conspiracies to stave off the monotony of Lee's pitches for his company's products and promises that listeners that become franchised dealers will soon gain a fortune.

Dennis Lee is significant not only as a modern wonder show operator but also as a member of a fringe science movement: the contemporary free energy movement dedicated to finding clean and virtually limitless energy sources. This movement had its roots in the utopian energy efforts of Tesla and in the pitches of 1950s hucksters like Otis T. Carr; it also relates to the alternative energy interests that rose to prominence in the 1970s with the environmental and anti-nuclear movements.

Lee fits well into the company of such past dreamers, inventors, and hoaxers.[23] He has attributed the source of his mysterious motor to numerous possibilities including magnets, but he also has allied himself with the slightly more respectable contemporary cold fusion movement. This movement dates to 1989 when University of Utah chemist Stanley Pons and his colleague Martin Fleischmann, a world leader in the field of electrochemistry, reported at a press conference that they had developed an electrical method for achieving nuclear fusion reactions at room temperatures. Pons and Fleischmann announced that with their simple apparatus, which involved minimal electrical input, enormous heat was released. Their equipment included four electrolysis cells with palladium electrodes immersed in heavy water. With this simple, inexpensive apparatus, they announced they had fused hydrogen nuclei within the interstices of the palladium atoms and generated fusion reactions that emitted heat and nuclear energy. They suspected it could have powerful commercial applications.

The scientific and popular reaction was immense. With promised help from the Utah legislature, the University of Utah began plans for a multimillion-dollar National Cold Fusion Institute.[24] Thousands of scientists and enthusiasts began to tinker with their own tabletop fusion kits. Early confirmations of anomalous energy emissions eventually led to new studies and retractions of many of the confirmations, as researchers refined methods for measuring the actual output of heat, radioactive particles, and rays. "Hot" fusion experts insisted that the recorded emissions, if accurate, were far too low for fusion reactions and rejected other attempts to theorize anomalous heat output. Cold fusion largely lost legitimacy, as science, in 1990, one year after its "discovery." Yet throughout the 1990s interest in free energy of the cold fusion variety remained.

Scientists who openly continued to conduct such research were stigmatized. As an alternative, many researchers continued on the sly or turned to the public realm, allying themselves with the amateurs who are often called "cold fusioneers." Together, they run their own conferences and have developed popular websites and journals, such as *New Energy News, Infinite Energy Magazine,* and the *Journal of New Energy.*

These researchers have formed their own loose alliance of dedicated amateurs, after-hours scientists, engineers, and futurists. Free energy advocates also are generally welcomed both in New Age circles and in fundamentalist Christian groups, by those whose millennialist assumptions make such miracle devices the norm rather than the unexpected.[25] These assumptions are inscribed haphazardly in the Institute for New Energy website—an encyclopedic mass of links to web postings and articles ranging from the highly technical to the loosely metaphysical. Many of the postings dispute ruling paradigms of science and revel, for example, in questioning the principles of thermodynamics, while exploring antigravity devices, low-energy nuclear reactions, transmutation, and perpetual motion. Others insist, with some scientific basis, that the "ether" of earlier physics tallies with the quantum theory premise that space is not empty but contains quantum-level fluctuations, which, free energy researchers maintain, might be harvested like wind or solar power. The postings also promote a rebellious attitude toward a sinister status quo that encompasses the political, scientific, and business establishment. A scattering of reports of espionage and sabotaged projects also links this website to a strong current of similar folklore. Many cold fusioneers suspect that they, having neared a new scientific secret that will allow for miraculous technology, have been stopped and oppressed by economic interests.

Borrowing from the free energy community's quasi-scientific status, Dennis Lee also has argued that "free energy" and the transmutation of radioactive materials to inert materials occur in cold fusion cells. Like many of the cold fusioneers, he has portrayed himself as a modern alchemist, bringing spirituality to technological research. Lee also has relied on a conspiracy argument of the "If authorities are calling me a fraud and out to get me, then I must be onto something" variety to boost his appeal. His lectures and web broadcasts also point to the obvious truth that the passion or cognitive pathways opened by wonder have long been a part of American commercial and sales culture.

Lee is a far better live performer on stage than on radio. The stage for the Dennis Lee show at the church in Austin was filled with apparatus such as engines, a boiler rigged to run off of gas from a septic tank, a welding machine, a kitchen stove, a generator beneath a row of lightbulbs, and posters for products such as Fire Shaker and Sonic Bloom. Lying about were oddities that suggested Lee had an instinct for clowning, such as a cane with a bulbed bicycle horn attached. The audience was mainly middle-aged and white. There were also a few children, teenagers, and people of color scattered about the pews. Many of the men wore remarkably long beards and seemed quite serious. Many, like myself, held pens and notebooks.

Lee came on stage and asked, "Are you ready to have fun tonight?" He was a big man, with a Fu Manchu moustache, light sideburns, and an intense look as he sized up his audience, trying to decide if he would have any hecklers or "trouble." He immediately asked the audience if he could remove his jacket, gained their assent, and promised he would not take anything else off.

Lee seemed comfortable on stage, capable of both physical and verbal comedy; he has a droll voice with a wheedling quality and he enjoyed playing to his audience, frequently asking rhetorical questions like, "Does anyone remember Free Enterprise?" This brought shouts of "Yep!" or "Sure do!" He began by noting that he would be telling us what we did not know and "What it is we don't know in America will shock you." He complained about "good old boy politics" and corrupt big business. He informed the audience he had technology that could eradicate all forms of pollution in the United States. "What level of pollution is OK for the United States?" he asked. "None!" came the replies. He also said that at the end of the night he would make a job offer to everyone; at no cost to us we could earn more than our present salaries while working only part-time. He also let us know that his other forty-one performances of the 2001 tour had been in hotel conference rooms—not churches. He had no connection with Promiseland; however, he announced, "I am a Christian from the top of my head to the bottom of my toes. That's who I am." He admitted that this did not necessarily give anyone credibility in this day and age, yet he wanted it noted. The inventions he would reveal were the Lord's work.

Like someone doing the "Lord's work," he was attempting to convert audience members to a new worldview. Lee has been a salesman for many years and knows instinctively what American sales manuals long have been teaching apprentice salespeople and what evangelical pamphlets have been teaching leaders of revivals. A sale, like a religious conversion, is fraught with dangers. It requires an attention to setting, an intuitive reading of the audience, a presentation that blends logic and emotion, and a skillful "closing." The salesman and the evangelist not only must rely on a simple display of logic, but also must create a heightened emotional atmosphere. In his analysis of revivalism, *Lectures on Re-*

vivals of Religion (1835), Charles Grandison Finney remarked, "The state of the world is still such, and probably will be till the millennium is fully come, that religion must be mainly promoted by these excitements."[26]

If novels such as Sinclair Lewis's *Elmer Gantry* (1927) made the evangelical preacher as confidence trickster an American archetype, then the salesman who offers a revivalist show, like Lee, is yet another. Like an evangelist cast among sinners, Lee had a difficult task before him. To convince his audience that the "impossible" was not impossible, that "free energy" could exist, and that they would soon have such a device on their lawn required a powerful conversion process. To these ends Lee generated wonders and delight. In addition to offering wonders that might evoke miniature epistemological crises in audience members, Lee also relied on more worldly strategies to delight. He followed to the letter sales manual stratagems—to quote one from 1916, "The sale is made, let it be whispered, to the child in the man. . . . It is the child that dares, that ventures, and that loves the new and untried. . . . To put analytical reason off her guard by pleasing with simple reason is the aim of the logical presentation. When the seller has accomplished this, he may address the child."[27] Lee offered enough apparent technical knowledge to relax his audience, but appealed throughout, with humor, clowning, and spectacle, to the child in each of his spectators.

Early in his presentation, he held up a small pedestal, noting it was the sort that "a toy elephant might dance on." He set it down and spun a top on it. "Let's see how long that keeps spinning." This led to a monologue on how perpetual motion wasn't ridiculous if linked to a perpetual source, such as the motion of the earth around the sun. Here Lee was affirming Tesla's and other technologists' belief in the possibility of perpetual motion devices "of the second kind." Lee said, "Everyone in this room is sitting on the biggest mass known"—laughter ensued—"No, I'm not insulting anyone—you're sitting on the earth, which is moving seventy-eight thousand miles per hour. How does that make you feel? How long has it been moving? A long time. When will it stop? Not for a long time." These thoughts led to a brief side discussion of his company's "dietary aids," which he depended upon to keep his weight down; he held up a spoon with a hole in it, again to much laughter.

He then went on to explain how he wanted to put a generator on each of our houses, as this "was a dream God gave to me." The machine would offer 100 percent of our heat, hot water, air-conditioning, and electricity. It would also put out fifteen times more than the needed wattage. He would sell the surplus to the local power company to make his money and let us use the rest for free.

His first demonstration was designed as a game—he informed us that he could modify our cars to run on pickle juice. He brought his two assistants—one of whom wore a trucker's cap, the other a black T-shirt bearing the portrait of Nikola Tesla—on the stage to help him run a small "infernal combustion motor." He was going to prove that we could run the engine on anything as part of his pitch for the environmentally friendly and economically pleasing formula of 80 percent water and 20 percent gasoline.

Acting like three conspiratorial clowns, he and his assistants began to produce samples of various household products, and Lee urged the audience to take sips or sniffs to authenticate each ingredient. Into a jar they poured samples of Coca-Cola, water, "Hot as Hell" hot sauce, crude oil, Aqua-Velva, sugar, salt, pickle juice, Frappuccino, and urine, which he referred to as "technician's juice." This routine included many broad comedy moments, as when an assistant bravely tasted the hot sauce and had a delayed reaction of horror. Lee then told us with his modified engine, no pollution would be emitted from the exhaust pipe, and, in fact, the exhaust would be 97 percent oxygen and perfectly safe to breathe. They attached the jar of fuel to the apparatus, and after many pulls on the starter and a few engine starts and sputters, they got the engine running. Lee held a white handkerchief before the exhaust, showed it was still clean, then leaned down to breathe in the exhaust. Lee then told us his modified engine relied on a mysterious "reactor rod." Scientists had been "astounded" by this rod, and nobody knew how it worked. Fiber-optic photography showed that "a blue lightning storm" was going on while it worked. He spoke of how these rods could be installed on a car, a complicated process of "tuning" while having it face magnetic north, and making other seemingly magic adjustments. With such modified engines, we could save enormously on fuel bills by using a water/gasoline mixture. He also attempted to demonstrate how a lawn mower could "run on its own exhaust," though he admitted that physicists would tell you this was "impossible." During this demonstration, as with the last, the lawn mower frequently stalled and Lee, the impatient showman, finally told his assistants, with some disgust, "Take it away." Here Lee's acting out of exasperation gained the audience's trust despite the demonstration's failure.

Lee next demonstrated his plug-in "water gas" welding units. The gas was not the mixture of hydrogen and oxygen one would expect from the electrolysis of water but a mysterious "water gas" that does not explode but "implodes," and had the structure of H-O-H rather than the H-H-O that he incorrectly claimed was standard for water. Here he also added a new motif: the idea that organic nature has an anthropomorphic quality. Water gas had the remarkable homeopathic quality of adjusting its temperature to melt whatever substance you were working with. In this way it never burned your hand. He said he discussed this with a scientist at Brigham Young University who said, "Atomic reactions must be involved for no conduction of heat to occur." Water gas also left "no slag" on steel when cut, left water streaks on surfaces it cut, and could burn through any substance on this planet, including diamonds. He and his assistants proceeded to cut various pieces of metal and discuss the costs of using the "water gas" instead of acetylene. The cost for one of the plug-in welding units was approximately $1,200, but Lee would also be happy to sell the bottled gas as well.

His following exhibit involved the concept, popular in the cold fusion and free energy community, that the transmutation of radioactive elements to inert elements was possible through cold fusion processes. He told us that the federal government was putting our lives at risk. He described an aboveground nuclear waste storage facility in Richland, Washington, with brine circulating around spent rods to prevent spontaneous reactions, and how authorities were scrambling around to find salt mines to bury the waste in. This was all foolery, since Lee had "a machine to neutralize all radioactive waste into inert materials. We know they know that," he added, because he had demonstrated the device for two unnamed U.S. senators, one of whom responded favorably and was promptly voted out of office. Though scientists and the Department of Energy would disagree—"anyone from the Department of Energy here tonight? No? I always invite them"—it was possible to "transmutate" [*sic*] the nucleus of an atom. "The alchemists were right."

After demonstrating what background radiation sounded like with a Geiger counter, Lee had an assistant mix up a control sample and a solution of one gram of radioactive thorium along with 125 grams of water and an undisclosed amount of hydrochloric acid. These would be placed in his "pure zirconium" cooker, with its electrodes, for thirty minutes. With a radioactive gauge they would test the sample before and after. Lee told us what to expect: a lowered radioactive count and traces of titanium and copper and other metals would be in the solution—proof of the transmutation of the thorium. Though the most likely explanation would involve contaminated samples or electrodes, Lee insisted that this was out of the question because the cooker was 99.9 percent pure zirconium.

While the samples "cooked," Lee spoke about the conspiracy of the power companies to rip off consumers with inefficient meters and appliances that drew more current than needed. It was all a result of the "good old boy routine," the short script of which ran, "You lie, I'll swear to it." He then extolled the virtue of his company's numerous products, starting with a power regulator that would make sure machinery only drew the amount of energy needed, and he demonstrated the device on a small generator hooked up to an array of lightbulbs.

Further sales pitches for household products relied less on humor and instead played on the audience's desire for security. Earlier, he had made the audience uneasy about pollution, corporate conspiracies, and radioactive waste. Now he showed videos for a fire barrier spray and a fire shaker that put out kitchen fires that could otherwise swiftly spread through a home. He also revealed the "Bandit" alarm system for homes or stores, which took three seconds to fill a room with thick fog, and "Miracle Shield," which was an anti-graffiti liquid—a product that catered to the audience's fear of teen gang members. Next came an enzyme soil remover that "cleans up the environment instead of polluting it." Those "little bugs," he told us, "ate all the oil they could find, then cannibalized each other. Eventually you have only one giant bug left to battle." He then did some shadow boxing, as if trying to knock out a giant adversary, then grinned and said, "Just kidding."

After these videos, Lee returned our attention to the zirconium cooker; his assistant offered precise measurements to confirm that the radioactivity level had lowered. Lee then urged an audience member to bring a test tube of the transformed solution to a university lab where trace amounts of newly formed atoms of copper, titanium, and other metals would be found. Lee then continued his theme of the insecurity of life in America, alluding to the terrorist attack on the World Trade Center in Manhattan that had taken place several weeks prior to his show. This was the one point in the lecture where his impeccable sense for what his audience would be willing to hear failed somewhat—a brief shudder seemed to greet the allusion. He soldiered on, though, and solemnly intoned that jet planes could use water as fuel. This had been demonstrated irrefutably. A "hydrogen pulse separator" would turn the water to hydrogen fuel (apparently preferable, in this case, to "water gas"). If a plane loaded with fuel tanks of water ran into a tower, he asked, what would happen? "Far less damage."

He then moved on to describe how he had been harassed in Kentucky and forced off the stage, and took a passionate stand about returning with federal marshals to claim his right to free speech.[28] When the applause faded, he began his final pitch. It was over three hours into his performance, and as no hecklers had challenged him, he suspected, correctly, the docile crowd was now ready to accept his more absurd pronouncements. He told us that he believed he had found a perpetual energy source—and it was the magnet. "I believe," he said, "that energy flows into magnets." Most scientists, he admitted, would disagree.

He offered us a few clues as to how his device might work. At the front of the stage he earlier had set up a series of small magnet-laden windmills. Spinning one caused the others to spin in a haphazard, chaotic way that Lee enjoyed; he gave personalities to these mills and their quirks, underlining his identification of the magnet as a trustworthy friend, a reliable source of unlimited power. Next he demonstrated an unusual magnetic phenomenon known since the nineteenth century. He showed us a permanent magnet and a segment of wide copper pipe. He upended the pipe and set up a mirror so that a video camera could look down it. Then he had his assistant drop a magnet down the copper tube. The video screens showed it slowly falling, as it induced magnetic currents in the copper; it fell in what seemed slow motion. He then said that a stronger magnet would go even slower. We watched its slow, somewhat magical free fall through the short length of tube on the video screen. "My motor will use one hundred permanent magnets," he said.

He then said he was not demonstrating his free energy motor this tour because in 1999 he had demonstrated it to all comers. Using the magician's art of misdirection, he then switched the audience's attention from his miraculous but unseen "permanent magnet motor" to the generator that it would be attached to on our lawns, the "G-10." The G-10 was brought out on stage and hooked to a heater, and amid its thunder and fury, several spectators came down with devices to measure its efficiency. He assured us that the complete energy-producing unit would not make much noise on our lawn.

It was now close to 11:00 P.M. Lee's audience had been listening to him for four hours. They were willing to take on trust Lee's miraculous free energy machine and to instead test the efficiency of the G-10 generator. They apparently had faith in his "permanent magnet motor" and interest in his other products, like the "noiseless jackhammer" and the welding unit that used "water gas." All that was left for the final hour was to pitch products and dealerships. The incentive for takers would be personal wealth and a chance to help improve the environment. By moving our imaginations ahead to the day we would have the G-10 working for us, Lee had already psychologically closed the deal. Some spectators might conclude that even if Lee was running a racket, he was a tireless front man who could help them sell products to others and so gain commissions.

I left Promiseland worn down—able, almost, to believe in the miraculous. Lee seemed to me the embodiment of an American archetype, salesman, showman, tinkerer, and trickster. Lee preyed on his audience's fears and insecurities—painting verbal pictures with video supplements of houses going up in flames, stores burglarized, jet planes exploding, terrorist attacks, and a variety of ecological disasters. He offered an "alternative"—miracles that had some technical grounding, surrounded by the fog of magic and double-talk.

He gained his audience's trust by making religious appeals and by fanning distrust of big government, big business, and big science. He championed an alternative science, seemingly like that of the cold fusioneers, one able to challenge established truths. Lee offered his audience not just "free energy" but another form of "power." He was selling a vision of democracy, one that gave each audience member the authority to reject scientific expertise and shape his or her own worldview. He offered a miracle gadget on the lawn but, more important, he offered a cosmos that was not grim and mechanistic but full of possibilities. Like Francis Bacon before him, Lee was offering a "Great Instauration"—a restoration that linked the American Dream to the millennial Christian dream of the approaching reign of heaven on earth.

Grassroots Wonder and Salesmanship

Ortner the evangelist, Lee the libertarian inventor with strong Christian convictions, and Olszewski the light therapy inventor all offer variants on the wonder show. These performers and their shows reveal a clear lineage to the eighteenth- and nineteenth-century performers examined earlier in this work. Olszewski was a modern counterpart to Charles Came, who toured in the mid–nineteenth century. Both offered miraculous healing technology, magic lantern shows, and a therapeutic worldview. Lee, with his mixture of technical knowledge and fancy, could be likened to Katterfelto and his dubious promotions.

Among these showpeople, as with performers in allied traditions such as the circus or the medicine show, traditions are handed down orally. Benjamin Silliman trained his son Benjamin Silliman Jr. to continue his financially rewarding natural philosophy lectures in the 1840s. When Charles Came retired after the Civil War, his son-in-law took over the wonder show business and toured upstate New York for several more decades with Came's equipment. And when Irwin Moon gave up lecturing for filmmaking in the 1940s, he trained his son to perform Sermons of Science, then found another successor, George Speak, who passed on the tradition to others, including Dean Ortner.

Ortner, Lee, and Olszewski all had widely divergent worldviews but shared common ground. All three believed in "the wonders of science" and held the belief that technology and science were here to benefit humanity. Common to these showmen, also, was the view that a person was not just a "walking beefsteak" but a spiritual being. Proponents of New Age or metaphysical belief systems, like Olszewski, and of fundamentalist Christianity, like Ortner and Lee, had in common a strong belief in the spiritual realm and the afterlife. They either attempted to convey these beliefs to their audience or to woo an ideal spectator that shared in them. All three—Olszewski and Lee, the salesmen, and Ortner, the evangelist—offered variants of the wonder show in which technology is presented as an aid to one's higher calling. And if all three also offer a critique of the status quo, it is made in the name of a higher spirituality or ethic—premised on a "betrayal" of the scientific project's millennial promise.

In these forums, wonder is used to promote sales and to announce utopian aims and, often, spiritual agendas. Dr. MegaVolt, dancing on a flatbed truck in the deserts of Nevada, helped to bind the quasi-utopian quasi-anarchist community that cohered in the desert during the week-long Burning Man arts festival. Ortner, like Silliman before him, harnessed wonder and science to the wider goal of advancing Christianity in America. Olszewski dedicated his efforts to the therapeutic goals of the New Age movement; Dennis Lee offered a utopian future where the disenfranchised reinherited their political centrality.

While Ortner and Olszewski appear to have a genuine desire to help their audiences and to demonstrate scientific principles, Lee is more slippery to categorize. An evaluation of his act points out the limitations set at the beginning of this work—that is, to consider the cultural dynamics of these shows outside the standard dialogue about the moral dangers of pseudoscience. Yet Lee's legal entanglements and his apparent efforts to prey upon his audience's fears and desires make it difficult to take a thoroughly objective stance. He is an entertainer, seems reasonably knowledgeable technically, and has some "genuine" products to sell, and yet he also is attempting to sell the miracle of "free energy" to which he has no genuine access. Like the rainmakers who toured the drought lands of Kansas one hundred years ago, he was out to make a buck and would evoke the technological sublime to do so.

Yet, the automatic condemnation of wonder as a simple handmaiden to superstition and fraud is too easy. Wonder is a theatrical strategy that can be harnessed to both immature and more mature worldviews, whether scientific or religious. It is debatable whether the "scientific" have a monopoly on the mature worldview; even Margaret Storm, who might be categorized as an "outsider artist," conceivably mentally deranged, successfully argued that the nuclear arms race born of technocratic impulses reflected less than rational and mature social governance. Further, the condemnation of wonder, which often seems a primary goal of the skeptical organizations of the West, is as doomed as any effort to outlaw myth.

Despite its possible dangers in abetting swindles, the wonder show harbors a critique of Enlightenment values and offers an alternative to a vision of the cosmos as rational, complex, and ultimately meaningless. These shows as well lampoon the tendency in Western culture to make scientists and those who consider themselves "scientific" into ultimate arbiters of truth. The widening distance between the lay public and the technical elite has made of scientists an "exotic other," and as such, the wonder show can be viewed as an attempt to reclaim and mimic the scientific project, offering dramatic narratives that help to explain and place this "other" culturally.[29]

Pseudoscience Revisited

Although a solid argument can be made separating wonder from superstition, it would be misleading to deny that the passion of wonder, which until the eighteenth century intellectuals harmonized with the scientific impulse, in the twentieth century often had become a species of cheap thrill for the dull and jaded, most suited as the bedfellow of pseudoscience. The passion that natural philosophers and connoisseurs savored in wonder cabinets often devolved into the gaping astonishment of the carnival midway.

Yet even in the twentieth century there still was room for wonder as the handmaiden of a "mature" religious or scientific vision. For example, scientist and author Carl Sagan, in almost solitary fashion among his fellow skeptics, stressed the importance of shows of wonder to fuel the scientific impulse. In the preface to *The Demon-Haunted World* (1995), Sagan recalled as a child visiting the 1939 World's Fair in New York City and being thrilled by displays—likely the work of GM's Kettering—that announced "See Sound!" and "Hear Light!"[30] Sagan insisted that his fascination with these World's Fair displays prompted his interest in science. He argued that scientists needed to combine an openness to new concepts no matter how outlandish, or the capacity to wonder, along with the "ruthless scrutiny of all ideas."[31] In the evocatively titled chapter "The Marriage of Skepticism and Wonder," among other matters, Sagan questioned whether an unlikely belief that causes more good than harm should be discouraged or not, and ultimately recommended that educators stimulate a "prudent balance" between the two forces to make good scientists and citizens.[32]

As with other skeptical accounts, however, Sagan's book was a warning about the rise of pseudoscience. He even offered the formula that "pseudoscience is embraced, it might be argued, in exact proportion as real science is misunderstood."[33] Sagan also insisted that as scientific authority diminished in a society—as in post-Soviet Russia—pseudoscience, cults, and other noxious forms of thought would fill in the gap like weeds.[34] The need for rationality, critical thinking, and social order appeared preeminent to Sagan. As such, he refused to appreciate pseudoscience on aesthetic grounds or as an expression of healthy diversity. Wonder shows like those of Lee would simply appall him, and the creationist agenda hidden beneath the surface of Ortner's Wonders of Science show would make him cringe. Wonder, as in earlier eras, had to be allied with critical thinking.

Critiques like those of Sagan placed the capacity for wonder into a larger debate about the dangers of pseudoscience. On one side were Sagan and other champions of scientific thinking who fretted about an apparent rise in superstition and decline in critical thinking; on the other side were postmodern academics who saw value in a democracy of ideas and in grassroots challenges to scientific authority. If the skeptics served as the pseudoscientist's prosecutors, the postmodern academics worked to defend the accused: whether psychic mediums, creationist scientists, parapsychologists, cultists, UFO abductees, or the inhabitants of science's fringe, such as wonder showmen Olszewski and Lee. Performers and reformers who appear to straddle the divide between science and religion have been easily targeted as promoters of dangerous pseudoscience. And, as in the nineteenth century, other performers, the guild of magicians, have been prepared to aid in such debunking.

Dennis Lee, in particular—and with good reason—has drawn the wrath of researchers dedicated to debunking pseudoscience. Although there are many such groups, the best known is the Committee for the Scientific Investigation of Claims of the Paranormal (CSICOP), an alliance of scientists, academics, and journalists that has actively debunked fringe science, creationist science, and psychic research. It was founded in 1976, when the Christian fundamentalist challenge to evolution was on the rise; in that era public interest in psychic performers such as Uri Geller also pointed to an occult revival and a public taste for the irrational. CSICOP has since also found numerous targets in New Age practices, products, and beliefs.

True to the efforts of magicians like John Nevil Maskelyne in the nineteenth century and Houdini and Dunninger in the early twentieth century, CSICOP has included the magician James Randall, aka the Amazing Randi, as one of its most active members. Randi's efforts in the 1970s to debunk psychic "spoon-bending" performer Uri Geller helped gain him an audience and enabled Randi to position himself as a crusader for the liberal, secular, scientific, "commonsense" view of reality that skeptics embrace. Just as Houdini offered cash prizes to Spiritualists who could prove their abilities were real, the Amazing Randi has offered a "One Million Dollar Paranormal Challenge" to anyone who can demonstrate "any psychic, supernatural, or paranormal ability of any kind under satisfactory observing conditions."[35] Beneath the area in the application form where the applicant must place his signature was a note from Randi recommending that the applicant try a few trial "double-blind" tests before applying. He cautioned, "Please be advised that several claimants have suffered great personal embarrassment after failing these tests."[36] Eric Krieg, a founding member of the Philadelphia Association of Critical Thinking, a skeptic organization, is an engineer who has dedicated much energy to debunking Dennis Lee, and has mounted a similar challenge, offering ten thousand dollars to anyone who can demonstrate a working free energy or "over unity" device.[37]

CSICOP, Randi, Krieg, and other skeptics engage in the "boundary maintenance" that sociologists argue is crucial to the scientific project. Yet philosophers of science have long pointed out that a schism exists between boundary maintenance that relies strictly on methodological definitions of science and those efforts that consider ontology—that is, science's relation to basic beliefs about the structure of reality.[38] An ontological bias common to many scientists is that of a mechanistic and materialistic universe—that is, one that "doesn't care." Such a bias, for example, made it difficult for many scientists and academics to take J. B. Rhine's methodologically respectable efforts to prove ESP seriously. The psychic forces Rhine wished to isolate simply did not mesh with the mechanistic worldview. Rhine's efforts were analogous to a proud bridegroom bringing home a young woman from the wrong side of the tracks. Despite the engrained sense that science is "without presupposition," scientists often rely on biases to identify and stigmatize deviant science or pseudoscience. At least one portion of this decision-making process is "instinctive"; judgments about what is or is not a sound basis for science or appropriate research will partially depend on the reflexes and tastes of the elite members of the community.[39]

While not all scientists of necessity share the ontological assumption of a mechanistic and materialistic universe—a vision threatened by quantum mechanics[40]—it is fair to say that most of the members of CSICOP would adhere to such a vision. For example, journalists frequently call upon magician James Randi to debunk cultists and pseudoscientists. In Randi's effort to debunk "near death" experiences, he remarked that he also has had two near death experiences. During these experiences, he also saw a tunnel of light. However, he concluded that this was simply a case of wish fulfillment, a product of his brain's last chemical frenzies. "Just because you see something doesn't mean there's a real tunnel or a real light. . . . It's a physiological phenomenon, not a spiritual phenomenon. It's what happens when the nervous system begins to relax."[41] Yet from a subjective point of view he had no real way to distinguish the two. His explanation has no firmer basis in truth than that of the "near death" theorizers. Others, with different ontological assumptions, would take such experiences as evidence pointing to the possibility of an afterlife.

CSICOP members often justify their policing of science's boundaries as a ground-clearing effort that allows scientists to dedicate valuable time instead to research, but certainly some scientists, as did Carl Sagan, have promoted the skeptical cause. Another, Robert L. Park, a physicist and director of the American Physical Society's public affairs office in Washington, D.C., joined the skeptical movement with great glee and gusto when he stitched together a number of his editorials denouncing pseudoscience into the book *Voodoo Science: The Road from Fraud to Foolishness* (2000). He dedicated chapters to ridiculing the free energy movement, Dennis Lee's schemes, perpetual motion, alternative medicine, channeling, and other fads that, in his mind, were mere superstition, thinly veiled.

Park put a twist on Sagan's assertion that pseudoscience rises when scientific thinking wanes. Several times in his book Park asked whether, in some way, scientists in fact were to blame for the proliferation of pseudoscience today—particularly those elements of it that can be found in the New Age movement. He also noted that public hostility toward scientific authority was on the rise; if science's involvement with government and the military was regarded as patriotic during World War II, the youth movement and the student protests of the Vietnam War era as well as the nuclear protest movement had changed these positive associations. Likewise, possibly with mock humility, he asked: "Did we set people up for this? In our eagerness to share the excitement of discovery, have scientists conveyed the message that the universe is so strange that anything is possible?"[42]

Park's query skirted the edge of a very important truth—too often scientists and defenders of the "scientific" have allowed their enterprise to appear not simply "strange" but as having its own sacred purpose. Yet Park ignored much of this process of "top down" science mystification to instead scold the public and aberrant scientists who justify varieties of the mystical in scientific terms. Far from aiding science's advance, Bacon's "strange facts" were best left at science's periphery. For similar reasons, Freud did not originally report his belief in telepathy, because to do so, with its mechanism unexplained, he worried, would be to encourage occultists to raise a ladder on which to "rise above science."[43] Yet scientists who have become interested in the paranormal will occasionally argue that the materialist assumption is simply invalid. Contemporary physicist Hal Puthoff, for example, noted that his often-ridiculed research in parapsychology at the Stanford Research Institute left him, nevertheless, with the conviction that the truth of the paranormal "must be taken into account in any attempt to develop an unbiased picture of the structure of reality."[44]

Such refusals to disbar phenomena once called occult—or to actively court such phenomena—also align with the postmodernist academics' view of the power struggle to control knowledge and information. If avant-garde scholars once preferred to ally themselves with science, many in the late twentieth century have chosen to battle against science's hegemonic hold on "truth" and to do so often have embraced "superstition." As with the skeptics, they have claimed a "progressive" attitude. They have argued that it was more progressive to permit the noxious flowers of pseudoscience to flourish than it was to attempt to stamp them out in the name of a higher moral purpose. One could also read these academics' efforts as a somewhat petulant attempt to regain authority in the name of the humanities.

Postmodern theorists have argued that no one person or group, any longer, can proclaim themselves the ultimate authority on truth; meaning no longer had a center; "reality" itself was malleable. Some theorists, tending to extreme relativism and pessimism, have lamented that critical thinking can at best help us navigate through the endless hall of mirrors that late capitalism has erected. According to thinkers such as Fredric Jameson and Jean Baudrillard, a media-saturated society has encouraged a new "depthlessness" and an epistemological "collapse of the real."[45] The liberal consensus that once permitted "a center to hold" has vanished. While some theorists, such as Jameson, offer this in the spirit of a neo-Marxist critique of capitalism, others, such as Baudrillard, imply that an irreversible epistemological shift has occurred. Political theorist Jodi Dean related the shift not only to "late capitalism" but also to the innovation of cyberspace. She argued that all claims to truth—like those posited on countless websites or in countless academic journals—rely on tissues of evidence much like those that hold together conspiracy theories. A collapse of the "liberal consensus" is the result.[46] Unlike Park, these academics do not bemoan the passing of the golden age of World War II, when scientists and physicists were universally hailed as authorities with a monopoly hold on the one true, liberal vision of reality.

Pressing their own case, CSICOP members believe their efforts will remove noxious beliefs from the public sphere and help genuine critical thinking to flourish.[47] The philosophy of such skeptics resembles the ideology of republicanism promoted by the elites of the eighteenth and nineteenth centuries. This ideology insisted that a democracy could only flourish with a virtuous, educated public. In melodramas of the early republic that played out this ideology, the virtuous American farm girl falls prey to foreign or urbanized seducers whose base motives are ultimately revealed.[48] According to the updated version of this melodrama, the virtuous public should not give in to the seductive maneuvers of pseudoscientists, postmodernists, or New Age charlatans with "foreign"—i.e., unscientific—heritage. In both cases it was the duty of an educated elite to tend to the education and moral training of the masses.

Offering an opposing viewpoint, academics like Andrew Ross, Sandra Harding, and Jodi Dean have been concerned less with "public virtue" and more with encouraging a democracy of ideas.[49] Though they have not necessarily embraced pseudoscience in itself, such scholars have embraced it as a symptom of dissent. Scientists for too long have had a lock hold on the "truth." Opposition to scientific pronouncements on matters such as UFOs or alternative medical treatments imply a public wishing to exercise its independence and shake off the controls of experts. Presumably, the opposing skeptics are attempting to enforce a "top-down," elitist model of social control.[50] The postmodernists have also argued that skeptics such as Sagan were naïve to assume that the "scientific" worldview was automatically progressive. Even if the democratic sharing of ideas is a central value in the scientific community, scientific and technological research has flourished under all sorts of patronage, including that of Christian fundamentalists, nineteenth-century racist elites, twentieth-century totalitarian regimes, and military manufacturers who place profits above progressive democratic ideals.[51]

Dean, Ross, and others were intent on helping the public regain political power even at the risk of allowing "superstition" to flourish. The skeptics were intent on restoring rationalism, and, to scientists, the authority that was their due. Like the alien visitor Klaatu in the film *The Day the Earth Stood Still,* the skeptics concluded that scientific thinking was our last best hope. Skeptical critics longed to see scientists restored to social prestige, a condition tarnished during the 1960s countercultural movements with their vilification of scientists and technocratic rule.

Reestablishing the cultural authority of science and scientists was high on Park's agenda in his book *Voodoo Science.* Park argued that to regain cultural authority, the scientist must follow a strategy quite different from that which led to scientists' eager participation in the popularization of science at the 1933 Chicago exposition. Park wrote that scientists "are eager to tell people what it's like on the frontier. They want to talk about neutrino oscillators, Higgs bosons, cosmic inflation, and quantum weirdness—the things that excite them." However, they must learn to stop "pandering to the public's appetite for the 'spooky' part of science," or the public will assume "anything is possible."[52]

The appropriate wonder show formula to someone like Park was for the public to admire the combined mental processes of scientists; that is, if we need a vision of human powers advancing alongside technology, we should be satisfied with the advancing body of scientific knowledge. This, too, could provide a thrill. But if wonder was left to amateurs who might evoke scientifically imprecise mappings of the universe, Park argued, we all would suffer the consequences of the return of "superstition."

Park concluded his book by describing an IMAX film, *Cosmic Voyage,* which he attended in 1996 at the Air and Space Museum in Washington, D.C. He described how the film's "'cosmic zoom' hurled viewers to the outer limits of the universe, plunged them down to the domain of the quark, and sent them tumbling back through billions of years."[53] It sounds like a production that Charles Came, with his magic lantern slides of protozoa and the solar system in the 1840s, would have approved, as would Irwin A. Moon, whose films of the 1950s also relied on contrasting microcosmic and macrocosmic dimensions. Park was profoundly stirred by the film. He particularly admired the fact that the production did not offer spectators the comfort that the universe "cares about them." Yet it did provide him with solace, perhaps akin to that of Sagan, who remarked, "When we recognize our place in an immensity of light-years and in the passage of ages . . . then that soaring feeling, that sense of elation, is surely spiritual."[54]

Yet Park and Sagan were "experts"; how would a layperson respond to the IMAX film—perhaps with the inappropriate, irrational assurance that the universe cared? Park spoke to his secretary, who had also attended the film, to determine the reactions of the layperson. He learned that she was at first terrified, but left the theater filled with a sense of "wonder." She realized, as did he, that the real source of wonder was the reach of human knowledge. The ultimate wonder was that humanity—though merely "self-replicating specks of matter . . . have managed to figure all this out."[55] As with the stage magicians of the turn of the twentieth century, Park sought to drain wonder from "objects" or "phenomena" and relocate it in human craft, skill, and intellectual reach.[56]

Within the boundaries of the standard skeptical argument against superstition, Park's tone seems reasonable and his demand that scientists stop pandering of value. However, the standard skeptical argument overlooks the historically constructed mystique of science. Ignoring this larger version of pandering, Park easily moves on to critique an infantile public longing for a universe that "cares." Such perversity, he believes, must give way to a deeper satisfaction with the tales scientists tell. The public cannot be a part of the discovery process, nor be permitted to browse among different cosmologies or belief systems. Instead, according to Park and other skeptics, we must rely on scientists to choreograph the wonder show; in turn, they will continue to allow us front row seats.

The assumption that "the universe doesn't care" might be appropriate to good science, but it remains vaguely unsettling and at odds with the belief that science aids progress. In a universe that "doesn't care," the common philosophical prescription is for "endurance" not "progress." Progress is an idea in search of a mythic framework to anchor its meaning. Hidden beneath the scientific enterprise remain traces of the millennialist impulse—with science, as in Roger Bacon's dream, helping humanity toward redemption or toward quasi-divine status in what may be at best an orderly but absurd universe. The tragic worldview that many science defenders believe is its necessary grounds inevitably clashes with the enterprise's utopian promise—encouraging old ghosts to rise again.

BIBLIOGRAPHICAL ESSAY

After spending several years studying how-to pamphlets on salesmanship, hypnotism, mind reading, and magic, as well as books detailing assorted tricks and illusions and performers' denouncements of one another as frauds, it is finally my turn to "unmask"—that is, to acknowledge the sources of my scholarship and inspiration. This essay surveys scholarly works that have themes in common with this book and evaluates research sources, relying on the chapter plan as its ordering framework.

As the genre here defined as "wonder shows" has no previous histories, I will mention first a few texts that have close thematic ties. In order to connect the familiar American studies notion of the "technological sublime" to wonder, this book is deeply indebted to Lorraine Daston and Katharine Park's *Wonders and the Order of Nature* (New York: Zone Books, 1998). This ambitious history of science treated the cultural implications of wonder from the medieval period through the scientific revolution and conveniently stopped where my book begins—with the "vulgarization" of wonder.

Several books have linked science to showmanship and mystification. David F. Noble's *The Religion of Technology: The Divinity of Man and the Spirit of Invention* (New York: Alfred A. Knopf, 1997) is a short work that, in broad strokes, established the central thesis that technologists and promoters of the scientific project have often assigned a millennial role to the technical fields. In the process a "religion" of technology has evolved. Richard Altick's *The Shows of London* (Cambridge: Harvard University Press, 1978) helped turn scholarly attention to popular theater. It canvassed the nontraditional theatrical performances offered in London in the late eighteenth and early nineteenth centuries; many of the shows involved entertaining scientific or pseudoscientific demonstrations. John C. Burnham's *How Superstition Won and Science Lost* (New Brunswick, N.J.: Rutgers University Press, 1987) examined how science has been popularized in America in the past two centuries. Burnham's work is a good introduction to the historical debates about science and superstition and the way therapeutic culture has shaped science journalism and popular thought. More recently, Jeffrey Sconce's *Haunted Media: Electronic Presence from Telegraphy to Television* (Durham, N.C.: Duke University Press, 2000) examined the public's tendency, historically, to connect new media forms to the supernatural.

Science as a species of performance has been explored in many of Simon Schaffer's essays and books including one coauthored with Steven Shapin, *Leviathan and the Air-Pump: Hobbes, Boyle, and the Experimental Life* (Princeton, N.J.: Princeton University Press, 1985). In a similar effort, Iwan Morus, in *Frankenstein's Children: Electricity, Exhibition, and Experiment in Early Nineteenth-Century London* (Princeton, N.J.: Princeton University Press, 1998) examined cultural hierarchy within the electrical trade and its expression in electrical exhibitions.

The chapters that examine show business and science in the early republic and antebellum periods owe much to scholarship on museum operators Charles Willson Peale and P. T. Barnum. For my look at the early history of museums I relied on two books about early museum curator Charles Willson Peale: Charles Coleman Sellers, *Mr. Peale's Museum* (New York: W. W. Norton, 1980), and David Brigham, *Public Culture in the Early Republic: Peale's Museum and its Audience* (Washington, D.C.: Smithsonian Institution Press, 1995). For a look at Barnum, Neil Harris, *Humbug: the Art of P. T. Barnum* (Boston: Little, Brown, 1973), is still an excellent point of departure, as is Bluford Adams, *E Pluribus Barnum* (Minneapolis: University of Minnesota Press, 1997). Recent works by James W. Cook, *The Arts of Deception* (Cambridge: Harvard University Press, 2001), and Benjamin Reiss, *The Showman and the Slave* (Cambridge: Harvard University Press, 2001) also add to the growing field of Barnum studies. For the anthropological analysis of performance, this book has relied on a thematic strand from Rogan Taylor, *The Death and Resurrection Show* (London: Anthony Blond, 1985), which argued that modern show business is a historical outgrowth of shamanism. The dramatic power of the "death and resurrection" theme is clearly offered in many wonder shows. In a parallel vein, anthropologist Victor Turner, in *From Ritual to Theater* (New York: Performing Arts Journal Publications, 1982), argued that tribal ceremonies evoked a "liminal" or sacred threshold for participants to cross, while modern theater evokes a quasi-liminal or "liminoid" space for audiences.

For an understanding of antebellum science, American culture, and intellectual currents, an excellent introduction is Herbert Hovenkamp, *Science and Religion in America, 1800–1860* (Philadelphia: University of Pennsylvania Press, 1978). This era and its hybridization of science and religion have been revisited in Craig James Hazen, *The Village Enlightenment in America: Popular Religion and Science in the Nineteenth Century* (Urbana: University of Illinois Press, 2000).

The staff at the Smithsonian Museum led me through the archival material available on Charles Came, the nineteenth-century showman highlighted in the first chapter. I was fortunate to have the help of curator Roger Sherman, who also let me examine his office files on Came. Came's letters are available at the National Museum of American History archives (NMAH), in Washington, D.C., and much of his apparatus is also in storage at the Smithsonian Institution; queries will eventually place you on a shuttle bus from the NMAH to the storage facility that houses, among other Came paraphernalia, his foot warmer, orrery, posters, healing crystal, and electrostatic devices. I placed Came in the medicine show context with the help of Brooks McNamara's colorful history of the medicine show, *Step Right Up* (Jackson: University Press of Mississippi, 1995 [1975]), Andrea Stulman Dennett, *Weird and Wonderful: The Dime Museum in America* (New York: New York University Press, 1997), and several histories of medicine in America, including J. G. Burrow, *Organized Medicine in the Progressive Era* (Baltimore: Johns Hopkins University Press, 1977), and J. H. Cassedy, *Medicine in America: A Short History* (Baltimore: Johns Hopkins University Press, 1991). Microfilm reels of the *New York Sun* of that era also revealed how deeply the penny press relied on revenues from patent medicine and "irregular" medical advertisements.

Electricity

My chapters focusing on the early electrical industry relied on several overviews. The cultural implications of electricity were carefully explored in David E. Nye's *Electrifying America: the Social Meanings of a New Technology* (Cambridge: MIT Press, 1990). Nye's look at how electricity was exalted and the technological sublime presented at World's Fairs is especially revealing. A related work, Caroline Marvin, *When Old Technologies Were New* (New York: Oxford University Press, 1988), looked at the electrical industry workers' promotions and perception of their own labors. Thomas P. Hughes, *Networks of Power* (Baltimore: Johns Hopkins

University Press, 1983), detailed the history of the power industry and examined the "war of the currents." Carolyn Thomas de la Pena, *The Body Electric: How Strange Machines Built the Modern America* (New York: New York University Press, 2003), explored the turn-of-the-century public's fascination with electrical healing.

Useful bibliographies of electrical history include Joyce E. Bedi, Ronald R. Kline, and Craig Semsel, *Sources in Electrical History: Archives and Manuscript Collections in U.S. Repositories* (New York: IEEE Center for the History of Electrical Engineering, 1989); and Judith A. Overmier and John Edward Senior, *Books and Manuscripts of the Bakken* (Metuchen, N.J.: Scarecrow Press, 1992), which is a catalogue of books about electrical healing, animal magnetism, mesmerism, and allied fields, all of which, along with a historical collection of electrical devices, can be found at The Bakken: A Library and Museum of Electricity in Life, in Minneapolis.

My look at electrical inventors involved an examination of the publicity crazes surrounding Nikola Tesla, Thomas Edison, and Charles P. Steinmetz at the turn of the twentieth century. For a book that looks at the historical underpinnings of the "mad scientist" archetype in popular culture, see Roslynn D. Haynes, *From Faust to Strangelove: Representations of the Scientist in Western Literature* (Baltimore: Johns Hopkins University Press, 1994). Perhaps the prime example of this archetype is Tesla. The cult that continues to grow around Tesla makes him a figure that bridges the "wonder show" circa 1890 with that of today. Tesla biographies include John J. O'Neill, *Prodigal Genius: The Life Story of Nikola Tesla* (New York: Ives Washburn, 1944), and Margaret Cheney, *Tesla: Man Out of Time* (Englewood Cliffs, N.J.: Prentice-Hall, 1981); the most recent and least hagiographic treatment of Tesla is in Marc J. Seifer's *Wizard: The Life and Times of Nikola Tesla* (Secaucus, N.J.: Carol Publishing Group, 1996). In defense of O'Neill's early study, his notion that Tesla engineered his own persona of "superman" resembles the vein of scholarship of the past few decades that examines how celebrities like Benjamin Franklin and Thomas Edison have crafted their identities for public consumption.

As most of Tesla's papers and effects were returned to Yugoslavia following his death in 1943, there is little in the way of primary materials on Tesla to be found in the United States. The Kenneth Swezey Papers at the National Museum of American History (NMAH) are somewhat useful, as Swezey, a science journalist, was long a fan of Tesla and the primary instigator of the celebrations of Tesla's seventy-fifth anniversary in 1931, as well as his one hundredth anniversary in 1956. These papers include Swezey's correspondence with other Tesla enthusiasts and a few letters from Tesla, as well as clippings about Tesla and Tesla imitators, copies of Swezey's articles heralding Tesla as a forgotten genius, samplings of Tesla's stationery, and such memorabilia as one of the meticulous inventor's white dinner gloves.

The Edison Papers, an enormous collection, is now on-line. At the time I was researching, only the first few pages of some of the lengthier manuscripts were available electronically. While biographies of Edison are many, I relied on the most recent, Paul Israel, *Edison: A Life of Invention* (New York: John Wiley, 1998); Israel suggested, along the lines of O'Neill's study of Tesla, that Edison's public persona was one of his greatest inventions. The electrical engineer Steinmetz is the subject of a biography that deftly handles both technical matters and cultural history, Ronald R. Kline, *Steinmetz: Engineer and Socialist* (Baltimore: Johns Hopkins University Press, 1992). Following Caroline Marvin's lead, I also browsed through decade-long runs of such electrical trade journals and popular science journals as *Electrical Review, American Electrician, The Electrician, Electrical World,* and Hugo Gernsback's *Science and Invention*. These periodicals are a rich source of material charting the rise of the electrical industry and of "electrical culture" at large.

Hypnotism

Research for the chapters on stage hypnotism, stage magic, and mind reading overlapped, as the same performer might perform hypnotism one week and then mind reading or magic the next; there were, however, plenty of hypnotic specialists. The literature on the history of mesmerism and hypnotism has been steadily growing. A good starting point is Adam Crabtree, *Animal Magnetism, Early Hypnotism, and Psychical Research, 1766–1925: An Annotated Bibliography* (White Plains, N.Y.: Kraus International Publications, 1988). Solid secondary sources include Robert Darnton, *Mesmerism and the End of the Enlightenment in France* (Cambridge: Harvard University Press, 1968), Robert C. Fuller, *Mesmerism and the American Cure of Souls* (Philadelphia: University of Pennsylvania Press, 1982), Alan Gauld, *A History of Hypnosis* (Cambridge: Cambridge University Press, 1992), and Alison Winter, *Mesmerized: Powers of Mind in Victorian England* (Chicago: University of Chicago Press, 1998). Jean-Roch Laurence and Campbell Perry, *Hypnosis, Will and Memory* (New York: Guilford Press, 1988), also provided insight into historical efforts to curb the use of hypnotism. The turn-of-the-century performer Professor Leonidas's *Stage Hypnotism* (Chicago: Bureau of Stage Hypnotism, 1901) can be found in many rare book archives, including the Harry Ransom Humanities Research Center (HRHRC) at the University of Texas, and the Library of Congress—it is a fascinating, book-length work that offers a novelistic impression of a hypnotist on the road plying his trade. The run of Sydney Flower's *Hypnotic Magazine,* which can be found at the Library of Congress and in other collections, offered superb insights into the anti-hypnotism campaign of the Progressive Era. Also extremely helpful to this chapter were Houdini's bound collection of pamphlets on hypnotism at the Library of Congress, and Houdini's scrapbook clippings about his friend, the hypnotist and "bloodless surgeon," Walford Bodie. The HRHRC, which has a great deal of Houdini's correspondence, also includes letters from Bodie to Houdini begging for the original electric chair, which Houdini had acquired, so that Bodie could use it as a stage prop. Houdini eventually relented.

Magic

Until recently, few academics had treated stage magic, but that neglect is fading. While Altick's *Shows of London* led the way, Kenneth Silverman started a more recent revival with *Houdini!!!* (New York: HarperCollins, 1996). More recently, Simon During, in *Modern Enchantments: The Cultural Power of Secular Magic* (Cambridge: Harvard University Press, 2002), has offered a major history of stage magic and its cultural implications. Along with broader arguments about the impact of stage magic on literature, intellectual trends, and consumer culture, During provided a superb micro-history of stage magic and its development in the social setting of London. James Cook's monograph on magic in nineteenth-century America, mentioned earlier in relation to Barnum, is another recent addition to this literature.

Many magicians also have shown a taste for scholarship: Houdini wrote several works and commissioned H. J. Moulton's *Houdini's History of Magic in Boston, 1792–1915* (Glenwood, Ill.: Meyerbooks, 1983), a fine sourcebook that tracks the variety acts that visited that city in the nineteenth century. Milbourne Christopher and Maurine Christopher, *The Illustrated History of Magic* (Portsmouth, N.H.: Heinemann, 1996) is a valuable history, lavishly illustrated, while Ricky Jay, *Learned Pigs & Fireproof Women* (New York: Warner Books, 1986), first led me to the raffish character of Walford Bodie.

In this work, Houdini is treated as a primary actor in the anti-Spiritualist movement. While biographies of Houdini are many, I relied primarily on Raymund Fitzsimons, *Death and the Magician: The Mystery of Houdini* (New York: Atheneum, 1981), and Silverman's text, which discussed Houdini's anti-Spiritualism. John Kasson's *Houdini, Tarzan, and the Perfect Man* (New York: Hill and Wang, 2001) added an analysis of the gender tensions in Houdini's performances. For primary resources on Houdini, I relied on the Houdini collection at the Harry Ransom Humanities Research Center (HRHRC) at the University of Texas at Austin and the Harry Houdini Collection at the Rare Book and Special Collections division at the Library of Congress. More Houdini material can be found at the Houdini Historical Center in Appleton, Wisconsin. The HRHRC collection includes numerous files of clippings, playbills, posters, photographs, and correspondence. The HRHRC also houses the extensive Arthur Conan Doyle collection, which includes, among other materials, a large trove of books on Spiritualism, correspondence that includes letters from Beatrice Houdini to Conan Doyle, letters from Spiritualist Le Roi Crandon to Conan Doyle, and Conan Doyle's annotated copy of Houdini's hatchet job on Spiritualism, *A Magician Among Spirits* (New York: Harper and Brothers, 1924). The Houdini Collection at the Library of Congress is enormous. Numerous volumes of Houdini's scrapbooks there are now in poor condition, and researchers may instead be asked to look at microfilm versions, which also can be ordered via interlibrary loan. The Library of Congress has several other magic collections in addition to its Houdini Collection. Their collections of stage magic equipment catalogues were a valuable aid to my thinking about the history of stage magic.

Perhaps my favorite research experience involved a trip to the Magic Castle, in Hollywood. This institution serves both as a training ground for amateur stage magicians and as a showcase of established talent. Its library, designed as a resource from which magicians can draw ideas for stage acts, has a small but useful collection of historical pamphlets on magic, hypnosis, and mind reading. As with other guests, scholars visiting the Magic Castle are required to say "open sesame" to enter the building. While in the library looking at turn-of-the-century hypnotism pamphlets, I enjoyed watching the librarian on duty demonstrate to other guests various trick shuffles and cuts. The sound of riffling cards and applause from distant theaters created an appealing diversion for a researcher used to hushed rooms and Tesla's fluorescent lights.

Spiritualism and Parapsychology

The literature of Spiritualism is vast. Useful secondary sources include Alan Gauld, *The Founders of Psychical Research* (New York: Schocken Books, 1968), and R. Laurence Moore, *In Search of White Crows* (New

York: Oxford University Press, 1977). The feminist study of Spiritualism has been an important scholarly development. Examples of such work include Ann Braude, *Radical Spirits: Spiritualism & Women's Rights in Nineteenth Century America* (Boston: Beacon Press, 1989), and Alex Owens, *The Darkened Room: Women, Power and Spiritualism in Late Victorian England* (Philadelphia: University of Pennsylvania Press, 1990). Also delving into the connection between Spiritualism and feminism is Mary Gabriel, *Notorious Victoria* (Chapel Hill: Algonquin Books of Chapel Hill, 1998), a biography of Victoria Woodhull, a nineteenth-century Spiritualist, feminist, free love advocate, stockbroker, and presidential candidate.

My look at mind reading interwove variety performance with the history of Spiritualism, psychic research, and parapsychology. Seymour H. Mauskopf and Michael R. McVaugh, *The Elusive Science* (Baltimore: Johns Hopkins University Press, 1980), is an excellent history of psychic research and particularly of Joseph Rhine's role in the founding of parapsychology. Gauld's study of the Society for Psychical Research is also important. I also found useful two anthologies, William James, *Essays in Psychical Research* (Cambridge: Harvard University Press, 1986), and George Devereux, ed., *Psychoanalysis and the Occult* (New York: International Universities Press, 1979 [1953]). The theater collection at the HRHRC includes clipping folders for numerous stage magicians and "mystic vaudevillians." There I found much of my material on the mind readers the Zancigs, S. S. Baldwin, and the celebrated muscle reader Washington Irving Bishop. The Library of Congress also has Houdini's bound volumes of pamphlets about mind reading.

Gee-Whiz Science

Those interested in the history of World's Fairs should begin with Robert Rydell's texts, *All the World's a Fair: Visions of Empire at American International Expositions, 1876–1916* (Chicago: University of Chicago Press, 1984), and *World of Fairs: The Century-of-Progress Expositions* (Chicago: University of Chicago Press, 1993). A handy single volume on World's Fairs is Rydell, John E. Findling, and Kimberly D. Pelle, *Fair America* (Washington, D.C.: Smithsonian Institution Press, 2000). John E. Findling, *Chicago's Great World Fairs* (Manchester: Manchester University Press, 1994), concentrated on the World's Columbian Exposition and the Century of Progress Exposition. Roland Marchand's *Creating the Corporate Soul* (Berkeley: University of California Press, 1998) treated at length the history of World's Fairs, and David E. Nye explored the architecture and electrical displays of World's Fairs in *Electrifying America*. For primary holdings on World's Fairs, the Smithsonian Institution has large reserves. The University of Illinois at Chicago (UIC) houses the papers that the Century of Progress management generated during that 1933–1934 World's Fair; this archive includes extensive collections of newspaper and magazine clippings, pamphlets, management correspondence, press releases, and scrapbooks.

James Gilbert's *Redeeming Culture: American Religion in an Age of Science* (Chicago: University of Chicago Press, 1997) treated the career of the million-volt evangelist Irwin Moon at length while probing the connections between evangelical Christianity, the armed forces, and the larger culture of the cold war. The Moody Bible Institute Library in Chicago has three boxes of material about Irwin Moon and his Sermons from Science. The folders, haphazard in arrangement, consist mainly of clippings of reviews of his lectures, press releases, publicity photographs, and internal documents meant for Moody officials and for other Sermons from Science performers. Roland Marchand's *Creating the Corporate Soul* has an excellent account of General Motors' Parade of Progress of the 1930s. Biographies of General Motors' Charles "Boss" Kettering include Stuart Leslie, *Boss Kettering* (New York: Columbia University Press, 1983). One archive I did not get the chance to visit is the GMI Institute in Flint, Michigan, which has documentation of GM's Parade of Progress in its Charles F. Kettering Papers.

UFOs and the Occult

The literature of UFOs is a thicket that only the stalwart should enter. George M. Eberhart has provided an excellent two-volume bibliography, *UFOs and the Extraterrestrial Contact Movement* (Metuchen, N.J.: Scarecrow Press, 1986). Another serviceable bibliography of earlier literature is Lynn Catoe, *UFO's and Related Subjects: An Annotated Bibliography* (Washington, D.C.: United States Government Printing Office, 1969). The academic literature of UFO culture is slowly growing. Leon Festinger, Henry Riecken, and Stanley Schachter, in *When Prophecy Fails* (New York: Harper Torchbooks, 1964 [1956]), treated the sociology of one 1950s flying saucer cult. David M. Jacobs, *The Flying Saucer Controversy in America* (Bloomington: Indiana University Press, 1975), provided the first academic history of this subject. Another history of UFOs is Curtis Peebles, *Watch the Skies! A Chronicle of the Flying Saucer Myth* (Washington, D.C.: Smithsonian Institution Press, 1994). Walter McDougall provided an excellent history of the United States' space program in *The Heavens and the Earth: A Political History of the Space Age* (New York: Basic Books, 1985). Jodi Dean, *Aliens in America* (Ithaca: Cornell University Press, 1998), treated the cultural and political ramifications of interest in UFOs, aliens, abductions, and conspiracy theory. An earlier entry, William Sims Bainbridge, *The Spaceflight Revolution: a Sociological Study* (New York: John Wiley, 1976), is an important though haphazard study of the pop religion underpinnings of the rocketry movement.

The flying saucer cults emerged from America's occult underground, much of which can be traced back to 1875, the year that Helena Blavatsky, formerly a Spiritualist medium, founded the Theosophical Society, recruiting many prominent citizens as members, including Thomas Edison. For a discussion of the Theosophy movement that avoids the impenetrable jargon that members rely on in their own accounts, see Bruce F. Campbell, *Ancient Wisdom Revived: A History of the Theosophical Movement* (Berkeley: University of California Press, 1980). Also somewhat useful is James Webb, *The Occult Underground* (LaSalle, Ill.: Open Court, 1974). For a more recent scholarly study of the occult, see Dan Burton and David Grandy, *Magic, Mystery, and Science: The Occult in Western Civilization* (Bloomington: Indiana University Press, 2004). The authors insist that "magic" be placed alongside "science" and "religion" as a third major stream in human culture and thought. They offer a solid survey of the history of the occult from ancient Egypt though the New Age movement of the twentieth and twenty-first centuries.

Carl Jung was one of the first to put a religious spin on UFO culture in *Flying Saucers: A Modern Myth of Things Seen in the Skies* (Princeton, N.J.: Princeton University Press, 1978 [1959]). Two more recent examples are James R. Lewis, ed., *The Gods Have Landed: New Religions from Other Worlds* (Albany: State University of New York Press, 1995), and Brenda Denzler, *The Lure of the Edge: Scientific Passions, Religious Beliefs, and the Pursuit of UFOs* (Berkeley: University of California Press, 2001). Encyclopedias such as Margaret Sachs, *The UFO Encyclopedia* (New York: G. P. Putnam's Sons, 1980), and Ron Story, ed., *Encyclopedia of Ufos* (Garden City, N.Y.: Doubleday, 1980), are useful for exploring the literature of UFO enthusiasts. Of the 1950s contactee narratives, I would second Jung's vote that the most compelling is Orfeo Angelucci's *The Secret of the Saucers* (Stevens Point, Wis.: Amherst Press, 1955). Margaret Storm's curiosity, *The Return of the Dove* (Baltimore: privately printed, 1959), helped me connect the UFO cults of the 1950s to earlier occultist strands in American culture. This book also shed some light on why Tesla has become such a cult figure for these communities, as well as for the alternative rock community, and right-wing and libertarian groups.

Skeptics All

While most skeptic authors abhor pseudoscience, those of good humor, such as Martin Gardner, at times display an aficionado's appreciation for the nuances of such theories. A good entry point to the skeptic literature is Gardner's *Fads and Fallacies in the Name of Science* (New York: Dover Publications, 1957). Other entries include Carl Sagan, *Demon-Haunted World* (New York: Random House, 1995), Michael Shermer, *Why People Believe in Weird Things: Pseudoscience, Superstition, and Other Confusions of Our Time* (New York: W. H. Freeman, 1997), and Robert L. Park, *Voodoo Science* (Oxford: Oxford University Press, 2000). Numerous periodicals are also dedicated to the skeptical stance, including the *Skeptical Enquirer* and *Skeptic*. The Amazing Randi keeps his followers informed of his debunking activities on his website.

Efforts to "debunk" or parody science are harder to come by. The journal *Annals of Improbable Research* (AIR) offers some amusing send-ups of scientific research, but it is an insider critique that does not stray philosophically beyond the bounds of the scientific orthodoxy whose members are its primary audience. AIR hosts an annual Ig Nobel Awards ceremony in Cambridge, Massachusetts, honoring ridiculous-sounding research. Prize winners often demonstrate their award-winning work: the 2000 physics prize, for example, went to Sir Michael Berry and Andre Geim, "for using magnets to levitate a frog."

For those who prefer an "outsider" view of science, I recommend browsing Charles Fort, *The Books of Charles Fort* (New York: Henry Holt, 1941). A friend of Theodore Dreiser, Fort was a journalist who spent decades collecting clippings of strange phenomena such as showers of frogs or rains of slime and relied on these puzzling incidents to illustrate his theory that the ruling forces of the universe are capricious and far from interested in or bound to the latest scientific theories. Fort popularized many themes of the paranormal that later made their way into science fiction. Louis Kaplan's quasi-academic treatise, *The Damned Universe of Charles Fort* (Brooklyn: Autonomedia, 1993), written in the comic style of Fort, includes an essay that pays tribute to the master along with choice excerpts from Fort that illustrate his ruling ideas. The book is quite useful, as Fort's books, though full of insights and wit, quickly get tiresome. Another unabashed celebration of crackpots is Margaret

Nicholas, *The World's Greatest Cranks and Crackpots* (New York: Exeter Books, 1984 [1982]). The *Fortean Times* is a British periodical dedicated to a Fortean point of view, but it generally jettisons the "master's" wit and philosophical subtleties for a "Believe It or Not" approach, in a marketing sleight of hand that makes it far less entertaining than the title, cover, and cover price would lead the hopeful to believe.

NOTES

Preface

1. Alexis de Tocqueville, *Democracy in America* (New York: Doubleday, 1969), 404.
2. Jon Butler argued that interest in the occult and supernatural has long been a component of the American religious experience in *Awash in a Sea of Faith: Christianizing the American People* (Cambridge: Harvard University Press, 1990).

Introduction: Beyond the Z-Ray

Epigraphs: L. Frank Baum, *The Wonderful Wizard of Oz* (New York: Penguin Books, 1984 [1900]), 156; poster for Goodland, Kansas Fair, 1891, Kansas State Historical Society. See http://www.kshs.org/cool2/graphics (accessed January 20, 2003).

1. For a full discussion of Spring Garden, formerly Cox's Museum, and the entertainment milieu from which it emerged, see Richard Altick, *The Shows of London* (Cambridge: Harvard University Press, 1978), 69–76. Katterfelto is discussed in Altick and in Simon During, *Modern Enchantments: The Cultural Power of Secular Magic* (Cambridge: Harvard University Press, 2002), 247–252.
2. Clipping, Magicians Collection, b.32 "Katterfelto," Harry Ransom Humanities Resource Center (HRHRC), Austin, Texas.
3. Clipping, "Katterfelto," HRHRC. E.O. refers to "even-odd" and is similar to more modern variants such as Faro and Keno.
4. Cowper, "The Task," *Universal Magazine,* 1792, 420.
5. Clipping, Magicians Collection, b.32, "Katterfelto," HRHRC.
6. Clipping, "The Mirror," 1831, HRHRC.
7. During outlines this pedigree in *Modern Enchantments.*
8. During presents the stage magician as a secularized version of the natural magician and argues that the magician becomes an emblematic figure of modernity, the peddler of illusions who prefigures the media-saturated modern and postmodern worlds. In contrast, the more marginal performers that this book focuses on, performers who took on a less successful stage pose, are intriguing historically for suggesting tensions with modernity and for more clearly identifying themselves with genuine magic. This present study, to my mind, does not refute but complements During's study.
9. Stephen Greenblatt, "Resonance and Wonder," in *Exhibiting Cultures,* ed. Ivan

Karp and Steven D. Lavine (Washington D.C.: Smithsonian Institution Press, 1991), 42–56.

10. For a superb discussion of wonder and its historical development, see Lorraine Daston and Katharine Park, *Wonders and the Order of Nature* (New York: Zone Books, 1998).
11. Daston and Park, *Wonders,* 220–231.
12. Daston and Park, *Wonders,* 9–20.
13. Greenblatt, in "Resonance and Wonder," discusses how museum curators choose between display strategies that highlight the "resonance" or historic aura of objects and "wonder," which creates an immediate, ahistorical response to the displayed object.
14. For a discussion of the technological sublime, see Leo Marx, *The Machine in the Garden* (New York: Oxford University Press, 2000 [1964]), and David E. Nye's more recent *American Technological Sublime* (Cambridge: MIT Press, 1994). While Nye deftly traces the technological sublime back to the aesthetic sublime, he largely overlooks its connection to the fascination with wonder that originally aided the scientific revolution.
15. The argument that the technological sublime as seen at world's fairs has a hegemonic function is developed in Robert Rydell, *All the World's a Fair: Visions of Empire at American International Expositions, 1876–1916* (Chicago: University of Chicago Press, 1984); for the argument that the technological sublime has a democratic binding effect on American society, see Nye, *Technological Sublime.*
16. Henry Houdini and H. J. Moulton, *Houdini's History of Magic in Boston, 1792–1915* (Glenwood, Ill.: Meyerbooks, 1983), 11.
17. One earlier exception was the evening "philosophical" demonstration at Peale's Museum in Philadelphia of the 1820s with its many dazzling electrical effects including lightning flashes, dancing dolls, and a collapsing "thunder house." See Charles Coleman Sellers, *Charles Willson Peale,* vol. 2, *Later Life* (Philadelphia: American Philosophical Society, 1947), 354. As an even earlier precursor, Benjamin Franklin gave electrical entertainments much like those at Peale's Museum and hosted several electrical picnics on the Schuylkill River that featured electrocuted turkeys as the main course. Edward T. Canby, *History of Electricity* (New York: Hawthorn Books, 1963), 38.
18. See Robert Abzug, *Cosmos Crumbling* (New York: Oxford University Press, 1994), and Donald B. Meyer, *The Positive Thinkers: A Study of the American Quest for Health, Wealth and Personal Power from Mary Baker Eddy to Norman Vincent Peale* (Garden City, N.Y.: Doubleday, 1965).
19. See Herbert Hovenkamp, *Science and Religion in America: 1800–1860* (Philadelphia: University of Pennsylvania Press, 1978), 119–146.
20. For a description of the use of "challenges" and "imitations" in vaudeville, see Susan Glenn, "Give an Imitation of Me," *American Quarterly* 50, no. 1 (1998): 47–76.
21. For historical explorations of this penchant for realism and science in nineteenth-century America, see David Shi, *Facing Facts: Realism in American Thought and Culture* (New York: Oxford University Press, 1994), and Cecilia Tichi's *Shifting Gears* (Chapel Hill: University of North Carolina Press, 1987).
22. For an examination of the many "escapes," geographical, philosophical, and otherwise, that the middle and upper classes sought at the century's turn,

see T. J. Jackson Lears, *No Place of Grace* (Chicago: University of Chicago Press, 1994). Historians have pointed out that the taste for the exotic at the turn of the century neatly accompanied the rise of consumerism and the need to free the public from the straitjacket of Victorian morals to encourage spending. See, for example, William Leach, *Land of Desire* (New York: Pantheon Books, 1993).

23. See Charles Fort, *The Books of Charles Fort* (New York: Henry Holt, 1941). See also Louis Kaplan, *The Damned Universe of Charles Fort* (Brooklyn: Autonomedia, 1993).
24. An excellent early discussion of this trend can be found in Frederick J. Hoffman, *The Twenties: American Writing in the Postwar Decade* (New York: Viking Press, 1955 [1949]).
25. Arthur Conan Doyle, *The Professor Challenger Stories; Land of Mist* (London: John Murray, 1958), 340.
26. Cynthia Eagle Russett, *Darwin in America* (San Francisco: W. H. Freeman, 1976).
27. The changing public and scientific conception of evolution and its causes is sketched out well in Edward Larson's *Summer for the Gods* (Cambridge: Harvard University Press, 1997), 16–30.
28. Peter L. Berger, *The Sacred Canopy: Elements of a Sociological Theory of Religion* (New York: Doubleday, 1990), 25–28.
29. Millikan wrote prolifically on this subject. One example is his leaflet, *A Scientist Confesses His Faith* (Chicago: American Institute of Sacred Literature, 1927).
30. This thinking had its origins in the abbeys of medieval Europe, when scholars such as Roger Bacon equated advance in the mechanical arts with the restoration of humanity to the former glory of its "Adamic" powers. See David F. Noble, *The Religion of Technology: The Divinity of Man and the Spirit of Invention* (New York: Penguin, 1997), 21–34.
31. This book will not neatly distinguish magic from religion, since in the popular realm many hybrids of the two have emerged. John Butler's *Awash* argued that religious eclecticism, which includes mainstream Protestantism as well as folk beliefs and magic, has always been a component of American religious experience.
32. For an extended look at the X-ray fad of the turn of the century and its impact on modernist art, see Linda Dalrymple Henderson, "X Rays and the Quest for Invisible Reality in the Art of Kupka, Duchamp, and the Cubists," *Art Journal* 47, no. 4 (winter 1988): 323–340.
33. "The National Electrical Exhibit," *American Electrician*, May 1896, 4.
34. The producers achieved this optical effect by placing a plate of clear glass on the stage at a forty-five-degree angle to the subject. Lights dissolved on the subject while another light brightened on a painting of a skeleton on the wings of the stage that faced toward the plate glass. Viewers superimposed the skeleton's projected image over that of the upright subject. The lighting effects were reversed to return a vision of the subject in the shroud. See Albert Hopkins, *Magic: Stage Illusions and Scientific Diversions Including Trick Photography* (New York: Munn & Co., 1911), 55–58.
35. In "X Rays," Henderson made this point as well, connecting the X-ray fad of the turn of the century with occultism and the cubist movement in the art world.

Psychic researchers frequently insisted that the paranormal became more "normal" when placed along a continuum of strange events ranging from prophetic dreams to telepathy to ghost sightings and spirit communications.

36. Charles-Albert Reichen, *A History of Physics* (New York: Hawthorn Books, 1963), 78.
37. "Z-Rays," *American Electrician,* December 1896, 276.
38. R. W. Wood wrote of his exposure of Blondlot in "The n-Rays," *Nature,* September 29, 1904, 530–531.
39. Ray Whitcomb, "The Ultimate Ray," *Science and Invention,* August 1920, 449.
40. Clement Fezandié, "The Secret of the Sixth Sense," *Science and Invention,* April 1923, 1169. His prose bears a remarkable resemblance to the film scripts of later schlock auteur Edward D. Wood.
41. For the antebellum period as a "golden age" of popular science, see John C. Burnham, *How Superstition Won and Science Lost* (New Brunswick, N.J.: Rutgers University Press, 1987), and Roland Marchand, *Creating the Corporate Soul* (Berkeley: University of California Press, 1998), 249–311.

Chapter One: The Electric Wonder Show

1. These items were listed in William C. Richards, *A Day in the New York Crystal Palace* (New York: G. P. Putnam, 1853).
2. "The Crystal Palace—Opening of the Exhibition," *New York Times,* July 15, 1853, 1.
3. Ibid.
4. "Popularizing Science," *Nation* 4, no. 80 (January 10, 1867): 32.
5. Iwan Morus, *Frankenstein's Children: Electricity, Exhibition, and Experiment in Early Nineteenth-Century London* (Princeton, N.J.: Princeton University Press, 1998), xi, 10–12. Also see Simon Schaffer, "Natural Philosophy and Public Spectacle in the Eighteenth Century," *History of Science* 21 (1983): 1–43; and Bruno Latour, "Give Me a Laboratory and I Will Raise the World," in *Science Observed: Perspectives on the Social Study of Science,* ed. Karina D. Knorr-Cetina and Michael Mulkay (London: Sage Publications, 1983), 141–170. Drawing from his research on Louis Pasteur, Latour concluded that "science is politics pursued by other means" (168).
6. "Popularizing Science," 32.
7. Ellen Hickey Grayson, "Social Order and Psychological Disorder: Laughing Gas Demonstrations, 1800–1850," in *Freakery: Cultural Spectacles of the Extraordinary Body,* ed. Rosemarie Garland Thomson (New York: New York University Press, 1996), 108–120.
8. Morus, *Frankenstein,* 29.
9. Ibid., 70–98.
10. For a history of Peale's many enterprises, see Charles Coleman Sellers, *Mr. Peale's Museum* (New York: W. W. Norton, 1980).
11. Sellers, *Mr. Peale's Museum,* 243–244.
12. Both Sellers's book and David Brigham's later study of Peale have stressed the implicit moralizing to be found in Peale's displays with their emphasis on a stable social hierarchy. Peale thought of his museum as a "temple" and citadel of wisdom. See David Brigham, *Public Culture in the Early Republic: Peale's Museum and its Audience* (Washington, D.C.: Smithsonian Institution Press, 1995). In the 1830s and 1840s, rising rents, competition, and changing public tastes

led Rubens and Titian Peale to allow musical concerts, novelty acts, and even "freaks" such as a black albino to be displayed. With Peale's Museum as a model, it would appear that the republican education model had already fused with the gee-whiz entertainment mode in the Jacksonian era, foreshadowing the declension many historians have saved for the twentieth century.

13. For a discussion of Barnum's transformation from raffish confidence man to middle-class exemplar, see Bluford Adams, *E Pluribus Barnum* (Minneapolis: University of Minnesota Press, 1997).
14. "Our Crystal Palace," *Putnam's* 2, no. 8 (August 1853): 127.
15. Roger Sherman, "Charles Came, Itinerant Science Lecturer and His 'Splendid Apparatus.'" *Rittenhouse* 5, no. 4 (August 1991): 119–120.
16. As quoted in John C. Greene, "Protestantism, Science and American Enterprise: Benjamin Silliman," in *Benjamin Silliman and His Circle,* ed. Leonard G. Wilson (New York: Science History Publishing, 1979), 22.
17. Quoted in Greene, "Protestantism," 23.
18. Much of my information on Lardner I have drawn from Elizabeth P. Stewart's "Diffusion Is the Watchword," National Museum of American History Colloquium (NMAH), June 14, 2000.
19. In addition to my own research at the Charles Came Collection at the National Museum of American History, I have also relied on the following articles, collected by Roger Sherman of the NMAH about Came: Edmund S. Carpenter, "The Strange Case of Dr. Came and the Sleeping Man," *New York Folklore Quarterly* 5, no. 4 (winter 1949): 241–256; Joan Lynn Schild, "Dr. Came, The Lightning Man," *Rochester Historical Society Scrapbook* 1 (1950): 9–24; Leatrice M. Kemp, "Dr. Charles Came, M.D.: The Great Electrician and Successful Operator" (unpublished), December 1987; Roger Sherman, "Charles Came, Itinerant Science Lecturer, and his 'Splendid Apparatus,'" *Rittenhouse* 5, no. 4 (August 1991): 118–128; and Elizabeth Stewart, "Diffusion."
20. Charles Came to Cynthia Came, April 20, 1845, Charles Came Collection, NMAH.
21. Kemp, "Dr. Charles Came," 15.
22. Handbill, Item .513, Charles Came Collection, NMAH.
23. See Grayson, "Social Order," in *Freakery,* 108–120.
24. Item .562, Charles Came Collection, NMAH.
25. Item .563, NMAH.
26. Item 1992.3092.566, NMAH.
27. Handbill, Item .514, Charles Came Collection, NMAH.
28. Morus, *Frankenstein,* 235–236.
29. Samuel B. Smith, *The Scientific Examiner . . . Electro-Magnetism Radically Applied,* 1849, 1 (pamphlet), NMAH.
30. These advertisements are from the *New York Sun,* July 30, 1844. Came had just begun his career as showman and physician in upstate New York at this time.
31. Rogan Taylor, *The Death and Resurrection Show* (London: Anthony Blond, 1985).
32. Throughout his book, Taylor argued that all modern entertainment descended from the shaman's ritual. In so doing, he followed up Mircea Eliade's concluding sentence to his groundbreaking work on shamanism: "What a magnificent book remains to be written on the ecstatic 'sources' of epic and lyric poetry, on the prehistory of dramatic spectacles, and, in general, on the fabulous worlds discovered, explored, and described by the ancient shamans." See Mircea Eli-

ade, *Shamanism, Archaic Techniques of Ecstasy* (New York: Pantheon Books, 1964), 511.

33. Brooks McNamara, *Step Right Up* (Jackson: University Press of Mississippi, 1995 [1976]), 7.
34. Charles Came to Cynthia Came, July 9, 1850, Charles Came Collection, NMAH.
35. Charles Came to Cynthia Came, July 19, 1850, Charles Came Collection, NMAH.
36. Charles Came to Cynthia Came, December 6, 1852, Charles Came Collection, NMAH.
37. Charles Came to Cynthia Came, July 16, 1853, Charles Came Collection, NMAH.
38. Item .566, Charles Came Collection, NMAH.
39. Letter reproduced in Carpenter, "Strange Case," 255.
40. Charles Came to Cynthia Came, April 20, 1845, Charles Came Collection, NMAH.
41. Item .525, Charles Came Collection, NMAH.
42. Charles Came Folder, Charles Came Collection, NMAH.
43. Charles Came to Cynthia Came, March 7, 1850, Charles Came Collection, NMAH.
44. Charles Came to Cynthia Came, June 26, 1853, Charles Came Collection, NMAH.
45. Ibid.
46. Charles Came to Cynthia Came, July 24, 1853, Charles Came Collection, NMAH.
47. Charles Came to Cynthia Came, July 23, 1853, Charles Came Collection, NMAH.
48. Charles Came to Cynthia Came, July 24, 1853, Charles Came Collection, NMAH.
49. Charles Came to Cynthia Came, August 12, 1853, Charles Came Collection, NMAH.
50. Charles Came to Cynthia Came, August 8, 1853, Charles Came Collection, NMAH.
51. Charles Came to Cynthia Came, August 12, 1853, Charles Came Collection, NMAH.
52. Advertisement, *New York Herald,* September 17, 1853, 5:6. An annotation of this clipping, procured by Elizabeth Stewart, is in the files of Roger Sherman at the NMAH.
53. "Phrenological Character of Mr. Cornelius Vrooman Given at Fowler and Wells Phrenological Cabinet Clinton Hall," September 8, 1853, Item .569, Charles Came Collection, NMAH.
54. Charles Came to Cynthia Came, August 20, 1853, Charles Came Collection, NMAH.
55. "A Curious Case: The Man Who Has Slept Five Years," *New York Daily Times,* September 9, 1853, 2:5.
56. "Disgusting Exhibitions," *New York Daily Tribune,* September 22, 1853, 4:4.
57. Charles Came to Cynthia Came, September 29, 1853, Charles Came Collection, NMAH.
58. Ibid.
59. Charles Came to Cynthia Came, October 4, 1853, Charles Came Collection, NMAH.

60. Charles Came to Cynthia Came, October 11, 1853, Charles Came Collection, NMAH.
61. "Funeral of Dr. Charles Came age 75," *Rochester Daily Union and Advertiser*, November 7, 1881, transcription, n.p., Roger Sherman's file on Charles Came, NMAH.
62. Historians such as John C. Burnham in *How Superstition* and Roland Marchand in *Creating the Corporate Soul* have analyzed science popularizations to show how efforts in public education slowly faded from the republican ideals of the nineteenth century to gee-whiz displays in the early twentieth century.
63. Stewart also delineates the "complicated" intertwining of education and entertainment in nineteenth-century science demonstrations.
64. See James W. Cook, *The Arts of Deception: Playing with Fraud in the Age of Barnum* (Cambridge: Harvard University Press, 2001).

Chapter Two: The Techno-Wizard

1. From William J. Hammer, *Electrical Diablerie*, promotional pamphlet, Hammer Collection, NMAH.
2. See Burnham, *How Superstition*, 31–44.
3. Bruce J. Hunt, "Lines of Force, Swirls of Ether," in *From Energy to Information: Representation in Science and Technology, Art, and Literature*, ed. Bruce Clarke and Linda Dalrymple Henderson (Stanford: Stanford University Press, 2002), 99–113.
4. David Nye, *Electrifying America* (Cambridge: MIT Press, 1990), 161.
5. "Electrical Wonder Working," *Electrical World* 26, no. 6 (August 9, 1890): 81.
6. "Abuse of the Word Electric," *Electrical Review*, August 29, 1891, 4.
7. This theme bases Caroline Marvin's study of electrical journals, *When Old Technologies Were New* (New York: Oxford University Press, 1988).
8. "Sparks," *The Electrician* 2, no. 12 (December 1883): 389.
9. "The National Electrical Exhibit," *American Electrician* 8, no. 1 (May 1896): 2.
10. "Sparks," *The Electrician* 2, no. 3 (March 1883): 83.
11. Marvin, *When Old*, 137.
12. "An Electric Belle," *Electrical Review* 20, no. 15 (June 4, 1892): 192.
13. "Sparks," *The Electrician* 2, no. 8 (August 1883).
14. *Electrical Review* 19, no.7 (October 10, 1891): 1.
15. See Nye, *Electrifying*, 29–47, for a discussion of electricity at world's fairs.
16. "Cranks," *Electrical Review*, January 17, 1891, 251. This concept is reminiscent of the harrowing conclusion of Mark Twain's *A Connecticut Yankee in King Arthur's Court* (1889) when Hank Morgan, aka "The Boss," wires his encampment with electricity to slaughter hundreds of attacking knights aligned with his enemies Merlin and the church.
17. "R. L. Garner Will Introduce the Telephone and Phonograph Among the Monkeys of the African Forest," *Electrical World*, July 23, 1892, 53.
18. "Execution by Means of Electricity," *The Electrician* 2, no. 4 (September 1883): 283.
19. "Execution by Electricity," *Electrical World*, January 21, 1888, 25.
20. Thomas P. Hughes, "Harold P. Brown and the Executioner's Current," *Business History Review* 32, no. 2 (summer 1958): 143–165.
21. "Died for Science's Sake," *New York Times*, July 31, 1888, 8.
22. "Testimony of the Wizard," *New York Times*, July 24, 1889.

23. Thomas P. Hughes, *Networks of Power* (Baltimore: Johns Hopkins University Press, 1983), 108.
24. *Electrical Review,* February 20, 1892, 354.
25. "Alternating Currents of High Frequency," *Electrical Review,* May 30, 1891, 185.
26. Ibid.
27. "Tesla at Royal Institution," *Electrical Review* (London), February 12, 1892, 192.
28. "Brilliant Experiments," *Electrical Review,* June 4, 1892, 193.
29. "Nikola Tesla and His Wonderful Discoveries," *Electrical World* 21, no. 17 (April 29, 1893): 323.
30. *New York Sun,* July 22, 1894, 1.
31. "The New Wizard of the West," *Pearson's Magazine,* May 1899, 470–476.
32. John P. Barrett, *Electricity at the World's Columbian Exposition* (Chicago: R. R. Donnelly & Sons, 1894), 168.
33. *Photographs of the World's Fair* (Chicago: Werner Company, 1894), 25.
34. Barrett, *Electricity,* xi.
35. "Opening of the World's Fair," *Electrical World* 21, no. 12 (May 13, 1893): 364.
36. Henry Adams, *The Education of Henry Adams* (New York: Modern Library, 1931), 342.
37. Ibid., 343.
38. Ibid., 388.
39. Paul Israel, *Edison: A Life of Invention* (New York: John Wiley, 1998), 365–368.
40. William James, *The Varieties of Religious Experience* (New York: Modern Library, 1929), 245.
41. Nikola Tesla, "Making Your Imagination Work For You," *American Magazine,* April 1921, 62.
42. Ibid.
43. "Nikola Tesla Receives Edison Medal," *Electrical Review and Western Electrician,* May 26, 1917, 881.
44. Tesla, and Ben Johnston, ed., *My Inventions: The Autobiography of Nikola Tesla* (Williston, Vt.: Hart Brothers, 1982), 59–60.
45. Chauncey Montgomery McGovern, "The New Wizard of the West," *Pearson's Magazine,* May 1899, 470.
46. Ibid., 471.
47. *New York World,* July 22, 1894, 1.
48. Tesla to Westinghouse Electric and Manufacturing Company, February 24, 1899, Swezey Papers, NMAH.
49. Nikola Tesla, "The Problem of Increasing Human Energy," *Century Magazine,* June 1902, 175–178.
50. Ibid., 188–189.
51. Ibid., 184.
52. Pamphlet quoted in John J. O'Neill, *Prodigal Genius: The Life Story of Nikola Tesla* (New York: Ives Washburn, 1944), 255.
53. As quoted in Marc Seifer, *Wizard: The Life and Times of Nikola Tesla* (Secaucus, N.J.: Carol Publishing Group, 1996), 166.
54. O'Neill argued that Tesla envisioned the Long Island site as the base for his wireless communications system, but planned to base his power system at Niagara Falls. See *Prodigal Genius,* 252.
55. As quoted in Margaret Cheney, *Tesla: Master of Lightning* (New York: Barnes and Noble, 1999), 100.

56. Cheney, *Tesla,* 107.
57. David H. Childress and Nikola Tesla, *The Fantastic Inventions of Nikola Tesla* (Stella, Ill.: Adventures Unlimited, 1993), 247.
58. "Marconi's Plan for the World," *Technical World Magazine,* October 1912, 150.
59. Edward R. Hewitt to Kenneth Swezey, May 9, 1956, Swezey Papers, NMAH.
60. Ronald R. Kline, *Steinmetz: Engineer and Socialist* (Baltimore: Johns Hopkins University Press, 1992), 101.
61. In his *Theory and Calculation of Alternating Current Phenomena* (1897) and *Theoretical Elements of Electrical Engineering* (1900).
62. Arthur Goodrich, "Charles P. Steinmetz, Electrician," *World's Work,* June 1904, 4867–4869.
63. "Noted Expert, Dr. C. P. Steinmetz, Talks of the Future Wonders of Scientific Discovery and Ridicules Many Prophecies," *New York Times,* November 12, 1911, IV, 4.
64. Charles P. Steinmetz, "The Bolshevists Won't Get You—" *American Magazine,* April 1919, 11.
65. Charles P. Steinmetz, "Industrial Efficiency and Political Waste," *Harper's Monthly,* November, 1916, 928.
66. Mary Vanderpoel Hun, "Steinmetz," *The Forum,* February 1924, 235.
67. Kline, *Steinmetz,* 281.
68. "Modern Jove Hurls Lightning at Will—Million Horse Power Forked Tongues Crackle and Flash in Laboratory," *New York Times,* March 3, 1922, 1.
69. Kline, *Steinmetz,* 266–268.
70. Ibid., 283.
71. Edward Larson, *Summer for the Gods* (Cambridge: Harvard University Press, 1997).
72. Robert Millikan, *A Scientist Confesses His Faith* (pamphlet) (Chicago: American Institute of Sacred Literature, 1927), 5.
73. Ibid., 7.
74. Ibid.
75. Ibid., 26.
76. Ibid., 27.
77. "Sources of Power," *New York Times,* November 8, 1922, 14.
78. Ibid.
79. Charles P. Steinmetz, "Science and Religion," *Harper's Magazine,* February 1922, 296.
80. Ibid., 297.
81. "Steinmetz," *The Forum* 71 (May 1924): 690.
82. "A Hunchback Who Played with Thunderbolts," *Literary Digest* 79 (November 17, 1923).
83. John Dos Passos, *USA* (New York: Modern Library, 1937), 327–328.
84. Roland Marchand and Michael L. Smith, "Corporate Science on Display," in *Scientific Authority & Twentieth Century America,* ed. Ronald G. Waters (Baltimore: Johns Hopkins University Press, 1997), 161.
85. "Edison's Views on Immortality Criticized," *Current Literature,* December 1910, 644.
86. B. C. Forbes, "Edison Working on How to Communicate with the Next World," *American Magazine* 90, no. 4 (October 1920): 11.
87. "Edison's Religion," *Literary Digest,* November 7, 1931, 19.

88. Austin Lescarboura, "Edison's Views on Life and Death," *Scientific American* 123, no. 8 (October 30, 1920): 459–460.
89. Forbes, "Edison Working," 11.
90. Steinmetz, "Science and Religion," 302.

Chapter Three: The Hypnotist

1. See Robert Fuller, *Mesmerism and the American Cure of Souls* (Philadelphia: University of Pennsylvania Press, 1982); Alan Gauld, *A History of Hypnosis* (Cambridge: Cambridge University Press, 1992); Alison Winter, *Mesmerized: Powers of Mind in Victorian England* (Chicago: University of Chicago Press, 1998).
2. George W. Alden, *An Essay on Human Magnetism: and Its General Views and Principles, as it Remains in its Embryo* (Columbus, Ohio: privately published, 1846), 7. From Harry Houdini, *Mind Reading Pamphlets,* vol. 3, Rare Book and Special Collections, Library of Congress.
3. Winter, *Mesmerized,* 53–54.
4. Robert Darnton, *Mesmerism and the End of the Enlightenment in France* (Cambridge: Harvard University Press, 1968), 58.
5. Gauld, *History,* 185.
6. Fuller, *Mesmerism,* 21.
7. John B. Dods, *The Philosophy of Electro Biology* (New York: Da Capo Press, 1982 [1850]), 17.
8. David Reese, *Humbugs of New-York* (reprint) (Freeport, N.Y.: Books for Libraries Press, 1971 [1838]), 49.
9. Timothy Shay, *Agnes; or, The Possessed. A Revelation of Mesmerism.* (Philadelphia: T. B. Peterson, 1848), 3.
10. Shay, *Agnes,* 85.
11. Shay, *Agnes,* 103.
12. Winter, *Mesmerized,* 60–78.
13. The divide between "mesmerism" and "hypnotism," though somewhat murky, is usually traced to the studies of British physician James Braid in the mid–nineteenth century; Braid and others rejected the assumption that the mesmerist was transmitting a subtle fluid to the patient; he preferred the name hypnotism and the assumption that a state of mind was being encouraged.
14. Kennedy Brothers, *Handbook on Mesmerism and Hypnotism* (New York: Benedict Publishing, 1883), 24.
15. William James, *The Principles of Psychology,* vol. 2 (New York: Dover Publications, 1950 [1890]), 595.
16. Kenneth Silverman, *Houdini!!!* (New York: HarperCollins, 1996), 16.
17. G. Odell, *Annals of the New York Stage,* vol. 14 (New York: Columbia, 1945), 674.
18. *Amusement Bulletin* (Philadelphia), October 31, 1892.
19. Robert Bogdan, *Freak Show: Presenting Human Oddities for Amusement and Profit* (Chicago: University of Chicago Press, 1988), 104–111.
20. X. Sage, and T. Adkin, *Scenes in Hypnotism and How To Produce Them* (Rochester: New York State Publishing, 1900), not paged. Magic Castle library, Los Angeles.
21. Regarding Mikhael Bakhtin's discussion of the "carnivalesque" and the festive grotesque style, stage hypnotists evoked displays of what Bakhtin might categorize as the later and less salutary "gothic" or romantic grotesque style. See Bakhtin, *Rabelais and His World* (Cambridge: MIT Press, 1968), 37.

22. Nathaniel Hawthorne, *Nathaniel Hawthorne Novels* (New York: Library of America, 1983 [1852]), 635.
23. L. A. Harraden, *Hypnotic Exhibitions* (Jackson, Mich.: privately published, 1900), 21–22.
24. Joe Laurie, *Vaudeville from the Honky Tonks to the Palace* (New York: Holt, 1953), 110–111.
25. Playbill, "The Great Newmann," Library of Congress, Rare Book and Special Collections.
26. Ricky Jay, *Learned Pigs & Fireproof Women* (New York: Warner Books, 1986), 137.
27. Miles Orvell documented this fascination with imitation and exposure in *The Real Thing* (Chapel Hill: University of North Carolina Press, 1989).
28. "Bodic Force" punned on the "Odic Force" that Karl, Baron von Reichenbach argued could explain mesmerism and other paranormal abilities.
29. Walford Bodie, *The Bodie Book* (London: Caxton Press, 1905).
30. Marc J. Seifer, *Wizard: The Life and Times of Nikola Tesla* (Secaucus, N.J.: Carol Publishing Group, 1996), 113.
31. Victor Turner, *From Ritual to Theater* (New York: Performing Arts Journal Publications, 1982), 20–59.
32. Walford Bodie, *The Bodie Book* (London: Caxton Press, 1905), 35.
33. See Silverman, *Houdini!!!*, and John Kasson, *Houdini, Tarzan and the Perfect Man* (New York: Hill and Wang, 2001).
34. Seifer, *Wizard*, 91.
35. For a study of the cultural implications of these electrical health devices, see Carolyn Thomas de la Peña, *The Body Electric: How Strange Machines Built the Modern Age* (New York: New York University Press, 2003).
36. Houdini and Bodie had met in Europe in 1909 when both were successful performers. In 1920, their fortunes had changed. In a series of letters to Houdini, the down-on-his-luck Bodie begged Houdini to send him the electric chair as a present. Houdini eventually complied. Bodie to Houdini, April 8, 1920; April 22, 1920; April 30, 1920; January 17, 1921. Harry Ransom Humanities Research Center (HRHRC).
37. J. F. Burrows, *The Secrets of Stage Hypnotism; Stage Electricity: and Bloodless Surgery* (London: The Magician Ltd., 1912), 18.
38. Cited in Jay, *Learned Pigs*, 130.
39. "Students Rag 'Dr.' Bodie. Wild Riot Scenes in a Glasgow Music Hall," *Lloyd's Weekly News*, November 14, 1909, not paged. Houdini "Hypnotism" scrapbook, Library of Congress, Rare Book and Special Collections.
40. Bodie, *Bodie Book*, 37.
41. Bodie, for instance, introduced a singing assistant, Mystic Marie, at this time. Sheet music for "The Bodie Hypnotic Waltz," dedicated to Mystic Marie, the "real Trilby," was available in London. See Jay, *Learned Pigs*, 141.
42. Kennedy Brothers, *Handbook*, 19.
43 Harry Arons, *Master Course in Hypnotism* (Newark: Power Publishers, 1948), 7.
44. J. R. Oldfield, "Transatlanticism, Slavery, and Race," *American Literary History* 14, no. 1 (March 2002): 136–137.
45. William Davies King, "Real Light on Real Darkness in Performance: H. E. Lewis and Electrobiology," conference paper presented at Mid-America Theatre Conference, March 5, 2004, Chicago.

46. Albert Cavendish, *How to Become a Mesmerist* (London: Scientific Publishing, circa 1890), 12.
47. Professor Leonidas, *Stage Hypnotism* (Chicago: Bureau of Stage Hypnotism, 1901), 7.
48. Exploring other identity boundaries, hypnotists might set up a "freak show" on stage with different subjects, or a "minstrel show" and encourage white audience members to "play" at being black.
49. Leonidas, *Stage,* 10.
50. Ibid., 37–39.
51. George Newmann Papers, "Miscellania Hypnotica," compiled 1928, clipping not dated. Library of Congress, Rare Book and Special Collections.
52. The professor was not humbugging the crowd. In the nineteenth century, many surgeons and dentists did successfully perform surgery using no other anesthetic than hypnotism. Winter argues that orthodox physicians began to experiment with chloroform and ether primarily to displace mesmerism as a means of pain management.
53. Leonidas, *Stage,* 17.
54. Jean-Roch Laurence and Campbell Perry, *Hypnosis, Will and Memory* (New York: Guilford Press, 1988), 218.
55. J. H. Cassedy, *Medicine in America: A Short History* (Baltimore: Johns Hopkins University Press, 1991), 91.
56. J. G. Burrow, *Organized Medicine in the Progressive Era* (Baltimore: Johns Hopkins University Press, 1977), 117.
57. Laurence and Perry, *Hypnosis,* 270.
58. "Dangers of Hypnotism," *New York Times,* December 28, 1890, 4.
59. X. LaMotte Sage, *Hypnotism: As It Is* (Rochester: New York State Publishing, 1902), not paged. Magic Castle library, Los Angeles.
60. Sidney Flower, "Introduction," *Hypnotic Magazine* 1, no.1 (August 1896): 1.
61. Sidney Flower, "'How to Hypnotize' Reviewed," *Hypnotic Magazine* 1, no.5 (December 1896): 284.
62. "The Death of Spurgeon Young," *Hypnotic Magazine* 2, no. 6 (June 1987): 291. In this era, although "scientific" defenses of racism were common, to their credit none of the contributors seemed concerned with Spurgeon Young's race.
63. Ibid., 293–294.
64. Ibid., 287–306.
65. Ibid., 302. As a curious sidelight, it appears that Dr. J. D. Buck of Cincinnati was also the acting president of the American Theosophical Society and an admirer of a fraudulent inventor, John W. Keely. A mildly satirical news column quoted Buck in praise of the inventor's new engine said to run on a "vibratory" method. See "Is Keely the New Mahatma?" *New York Times,* April 29, 1896, 4.
66. Ibid., 301.
67. Ibid., 302.
68. Sidney Flower, "'How to Hypnotize' Reviewed," 283–284.
69. P. H. McEwen, *Hypnotism Made Plain* (Fargo, N.D.: privately published, 1897), 13.
70. Ibid., 80.
71. L. A. Harraden, *Hypnotic Exhibitions* (Jackson, Mich.: privately published, 1900), 33.
72. "A Mesmerist's 'Horses,'" *New York Times,* February 26, 1885, 5.

73. Leonidas, *Stage,* 10.
74. Sage, *Hypnotism: As It Is,* not paged.
75. William Shaw, *How to Hypnotize and Mesmerize* (Chicago: privately published, 1896), 20–21.
76. George White, *Personal Magnetism, Telepathy and Hypnosis* (London: George Rutledge and Sons, 1907), 246.
77. Leonidas, *Stage,* 58.
78. Ibid., 19.
79. Ibid., 60.
80. *Stealing a Dinner* (1903), American Mutoscope and Biograph Company, G. W. "Billy" Blitzer, camera. Library of Congress, catalog no. 973. The catalog copy indicates this was not Leonidas's first film with his animal troupe. Also available online: http://memory.loc.gov/ammem/vshtml/vsfmlst.html.
81. "Animal Magnetism," *Electrical World,* January 28, 1888, 38.
82. "Hypnotism the Cure-All," *New York Times,* April 30, 1899, 12. The Columbia University professor highlighted in this article, D. Quackenbos, found his subjects at Charles Loring Brace's Newsboys' Lodging House in lower Manhattan. Quackenbos also cowrote with John Duncan *Hypnotism in Mental and Moral Culture* (New York: Harper & Brothers, 1903).
83. Worcester, McComb, Coriot, *Religion and Medicine: The Moral Control of Nervous Disorders* (New York: Moffet, Yard, 1908), 138.
84. In a letter to Houdini in 1913, Bodie complained bitterly when he initially was turned down for membership by the Magician's Club of London, presumably for his unsavory reputation.
85. Members of the positive thinking movement, which included Mary Baker Eddy, founder of Christian Science, had greater luck in linking hypnotism to mental life. In their systems, autosuggestion did not turn subjects into "grotesque displays" but instead provided mental relief and the promise of "abundance." See Donald B. Meyer, *The Positive Thinkers, a Study of the American Quest for Health, Wealth and Personal Power from Mary Baker Eddy to Norman Vincent Peale* (Garden City, N.Y.: Doubleday, 1965).
86. For an examination of the development of vaudeville, its standardized nature, and its founders' attempts to create and encourage a middle-class audience, see Robert W. Snyder, *The Voice of the City: Vaudeville and Popular Culture in New York* (New York: Oxford University Press, 1989).

Chapter Four: The Magician

1. Poster for "Mysterious Man," 1873. McManus Young Collection, Library of Congress.
2. R.G.A. Dolby, "Reflections on Deviant Science." In *On the Margins of Science,* ed. Roy Wallis (Staffordshire, England: University of Keele, 1979), 9–47.
3. Andrew Jackson Davis's first major work was the three-volume *The Great Harmonia, Concerning the Seven Great States* (Boston: Mussey, 1852). One of many studies that followed was *The Penetralia: Being Harmonial Answers to Important Questions,* 5th ed. (Boston: B. Marsh, 1866).
4. Anne Braude, *Radical Spirits: Spiritualism and Women's Rights in 19th Century America* (Boston: Beacon Press, 1989), 84–98.
5. Handbill reproduced in Milbourne Christopher and Maurine Christopher, *The Illustrated History of Magic* (Portsmouth, N.H.: Heinemann, 1996), 55.

6. Harry Houdini, *Houdini's History of Magic in Boston, 1792–1915,* facsimile ed. (Glenwood, Ill.: Meyerbooks, 1983), 11.
7. Ibid., 14.
8. Playbill, Steve Finer Catalogue 75, item 55, summer 1992.
9. "Spiritualism Jugglery—Curious Trial at Buffalo," *New York Times,* August 27, 1865, 2.
10. The same companies built equipment for both natural philosophers and stage magicians—and, as a consequence, reinforced the ties between these somewhat disparate communities.
11. "A New and Descriptive Catalogue . . ." Adams and Company, Boston, 1876. McManus Young Collection, Library of Congress.
12. See Burnham, *Why Superstition,* 19–44. Burnham argued that this tactic remained strong into the early twentieth century until corporate public relations departments and journalists began to again promote superstition in the form of isolated gee-whiz science facts.
13. Christopher, *Illustrated History,* 118–120.
14. "Sid Macaire's Descriptive Catalogue of Entirely New and Superior Wonders . . ." Chicago, circa 1885. McManus Young Collection, Library of Congress.
15. "Burlingame Catalogue," Chicago, circa 1887. McManus Young Collection, Library of Congress.
16. Houdini, *A History,* 88.
17. Ibid., 36, 42. This was either the prominent Spiritualist LaRoy Sunderland or an imitator. The "real" Sunderland was on record disapproving the sensationalizing of Spiritualism. See R. Laurence Moore, *In Search of White Crows* (New York: Oxford University Press, 1977), 17.
18. W. D. Leroy, "New Descriptive Catalogue . . . Anti-Spiritualistic Illusions," Boston, 1893. McManus Young Collection, Library of Congress.
19. Adams and Company, "A New and Descriptive Catalogue of Magical Apparatus," Boston, 1876. McManus Young Collection, Library of Congress.
20. "W. G. Magnuson catalogue," 1938. McManus Young Collection, Library of Congress.
21. Kenneth Silverman first made this point in his biography, *Houdini!!!,* 36–44.
22. Arthur Conan Doyle, *Professor Challenger Stories; The Land of the Mist* (London: John Murray, 1958), 363.
23. Houdini Collection, Box 11, Harry Ransom Humanities Research Center (HRHRC).
24. July 19, 1912. "Police" folder, Houdini Collection, cabinet 112, HRHRC. This cartoon referred to a 1912 case in which a corrupt New York City police lieutenant, Charles Becker, hired gangsters to shoot another gangster—Herman "Beansie" Rosenthal, believed to be an informant against Becker. One of the gangsters involved in the shooting, Big Jack Zelig, later was murdered before he could testify at police officer Becker's trial for murder; Becker was found guilty and electrocuted.
25. John Kasson has argued that the jail escape, which often began with an invasive medical examination of Houdini's naked body, including a prodding of his mouth and anal orifice, cast Houdini's eventual triumph as an act of reclaiming his masculinity. See John Kasson, *Houdini, Tarzan, and the Perfect Man* (New York: Hill and Wang, 2001).

26. Kenneth Silverman linked Houdini to the cult of the rigorous life in *Houdini*, 36–44. Kasson relied on the same insight as the starting premise for his more elaborate examination in *Houdini, Tarzan, and the Perfect Man.*
27. Walter S. Gibson, *The Original Houdini Scrapbook* (New York: Sterling Publishing, 1976), 39. One challenge that might elude any cultural analysis came from the Hogan Envelope Company of Chicago, as follows, "We believe a giant envelope can be made by us which will enclose Houdini and successfully prevent his escape" (32).
28. Beatrice Houdini to Arthur Conan Doyle, December 16, 1926, HRHRC.
29. See Taylor, *Death and Resurrection,* 144–154.
30. Raymund Fitzsimons, *Death and the Magician: The Mystery of Houdini* (New York: Atheneum, 1981), 98.
31. Beatrice Houdini to Arthur Conan Doyle, December 16, 1926, HRHRC.
32. Kenneth Silverman, 263–264.
33. Houdini, *A Magician Among the Spirits* (New York: Harper and Brothers, 1924), 154.
34. Ibid., 152.
35. The copy of *A Magician Among the Spirits* in the Sir Arthur Conan Doyle Collection at the Humanities Research Center is full of Conan Doyle's—sometimes lengthy—marginal comments.
36. Houdini, *A Magician,* 247.
37. *New York American,* August 15, 1925. Humanities Research Center, Houdini Collection, Box 11.
38. Crandon to Conan Doyle, July 30, 1924, HRHRC.
39. Crandon to Conan Doyle, May 2, 1924, HRHRC.
40. Crandon to Conan Doyle, May 13, 1924, HRHRC.
41. Crandon to Conan Doyle, September 22, 1924, HRHRC.
42. Crandon to Conan Doyle, June 6, 1924, HRHRC.
43. J. Malcolm Bird, *Margery the Medium* (Boston: Small, Maynard, 1925), is an ardent defense of Margery and her psychic abilities. Bird outbid Houdini in the pamphlet game, when he wrote this approximately five-hundred-page work.
44. Crandon to Conan Doyle, July 30, 1924, HRHRC.
45. Crandon to Conan Doyle, August 26, 1924, HRHRC.
46. *Boston Herald,* December 31, 1924. Library of Congress, American Memory website, "The American Variety Stage," Item 65 of 137.
47. Joseph Dunninger, *Dunninger's Secrets* (Secaucus, N.J.: Lyle Stuart, 1974), 243.
48. Harry Houdini, *Houdini Exposes the Tricks. . .* (New York, 1924), 6. Humanities Research Center, Box 11.
49. Crandon to Conan Doyle, October 24, 1924, HRHRC.
50. Playbill, Houdini Collection, Box 11, HRHRC.
51. Silverman, *Houdini!!!,* 407–409.
52. Beatrice Houdini to Doyle, June 28, 1928, HRHRC.
53. See, for example, David E. Shi, *Facing Facts: Realism in American Thought and Culture, 1850–1920* (New York: Oxford University Press, 1995).
54. See Orvell, *The Real Thing,* 20–23, 36–39.
55. Leach, *Land of Desire,* 246–260.
56. Cook, *Arts of Deception,* 163–213.

Chapter Five: The Mind Reader

1. J. B. Rhine and Louisa Rhine, "An Investigation of a 'Mind-Reading' Horse," *Journal of Abnormal and Social Psychology* 23, no. 4 (1929): 458.
2. Julius Zancig, *Two Minds with but a Single Thought* (London: Paul Naumann, 1907), 38.
3. In the mid–nineteenth century, to emphasize their progressive tendencies, many Spiritualist groups in America and England titled their newspapers *The Spiritual Telegraph*.
4. George M. Beard, "Physiology of Mind Reading," *Popular Science Monthly*, February 1877, 459.
5. Clipping, September 9, 1882. Baldwin file at HRHRC.
6. *Baldwin's Illustrated Butterfly*, 1889, 3, HRHRC.
7. Other show business pamphlets suggest a different lineage for Bishop, claiming he was an assistant to J. Randall Brown, or "the Celebrated Brown," a newsman turned muscle reader in 1873.
8. Program, February 27, 1887. Wallack's Theatre, New York City, HRHRC.
9. "Second Sight Explained" is a manuscript in the Bishop file at HRHRC.
10. As quoted in George M. Beard, *The Study of Trance, Muscle Reading and Allied Nervous Phenomena* (New York: George M. Beard, 1882), 33.
11. George Beard, *The Study of Trance*, 17. Electrical experimenter Michael Faraday had earlier relied on the thesis of unconscious muscular action to explain table-tipping and table-turning phenomena during Victorian séances.
12. Ibid., prefatory note.
13. Leonidas, *Stage Hypnotism*, 124.
14. Fear of premature burial was widespread in the nineteenth century and not just an aberration in the minds of Bishop's mother and Edgar Allan Poe. One electrical journal published illustrated plans for installing telegraph keys inside coffins so that those wrongfully buried could be rescued. See "Telegraphy from the Grave," *Electrical Review* 18, no. 21 (July 18, 1891): 279.
15. *New York Times*, June 4, 1889.
16. Bishop file, HRHRC.
17. Sid Macaire, *Mind Reading or Muscle Reading* (London: Simpkin, Marshall, 1905), 76.
18. Leonidas, *Stage Hypnotism*, 96.
19. Handbills number 7, number 8, Box 15, George Newmann Collection. Rare Book and Special Collections, Library of Congress.
20. Mark Twain, "Mental Telegraphy," *Harper's Magazine*, December 1891, 99.
21. Mark Twain, "Mental Telegraphy," 94.
22. Ibid., 101.
23. "The Prophet," *Electrical Review*, September 26, 1891, 66.
24. J. J. Carty, "Is Electricity Related to Nerve Force?" *Electrical Review*, February 20, 1892, 358.
25. Edwin J. Houston, "Cerebral Radiation," *Electrical Review*, June 4, 1892, 190.
26. "The Zancig Fever," *The Throne*, January 19, 1907, 404. Magician's Biography file, Julius Zancig, HRHRC.
27. "The Cleverest Music Hall 'Turn' On Record," *Daily Mail*, December 1, 1907, 6. Magician's Biography file, Julius Zancig, HRHRC.
28. Ibid.
29. *Daily Mail*, January 2, 1907. Magician's Biography file, Julius Zancig, HRHRC.

30. Julius Zancig, *Two Minds with but a Single Thought* (London: Paul Naumann, 1907), 39. Zancig quotes an unnamed reporter.
31. *Daily Mail,* January 2, 1907. Magician's Biography file, Julius Zancig, HRHRC.
32. "The Zancig Mystery," *Daily Mail,* December 27, 1906, 5. Magician's Biography file, Julius Zancig, HRHRC.
33. "The Zancig Fiasco," *The Throne,* February 23, 1907, 615. Magician's Biography file, Julius Zancig, HRHRC.
34. "Cleverest Music-Hall Turn," *Daily Mail,* December 1, 1906, 6. Magician's Biography file, Julius Zancig, HRHRC.
35. Julius Zancig, *Two Minds,* 35.
36. "The Zancig Mystery," *Daily Mail,* December 27, 1906, 5. Magician's Biography file, Julius Zancig, HRHRC.
37. Ibid.
38. *Daily Mail,* January 3, 1907. Magician's Biography file, Julius Zancig, HRHRC.
39. "The Queen and the Zancigs," *Daily Mail,* December 28, 1906, 5. Magician's Biography file, Julius Zancig, HRHRC.
40. Ibid.
41. "The Zancig Mystery," *Daily Mail,* December 27, 1906, 5. Magician's Biography file, Julius Zancig, HRHRC.
42. Julius Zancig, *Two Minds with but a Single Thought* (London: Paul Naumann, 1907), 10–13.
43. "The Cleverest Music Hall Turn," *Daily Mail,* December 1, 1906, 6. Magician's Biography file, Julius Zancig, HRHRC.
44. Zancig, *Two Minds,* 29.
45. W. G. Magnuson, who ran a magic catalogue company in the 1920s that catered specifically to dishonest psychics, indicated that Zancig was one of his customers. W. G. Magnuson to George Newmann, March 22, 1928. In George Newmann, *Spook Racketeer* (Minneapolis: privately published, 1943). Newmann Collection, Box 31, no. 41. Rare Book and Special Collections, Library of Congress.
46. Julius Zancig, *Crystal Gazing; The Unseen World* (Baltimore: J & M Ottenheimer, 1926), 9.
47. Ibid., 40.
48. For example, Larry Laudan in *Progress and Its Problems* (Berkeley: University of California Press, 1977), distinguished empirical problems from conceptual problems in science, and noted that one category of the latter could be labeled "worldview problems," when a scientific theory "is in conflict with any component of the prevalent *world view,*" 55.
49. R. Laurence Moore, *In Search of White Crows* (New York: Oxford University Press, 1977), 220.
50. Seymour H. Mauskopf and Michael R. McVaugh, in *The Elusive Science* (Baltimore: Johns Hopkins University Press, 1980), indicate that Freud was only a subscriber, but in his essay of 1925, "The Occult Significance of Dreams," Freud indicates that he was a member of both the American and English societies.
51. Alan Gauld, *The Founders of Psychical Research* (New York: Schocken Books, 1968), 138.
52. James also used this method of subjective empiricism to base his defense of the religious life in *Varieties of Religious Experience.*
53. Sigmund Freud, "Dreams and the Occult," in *Psychoanalysis and the Occult,* ed. George Devereaux (New York: International Universities Press, 1970, 1953), 95.

54. After his death, Hodgson became one of Piper's "controls," leading James to write a long inconclusive essay debating whether Hodgson was genuinely speaking through her. See William James, "Report on Mrs. Piper's Hodgson-Control," in William James, *Essays in Psychical Research* (Cambridge: Harvard University Press, 1986).
55. Devereaux, *Psychoanalysis,* preface, xii.
56. William James, "Telepathy," (1895): *Essays in Psychical Research,* 123.
57. William James, "Address of the President before the Psychical Research Society" (1896), in *Essays in Psychical Research,* 131. R. Laurence Moore relied on this speech for the title of his groundbreaking study of Spiritualism in America, *In Search of White Crows.*
58. J. B. Rhine, *Extra-Sensory Perception* (Boston: Bruce Humphries, 1964), 25.
59. Rhine, *Extra-Sensory Perception,* preface, xxviii.
60. Quoted in Mauskopf and McVaugh, *Elusive Science,* 79.
61. McDougall, a Lamarckian, sought to prove that learned behavior could be inherited. Rhine helped him with an experiment in which generations of rats "learned" and passed on the knowledge of leaving a tub of water by a dark unlighted passage rather than a lighted passage that gave an electric shock. For McDougall, Lamarckian principles implied the possibility of free will. See Mauskopf and McVaugh, *Elusive Science,* 82–84.
62. Rhine, *New Frontiers of the Mind* (New York: Oxford University Press, 1977), 106.
63. My own calculation, based on data in "Table XIX," *Extra-Sensory Perception,* 102.
64. Rhine, *Extra-Sensory Perception,* 117.
65. Rhine, *Extra-Sensory Perception,* 37.
66. Ibid., 163.
67. Ibid., 14.
68. Rhine, *New Frontiers,* 212.
69. From 1964 preface to *Extra-Sensory Perception,* xxxv.
70. Rhine, *Extra-Sensory Perception,* 222.
71. Rhine, *Extra-Sensory Perception,* 203–204. This quote was not a new addition to the 1964 edition, but can also be found in the original. See *Extra-Sensory Perception* (London: Faber and Faber Limited, 1935), 207.
72. J. B. Rhine, *New Frontiers,* 249.
73. Mauskopf and McVaugh, *Elusive Science,* 161–162.
74. Joseph Jastrow, "ESP, House of Cards," *American Scholar* 8, no. 1 (January 1939): 22.
75. B. F. Skinner, "Is Sense Necessary?" *Saturday Review,* October 9, 1937, 5.
76. Ibid., 5–6.
77. Dael L. Wolfle, "Extra Sensory Perception," *American Journal of Psychiatry* 94, no. 4 (January 1938): 947.
78. Wolfle, "Extra Sensory Perception," 943.
79. Harold O. Gulliksen, "Extra-Sensory Perception: What Is It?" *American Journal of Sociology* 43, no. 4 (January 1938): 623. Curiously, P. T. Barnum, to great success, had ballyhooed a "wild man" with the slogan "What-Is-It?" in the nineteenth century.
80. Ibid., 630.
81. Gulliksen, "Extra-Sensory Perception," 628.
82. Ibid., 629.
83. "Battle on Rhine," *Time,* April 11, 1938, 54.

84. Norbert Guterman, "Frontiers of Credulity," *New Republic,* November 17, 1937, 49.
85. Joseph Jastrow, "ESP, House of Cards," *American Scholar* (winter 1938–1939): 22.
86. "The Results of Our First Test of Telepathy," *Scientific American,* July 1933, 10.
87. "Our Second Test of Telepathy," *Scientific American,* September 1933, 108.
88. "Test for Telepathy," *Scientific American,* February 1934, 64.
89. Walter Franklin Prince, "Extra-Sensory Perception," *Scientific American,* July 1934, 5–7.
90. J. B. Rhine, "Telepathy and Clairvoyance in a Trance Medium," *Scientific American,* July 1935, 12–14.
91. "Telepathy Comes of Age," *Scientific American,* June 1937, 361.
92. J. B. Rhine, "ESP, What Precautions Are Being Taken . . . ?" *Scientific American,* June 1938, 328.
93. Edwin Teale, "Can We Read Each Other's Minds?" *Popular Science Monthly* 130, no. 3 (March 1937): 28. Dunninger insisted that he not only practiced on strangers in movie theaters, but also on elevator operators, occasionally saying the wrong number while thinking the right number, or not saying any number at all—invariably the operator would stop at the proper floor. Joseph Dunninger, *What's On Your Mind?* (Cleveland: World Publishing, 1944), 47–48.
94. Teale, 109.
95. Hugo Gernsback, "Can the Dead Be Reunited?" *Science and Invention,* April 1926, 1089.
96. Vernon W. Lemmon, "Extra-Sensory Perception," *Journal of Psychology* 4 (1937): 227.
97. Ibid., 230.
98. Ibid.
99. Ibid., 231.
100. Ibid., 232.
101. Ibid., 235.
102. Murphy listed the University of Colorado, Fordham University, New York University, Hunter College, Columbia University, and Tarkio College, and in investigations by six other groups outside the academy. Gardner Murphy, "Dr. Rhine and the Mind's Eye," *American Scholar* 7, no. 7 (spring 1938): 196.
103. Ibid, 200.
104. Ibid.
105. William James, "The Will to Believe," in *The American Intellectual Tradition,* vol. 2, ed. David A. Hollinger and Charles Capper (New York: Oxford University Press, 1989), 80–93.
106. William James, "Address of the President before the Psychical Research Society," 1895, in *Essays in Psychical Research,* 130.
107. Sigmund Freud, "Dreams and the Occult," in *Psychoanalysis,* 93.
108. Ibid., 108.
109. George M. Beard, "Physiology of Mind Reading," 472.
110. Freud, "Dreams and the Occult," 108.
111. The difficulty of establishing a grounds for the "falsifiability" of ESP, in philosopher Karl Popper's terms, suggested Rhine's difficulty in making it a "fit" scientific subject.
112. Charles Fort, *The Book of the Damned* (New York: Henry Holt, 1941 [1919]), 55.

113. Physicist Sheldon Goldstein has argued that Werner Heisenberg and Niels Bohr unnecessarily skewed quantum theory toward the extralogical. Presumably David Bohm's equations relieve quantum interactions of indeterminacy. Sheldon Goldstein, "Quantum Philosophy: The Flight from Reason in Science," in *The Flight from Science and Reason,* ed. Paul R. Gross, Norman Levitt, and Martin W. Lewis (New York: New York Academy of Sciences, 1996), 119–126.
114. J. B. Rhine, "Are We Psychic Beings?" *The Forum* 92, no. 6 (December 1934): 372.
115. J. B. Rhine, "Don't Fool Yourself," *The Forum* 94, no. 3 (September 1935): 189.
116. Joseph Dunninger, *Inside the Medium's Cabinet* (New York: David Kemp, 1935), 141.
117. Joseph Dunninger, *What's On Your Mind?* (Cleveland: World Publishing, 1944), 134–135.
118. Dunninger, as told to Walter Gibson, *Dunninger's Secrets* (Secaucus, N.J.: Lyle Stuart, 1974), 243.
119. Dunninger, *Dunninger's Secrets,* 64–66.
120. Such fascination tends to emerge along with new technologies. While Spiritualists "rapped" to the telegraph, mind readers "tuned into" partners as if they were radios. Artificial intelligence and robotics since have offered new millennial hopes, with many of the fields' founders looking forward to a "posthuman" future of super-intelligent artificial organisms.

Chapter Six: The Missionaries

1. See especially Warren Susman, *Culture as History: The Transformation of American Society in the Twentieth Century* (New York: Pantheon, 1984), and Leach, *Land of Desire.*
2. "Hall of Science Is Big Surprise of World's Fair," *Chicago News,* October 18, 1933. Clipping in Folder 14-332, Century of Progress Collection (CPC), Special Collections, University of Illinois, Chicago.
3. See Roland Marchand, *Creating the Corporate Soul* (Berkeley: University of California Press, 1998), especially chapter 7, "The Corporations Come to the Fair," 249–311.
4. See Roland Marchand and Michael L. Smith, "Corporate Science on Display," in *Scientific Authority and Twentieth Century America,* ed. Ronald G. Walters (Baltimore: Johns Hopkins University Press, 1997).
5. Edward Gibbons, *Floyd Gibbons, Your Headline Hunter* (New York: Exposition Press, 1953), 218–219.
6. John E. Findling, *Chicago's Great World's Fairs* (Manchester: Manchester University Press, 1994), 92–93.
7. The SAC's ability to gain the cooperation of the nation's top scientists and prompt their eager participation in radio talks and in fair exhibit design puts into question John C. Burnham's contention that "scientists who believed in science as a calling rather than an occupation tended increasingly to withdraw from popularizing during the twentieth century, leaving the field to media personnel and educators." See Burnham, *How Superstition,* 7.
8. Holland to Jewett, July 27, 1929. Folder 5-245, CPC.
9. Pamphlet, National Research Council Science Advisory Committee, October 1, 1929. Folder 5-266, CPC.

10. Maurice Holland, "Science Takes off the High Hat" (unpublished manuscript). Folder 5-248, CPC.
11. Press release, National Research Council, n.d. "Can 100,000,000 People Be Interested in the Life History of Science?" CPC, Special Collections.
12. National Research Council press releases, October 5, 1930, March 15, 1931. Folders 5-267, 5-268, CPC.
13. F. K. Richtmyer, "How Light Puts Electrons to Work." Press release announcing radio show, February 25, 1931. Folder 5-268, CPC.
14. Maurice Holland to Miss Martha McGrew, October 3, 1930. CPC.
15. Will Rogers, "Chicago's Great Will Rogers' Advice," *Herald and Examiner,* May 28, 1933, n.p. Folder 14-286, CPC.
16. H. D. Sanborn to C. W. Farrier et al. November 16, 1934. Folder 1-6177, CPC.
17. *Official Story and Encyclopedia of A Century of Progress* (Chicago: A Century of Progress Administration Building, 1933), 11.
18. Ibid.
19. Findling, *Chicago's Great World's Fairs,* 93.
20. "Preliminary Report of the Science Advisory Committee to Chicago Trustees," March 25, 1930, Folder 5-247, CPC, Special Collections.
21. R. P. Shaw to L. R. Lohr, December 21, 1929. Folder 5-246, CPC, Special Collections.
22. *World's Fair Weekly,* October 7, 1933, 22. Folder 16-153, CPC.
23. "Robot Lectures Like Gentleman; Insides Exposed," *Chicago Tribune,* May 31, 1933. Folder 14-287, CPC.
24. Hall of Science description in Edwin Teale, "History's Biggest Show," *Popular Science Monthly,* July 1933, 23–27, 91.
25. "Life Under the Microscope," *World's Fair Weekly,* July 22, 1933, 35. Folder 16-142, CPC.
26. Ibid.
27. "Hall of Science Is Revamped for this Year's Fair," *Chicago Tribune,* March 31, 1934. Folder 14-248, CPC.
28. Henry Crew, "The Human View of Science," *World's Fair Weekly,* May 13, 1933, 6–7. Folder 16-132, CPC, Special Collections.
29. Joseph Ator, "Hall of Science Coming Out in New 1934 Model," *Chicago Tribune,* May 23, 1934, n.p. Clipping, Folder 14-248, CPC, Special Collections.
30. "Science Bankrupt," *Chicago News,* January 2, 1934, n.p. Folder 14-339, CPC.
31. "10,000 Visitors Daily at Fair's Hall of Religion," *Chicago Tribune,* July 16, 1933. Folder 14-307, CPC.
32. "Science Forum Paints Future a Blaze of Hope," *Chicago Tribune,* May 26, 1934. CPC.
33. Ibid.
34. For a further look at the debate about the impact of technology on labor in the 1930s, see Amy Sue Bix, *Inventing Ourselves Out of Jobs* (Baltimore: Johns Hopkins University Press, 2000).
35. J. Parker Van Zandt and L. Rohe Walter, "King Customer at a Century of Progress," *Review of Reviews,* September 1934, 22.
36. Ibid., 23.
37. Ibid., 24.
38. Ibid., 26.

39. Ibid., 27.
40. From advertisement, *Official Guidebook of the Fair* (Chicago: Century of Progress Administration Building, 1933), 158.
41. Van Zandt and Walter, "King Customer," 24.
42. Van Zandt and Walter, "King Customer," 27.
43. *Official Guidebook of the Fair* (Chicago: Century of Progress Administration Building, 1933), 148.
44. Roland Marchand and Michael L. Smith, "Corporate Science on Display," 162.
45. G. E. Simons to E. Ross Bartley, October 14, 1933. Folder 1-6177, CPC.
46. "Here's a Magician Who Explains All His Tricks," *Chicago Tribune,* June 3, 1933. Folder 14-288, CPC.
47. Ibid.
48. *Magic Versus Science* (film). General Electric, 1932, Schenectady Museum.
49. John Hix, "Strange As It Seems," *Chicago Daily Times,* September 25, 1933, n.p. Folder 14-327, CPC.
50. In Paris in 1841 the magician Philippe presented a forerunner of this effect that relied on an electric circuit to light hydrogen jets in a theater. See Christopher, *Illustrated History,* 136.
51. G. E. Simons, press release, June 22, 1934. Folder 1-6177, CPC.
52. George E. Simons, "Lamps and Lighting," June 22, 1934. Folder 1-6177, CPC.
53. General Motors Corporation, press release, May 10, 1934. Folder 1-6212, CPC.
54. Joseph Ator, "Latest Magic of Scientists to Provide Thrills at Fair," *Chicago Tribune,* May 24, 1934, n.p. Folder 14-248, CPC.
55. Findling, *Chicago's Great World's Fairs,* 142.
56. E. Leichter, *Successful Selling* (New York: Funk and Wagnalls, 1916), 9, 53.
57. Charles F. Kettering and Allen Orth, *The New Necessity: The Culmination of a Century of Progress in Transportation* (Baltimore: Williams and Wilkins, 1932), 33.
58. Ibid., 219.
59. Thomas A. Boyd, *Professional Amateur: The Biography of Charles Franklin Kettering* (New York: Dutton, 1957), 218.
60. Advertisement reprinted in Marchand, *Creating the Corporate Soul,* 289.
61. Marchand, *Creating the Corporate Soul,* 283–291.
62. Tom M. Olson, "World's Fair at San Francisco Ends Forever," *Now,* circa 1940. Clipping, Irwin Moon folder, Moody Bible Institute Library (MBI).
63. F. O. McMillan to P.V. Jenness, October 29, 1938. Irwin Moon folder, MBI.
64. Poster, "Startling Scientific Spectacle," Moody Institute of Science (MIS) box, MBI.
65. "Atomic Structure of Matter Shown in Lecture," *Star-News,* May 24, 1944, n.p. Irwin Moon folder, MBI.
66. James Gilbert, *Redeeming Culture* (Chicago: University of Chicago Press, 1997), 123.
67. Personal interview with Margaret Moon. December 16, 2000.
68. Moon biography, n.d., MBI.
69. Charles J. Miller, "Sermons from Science," *Sunday School Times,* n.d., n.p., circa 1940. Clipping, MIS box, MBI.
70. Personal interview with Margaret Moon, December 16, 2000.
71. Miller, "Sermons."

72. Arnold Grunigen Jr., "Irwin Moon Presents His Startling Words," pamphlet, Irwin Moon collection, MBI.
73. Miller, "Sermons."
74. Press release, Moody Institute of Science, circa 1950s, for television series based on lectures and films, MBI.
75. "Evangelist in Scientist's Role Refutes," *Buffalo Courier,* October 19, 1938. Clipping, n.p., MBI.
76. "Facts of Faith," Moody Bible Institute pamphlet, circa 1950s.
77. Lecturer's Demonstrations Stir Auditorium," *Star-News,* May 23, 1944, n.p., MBI.
78. Press release, Moody Bible Institute, n.d., MIS box, MBI.
79. John Hix, "Strange As It Seems," Cushing, Okla., *Citizen,* October 16, 1941, n.p. Clipping, Irwin Moon File, MBI.
80. Poster for Grand Rapids, Michigan performance, May 2, circa 1944, MBI.
81. "Methuselah Lived On," *San Antonio Express,* November 18, 1941, n.p., MBI.
82. Advertisement, Portland, Oregon, *Journal,* November 23, 1940, n.p., MBI.
83. Mr. Lyons to Dr. Houghton, July 11, 1942, MIS box, MBI.
84. George Speake, "Informal Report from George Speake," April 30, 1958, MIS box, MBI.
85. As quoted in James F. Findlay Jr., *Dwight L. Moody: American Evangelist, 1837–1899* (Chicago: University of Chicago Press, 1969), 262.
86. Findlay, *Moody,* 266.
87. Publicity story, n.d., n.p. Begins "Crackling, blue-violet lightning . . ." 4, MBI.
88. M. D. Morrison to "Whom It May Concern," June 3, 1942, Camp Haan, California, MBI.
89. Martin L. Thomas to Noel O. Lyons, June 26, 1942, MBI.
90. Gilbert, *Redeeming Culture,* 126.
91. Gilbert provides an excellent account of Moon's ability to evangelize U.S. military troops, as well as a solid account of the development and impact of Moon's film projects. See *Redeeming Culture,* 121–145.
92. "Proposal for Christian Laboratory." See "Correspondence (1939–1946) Having to do with the Founding of Moody Institute of Science. From the files of F. Alton Everest," MIS box, MBI.
93. See Gilbert, *Redeeming Culture,* for a discussion of Everest's related American Science Affiliation, 147–169.
94. Gilbert, *Redeeming Culture,* 131.
95. For a history of "character guidance" in the military and a look at how Moon tailored his Sermons for Science for that context, see Gilbert, *Redeeming Culture,* 98–100, 130–144.
96. "Rev. Irwin Moon, 78, Science Film Producer" (obituary), *Chicago Tribune,* May 4, 1986, n.p. Clipping, Moon collection, MBI.
97. "Preface to Sermons from Science Outlines." n.d., MIS box, MBI.
98. Ibid., 9–10.
99. "A Young Lady Impersonates an Electrode." *Life,* September 3, 1951.
100. Bill Sharpe, "Sugar for the Science Pill," *The State,* January 12, 1952, 12–13. Swezey Papers, Archives Center, National Museum of American History.
101. See Paul Boyer, *By The Bomb's Early Light: American Thought and Culture at the Dawn of the Atomic Age* (New York: Pantheon, 1985), 109–130, 291–302.
102. Sharpe, "Sugar," 13.

Chapter Seven: Flying Saucers

1. David M. Jacobs, *The Flying Saucer Controversy in America* (Bloomington: Indiana University Press, 1975), 121.
2. Ron Story, ed., *Encyclopedia of Ufos* (Garden City, N.Y.: Doubleday, 1980), 381.
3. Hal Draper, "Afternoon with the Space People," *Harper's Magazine*, September 1960, 37–40.
4. For a photographic essay that covers flying saucer "vernacular" in roadside architecture and the pageants of flying saucer cults, see Douglas Curran, *In Advance of the Landing: Folk Concepts of Outer Space* (New York: Abbeville Press, 1985).
5. William Graebner, *The Age of Doubt* (Boston: Twayne Publishers, 1991), 21.
6. Gilbert, *Redeeming Culture*, 230.
7. Jodi Dean, *Aliens in America* (Ithaca: Cornell University Press, 1998), 11–21.
8. Carl G. Jung, *Flying Saucers: A Modern Myth of Things Seen in the Skies* (Princeton: Princeton University Press, 1978), 22.
9. See David Ketterer, *New Worlds for Old; the Apocalyptic Imagination, Science Fiction, and American Literature* (Garden City, N.Y.: Anchor Press, 1974). Fusing myth and science, the science fiction genre is the ultimate repository of the "populist science" impulse this work has examined.
10. Robert S. Ellwood Jr., "Religious Movements and UFOs," *Encyclopedia of UFOs*, 306–308.
11. George F. Dole, ed., *Emanuel Swedenborg: The Universal Human and Soul-Body Interaction* (New York: Paulist Press, 1984), 10–12.
12. Emanuel Swedenborg, *The Earths in Our Solar System* (New York: J. B. Lippincott, 1876 [1787]), 4.
13. Henry A. Gaston, *Mars Revealed* (San Francisco: A. L. Bancroft, 1880).
14. These two books are Leadbetter's *The Astral Plane: Its Scenery, Inhabitants and Phenomena* (London: Theosophical Publishing Society, 1895), and *The Devachanic Plane: Its Characteristics and Inhabitants* (London: Theosophical Publishing Society, 1896).
15. Many other such volumes are listed in George M. Eberhart's *UFOs and the Extraterrestrial Contact Movement: A Bibliography* (Metuchen, N.J.: Scarecrow Press, 1986).
16. William Sims Bainbridge, *The Spaceflight Revolution* (New York: John Wiley, 1976), 31.
17. Von Braun in a 1972 letter, cited in Noble, *Religion of Technology*, 126.
18. Paul Meehan, *Saucer Movies: A Ufological History of the Cinema* (Lanham, Md.: Scarecrow Press, 1991), 15.
19. According to Meehan, form-fitting outfits for astronauts first debuted in J. V. Leigh's 1919 silent film version of H. G. Wells's *The First Men in the Moon* (see Meehan, *Saucer Movies*, 21). The zipperless body suit highlights the space dwellers' angelic nature, presenting them as beings who apparently have moved beyond bodily functions.
20. Desmond Leslie and George Adamski, *Flying Saucers Have Landed* (New York: British Book Centre, 1953), 206.
21. Leslie and Adamski, *Flying Saucers*, 207.
22. Photographs Adamski took of this bell-shaped craft suggest the aliens shared a

1950s-earthling design aesthetic. One debunker has identified the photograph as representing the disassembled top of a Hoover vacuum cleaner.

23. Leslie and Adamski, *Flying Saucers,* 195.
24. Angelucci's visions that ensued after drinking from the goblet parallel those that Aldous Huxley attributed to hallucinogens; this suggests a parallel between the mescaline and LSD experimentation among intellectuals in the 1950s and the more lowbrow flying saucer movement; in his 1954 essay "Heaven and Hell," Huxley talked of the drug-induced encounter with the "non-human other" and descriptions of "stones of fire" and other such translucent materials during hallucinogenic raptures—imagery reminiscent of that with which contactees described flying saucer experiences. See "Heaven and Hell," in *The Doors of Perception* (New York: Harper and Row, 1990), 83–140. The LSD aristocracy also met with lowbrow contactee culture when Ken Kesey and his Merry Pranksters began to tout the *Urantia Book,* a cold war text of revealed wisdom.
25. Orfeo Angelucci, *The Secret of the Saucers* (Stevens Point, Wis.: Amherst Press, 1955), 20.
26. Angelucci, *Secret,* 31–32.
27. Ibid., 34.
28. Ibid., 103.
29. Ellwood, "Religious Movements," 306–308.
30. Story, *Encyclopedia,* 157.
31. Story, *Encyclopedia,* 156. Advertisement from *Los Angeles Mirror News,* July 22, 1960, n.p.
32. Margaret Storm, *Return of the Dove* (Baltimore: privately published, 1959), 2.
33. Storm, *Return,* 12.
34. Storm, *Return,* 14.
35. John J. O'Neill, *Prodigal Genius: The Life of Nikola Tesla* (New York: Ives Washburn, 1944), 302.
36. Storm, *Return,* 22.
37. Storm, *Return,* 22.
38. Storm, *Return,* 263.
39. Storm, *Return,* 57.
40. Charles Kettering, head of research at General Motors in the early twentieth century, probably had Tesla in mind when he lampooned an inventor who came to him with the design for a small, innovative dynamo, but when asked how it would avoid overheating scolded, "What are your research laboratories for? I can't think of all the good things." In Charles Kettering and Allen Orth, *The New Necessity* (Baltimore: Williams and Wilkins, 1932), 57–58. Tesla had designed a "bladeless turbine" relying on fluid dynamics and patented it in 1913, but it was not put into immediate production because of the problem of overheating. In 1920 he developed a gasoline powered model that he showed to various automobile manufacturers. See Cheney, *Tesla,* 109–115.
41. Storm, *Return,* 90.
42. Storm, *Return,* 107.
43. Storm, *Return,* 113.
44. Satirist and science fiction writer Karel Capek mapped out the connection between free energy and religion in his 1927 novel *The Absolute at Large.* In this tale, an inventor creates an atomic "Karburator" that not only produces unlim-

ited energy but religious ecstasy to those within its sphere of influence. Following the religious frenzies came schisms, economic and political havoc, and an orgy of warfare and killing. The Karburator, finally, is banned, and small cells of cultists that still worship the few remaining, hidden Karburators are subjected to police raids.

45. Perpetual motion scholar Arthur W.J.G. Ord-Hume argues that at least one such device "of the second kind" was developed in the 1760s, when James Cox developed a clock that was automatically rewound by a device triggered by changes in barometric pressure. See Ord-Hume, *Perpetual Motion: The History of an Obsession* (New York: St. Martin's Press, 1977), 11–24.
46. "Is Keely the New Mahatma?" *New York Times*, April 29, 1896, 4.
47. Keely still has disciples promoting a theory of "Sympathetic Vibratory Physics." The group's website includes a photograph of the shiny machines "Atlin" and "Symael" whose function is to promote love and a new spiritual science; see www.svpvril.com (accessed December 22, 2003).
48. Ord-Hume, *Perpetual Motion*, 139–150.
49. See Storm, *Return*, 72.
50. Storm, *Return*, 73.
51. Paris Flammonde, *The Age of Flying Saucers: Notes on a Projected History of Unidentified Flying Objects* (New York: Hawthorne Books, 1971), 128.
52. Storm, *Return*, 131.
53. Otis T. Carr advertisement for "Peaceful Atomic Energy," 1958. Swezey Collection, National Museum of American History (NMAH), Archives Center.
54. Otis T. Carr advertisement, 1958, NMAH.
55. Flammonde, *Age*, 129.
56. Draper, "Afternoon with the Space People," 37.
57. Flammonde, *Age*, 129.
58. The ambiguities of Storm's narrative were deepened to Nabokovian levels in 1972, when the other Tesla disciple that Storm ballyhooed, Canadian electrical engineer Arthur Matthews, published his own volume, *Wall of Light: Nikola Tesla and the Venusian Space Ship, the X-12* (Mokelumne Hill, Calif.: Health Research). Matthews insisted that his exhibition of his "Tesla-scope" in 1956 single-handedly sparked the renewed interest in the inventor—entirely ignoring the efforts of Kenneth Swezey and other Tesla Society organizers to mark the centennial celebration of Tesla's birth. Matthews also insisted he cowrote Storm's 1959 book *Return of the Dove*, and, even more curiously, he included the frontispiece portraits of Margaret Storm and John Storm that appeared in *Return*, but put his own name beneath that of the "John Storm" portrait. The extravagance of his claims make him an even more unreliable narrator than Storm; likewise, his prose lacks the wit of Storm's volume, suggesting that he had, at best, been her collaborator. At the very least, this was a very incestuous community.
59. OTC Enterprises Inc., "Space-O-Gram," April 1959. Swezey Collection, NMAH.
60. Flammonde, *Age*, 131.
61. Clipping, August 12, 1960. Swezey Collection, NMAH.
62. Otis T. Carr v. State of Oklahoma, No. A-12907, 1961 Okla. Crim. App. Lexis 128; Blue Sky L. Rep. (CCH) P70, 513, January 11, 1961.
63. Otis T. Carr advertisement, 1958, Swezey Papers, NMAH.
64. Draper, "Afternoon," 40.

Chapter Eight: The Many Gospels

1. I mean to distinguish the grassroots wonder "pitch" from mass culture forms that rely on special effects or that commonly tap fringe science—for example, tabloid newspapers, tabloid television, advertisements, and the many Hollywood movies and television shows that have explored the paranormal.
2. One year later, the website for the Q-Ray bracelets emphasized that they were sports enhancement devices and avoided any explanation of what the Q-Ray might be and any specific health claims. On July 22, 2003, the www.natures bracelets.com website featured testimonials from various athletes including William Kazmaier, "The World's Strongest Man," who said "Thanks to Q-Rays I feel like I can compete for my championship again. I can lift 25 lbs. more now!" Another website noted that a class action suit was under way in Cook County Circuit Court against the QT Inc. company of Elk Grove Village, Illinois, manufacturer of the devices, for encouraging sufferers of arthritis and other symptoms to rely on their bracelets rather than mainstream treatments. http://Bookstore.cee-ane.org/q_ray_bracelets.html (accessed July 22, 2003).
3. Personal interview with Greg Roberts, V.P. Whole Life Expo, October 9, 2001.
4. Judith Rosen, "Crossing the Boundaries; Regardless of its label, this increasingly mainstream category continues to broaden its subject base," *Publishers Weekly,* May 27, 2003, 25.
5. This figure comes from a New Age book publishing website: http://www.llewellyn.com/trade/newage.php (accessed February 2, 2004).
6. Dan Burton and David Grandy, *Magic, Mystery, and Science: the Occult in Western Civilization* (Bloomington: Indiana University Press, 2004), 318.
7. David Spangler and William Irwin Thompson, *Reimagination of the World: A Critique of the New Age, Science, and Popular Culture* (Santa Fe: Bear and Company, 1991), 57.
8. Contactees are to be found in abundance, however, at the Mutual UFO Network (MUFON) conferences.
9. Leaflet, "Crystal Tones" booth. Dallas Whole Life Exposition, September 21, 2001.
10. The *Jesusonian,* "Special Report," 1988.
11. *The Urantia Book, a Description* (pamphlet) (Chicago: The Urantia Foundation, n.d.). It would appear that the revelation preceded by two decades the publication of *The Urantia Book* in 1955.
12. The *Jesusonian,* 25.
13. *Transmission Meditation* (pamphlet) (North Hollywood: Tara Center, n.d.). Although the word "energy" has scientific connotations, the New Age use does not have the precision one would associate with the equations of thermodynamics. Instead "energy" closely approximates the nineteenth-century notion of animal magnetism, mesmeric fluid, or life energy—the Chinese concept of "chi" is also often evoked in this context. Andrew Ross discussed the use of the concepts of "energy" and the "natural" in the New Age movement in his *Strange Weather: Culture, Science, and Technology in the Age of Limits* (London: Verso, 1991), 15–74.
14. *Oxybliss* (pamphlet) (Mill Valley, Calif.: Oxygen Research Institute, LLC), 2.
15. According to one article, the cytochromes inside the mitochondria, when stimulated by light, encourage the mitochondria to produce more energy. Dan Drollette, "Can Light Hasten Healing in Space?" *Biophotonics International,* September/October 2000, 47.

16. Ibid., 46–49.
17. "NASA Shines a Healing Light on Wounds," *Houston Chronicle,* February 8, 2001, 12. See also Tiina Karu, "Photobiology of Low-Power Laser Effects," *Health Physics* 56, no. 5 (May 1989), 702.
18. Plastic surgeons have since begun to offer such treatments as an alternative to surgery. See Shari Roan, "Face-Lifts, One Cell at a Time," *Los Angeles Times,* May 26, 2003, part 6, 1. The article describes patients sitting before "blinking lights" for a few minutes, in a treatment reminiscent of that which the robot used to resurrect Klaatu in *The Day the Earth Stood Still.*
19. All David Olszewski quotes either from personal interview or his lecture of Sunday, September 23, 2001, at the Dallas Whole Life Expo.
20. Personal interview with Dean Ortner, December 15, 2000.
21. This description of Lee's show is based on the author's notes on Lee's October 22, 2001 appearance at Promiseland in Austin, Texas.
22. This home page also had contrasting advertisements for the "Million Mom March" against handguns and one for "Dangerous Books Online Bookstore," with a list that gave tips on how to open an offshore bank account, how to protect one's privacy, and how to travel internationally without a passport. Truth Radio home page. http://www.truthradio.com (accessed July 19, 2001).
23. Lee has run into trouble with the law in several states. One of his dealer's websites placed a positive spin on these problems, as follows: "In the states of Maine and Idaho, UCSA, BWT and ITEC have agreed with the Attorney Generals of those states that we will not *ever* promote, demonstrate or install proven or unproven Free Electricity generators or any device relating to electricity. *It is evident by the following, which is an exact quote from the agreement, that these states do not wish their citizens to have any device that produces free energy in their state either now or any time in the future.*" See http://www.ucsofa.com/announcement.htm (accessed September 19, 2003).
24. John Huizenga, *Cold Fusion: The Scientific Fiasco of the Century* (Rochester: University of Rochester Press, 1992), 23.
25. For a recent sociological study of cold fusion, see Bart Simon, *Undead Science: Science Studies and the Afterlife of Cold Fusion* (New Brunswick, N.J.: Rutgers University Press, 2002).
26. Charles Grandison Finney, "What a Revival of Religion Is," in *The American Intellectual Tradition,* vol. 1, ed. David A. Hollinger and C. Capper (New York: Oxford University Press, 1997), 202. The connection between salesmanship and evangelism is thoroughly explored in R. Laurence Moore, *Selling God: American Religion in the Marketplace of Culture* (New York: Oxford University Press, 1994).
27. E. Leichter, *Successful Selling,* 40.
28. According to an archived news release from the Kentucky Office of the Attorney General, Lee had been arrested during his sales seminar there. The charges were similar to those that free energy promoter Otis T. Carr faced when arrested in Oklahoma. Kentucky law "prohibits the sale of business opportunities without registering with the attorney general's office and posting a $75,000 bond." The attorney general's office cited Lee for three felony violations. See http://kyattorneygeneral.com/news/2001rel/026_9oct01.htm (accessed January 12, 2004).
29. In this way, the wonder showman can be seen as a side branch of the long-lived popular culture archetype of the "mad scientist." See Roslynn D. Haynes, *From*

Faust to Strangelove: Representations of the Scientist in Western Literature (Baltimore: Johns Hopkins University Press, 1994).

30. Carl Sagan, *The Demon-Haunted World: Science as a Candle in the Dark* (New York: Random House, 1995), xiii.
31. Ibid., 304.
32. Ibid., 295–306.
33. Ibid., 15.
34. Ibid., 17.
35. "One Million Dollar Paranormal Challenge," James Randi Educational Foundation website, www.randi.org/research/challenge.html (accessed September 9, 2001).
36. Ibid.
37. "Eric's Open Offer to Validate Claims of Free Energy," http://www.users.voicenet.com/~eric/freetest.html (accessed March 21, 2002).
38. For example, Larry Laudan in *Progress and Its Problems* distinguished empirical problems from conceptual problems in science and noted that one category of the latter could be labeled "worldview problems," when a scientific theory "is in conflict with any component of the prevalent *world view,*" 55.
39. See James McLenon, *Deviant Science* (Philadelphia: University of Pennsylvania Press, 1984), 2–3, and R.G.A. Dolby, "Reflections on Deviant Science," in *On the Margins of Science,* ed. Roy Wallis (Staffordshire, England: University of Keele, 1979), 9–47.
40. This explains why physicists recently have attacked the notion of "indeterminacy" that Bohr and Heisenberg once encouraged. See Goldstein, "Quantum Philosophy," in *The Flight,* 119–125.
41. Jerry Libonati, "Postcards from the Afterlife," *Austin American Statesman,* March 14, 2002, E4. Randi had become the authority journalists turned to to offer the debunker's view on virtually any paranormal phenomenon.
42. Robert L. Park, *Voodoo Science* (New York: Oxford University Press, 2000), vii. In his willingness to implicate scientists, he ignored the far more obvious commercial forces which have made both "gee-whiz" science and "wonder" as a sales device prominent and have altered the utopian character of the scientific project.
43. Sigmund Freud, "Psychoanalysis and Telepathy" (1921), in *Psychoanalysis and the Occult,* 58.
44. Hal E. Puthoff, "CIA-Initiated Remote Viewing Program at Stanford Research Institute," *Journal of Scientific Exploration* 10, no. 1 (1996): 76. Puthoff's critics, such as Martin Gardner, also complained of his participation in Scientology. Freeman Dyson, a respected physicist, has taken another tack; rejecting the "reductionist" view that all knowledge can be subsumed to the scientific, he has argued that although ESP and the paranormal are most likely not susceptible to scientific proof, they may yet be "real," established through different avenues of understanding. See Freeman Dyson, "Debunked! ESP, Telekinesis, Other Pseudoscience," *New York Review,* March 25, 2004, 4–6. Dyson's position, though welcome, marks yet another return to the philosophy of "separate spheres," which never seems to satisfy either side of the debate.
45. See Fredric Jameson, "Postmodernism or the Cultural Logic of Late Capitalism," *New Left Review* 146 (July–August 1984): 53–92. One forum in which Baudrillard announced the "collapse of the real" was an analysis of the works of

science fiction author Philip K. Dick. Jean Baudrillard, "Simulacra and Science Fiction," *Science Fiction Studies* 18 (1991): 309–313.

46. Jodi Dean, *Aliens in America: Conspiracy Cultures from Outerspace to Cyberspace* (Ithaca, N.Y.: Cornell University Press, 1998).

47. Their approach, however, at times has resembled the hucksterism of their "enemy." To give one example, Michael Shermer, the director of the Skeptics Society, sent free copies of his book *Why People Believe Weird Things* (New York: W. H. Freeman, 1997) to numerous academics in 2001, along with a copy of his *Skeptic* magazine and a glossy pamphlet called *The Baloney Detection Kit.* Included were "Eight Sample Syllabi: How to Teach a Course in Science and Pseudoscience." He also added narrative interest to his attack on pseudoscience by framing it as a conversion story. We learn that in his youth, as a competitive bicycle racer, he fell prey to any number of quack health regimens, before "seeing the light." See Shermer, 13–15.

48. For a history of the melodrama in America and its republican quotient, see David Grimsted, *Melodrama Unveiled* (Chicago: University of Chicago Press, 1968).

49. See Dean, *Aliens in America;* Sandra Harding, *The Science Question in Feminism* (Ithaca, N.Y.: Cornell University Press, 1986); Ross, *Strange Weather,* especially 1–74. Ross found himself at the center of an intellectual furor when, as a guest editor, he published physicist Alan D. Sokal's hoax, "Transgressing the Boundaries: Toward Transformative Hermeneutics of Quantum Gravity," *Social Text* 46/47 (spring 1996): 217–252. Sokal explained that his hoax was designed to critique the postmodernist embrace of intellectual relativism and attack the belief that there no longer was an objective reality to refer to or describe scientifically, as well as the postmodernists' reduction of all "reality" to linguistics. Sokal's hoax of the cultural studies stance toward science became fodder for a minibook boom, including editors of Lingua Franca, *The Sokal Hoax: The Sham that Shook the Academy* (Lincoln: University of Nebraska, 2000); Noretta Koertge, ed., *A House Built on Sand: Exposing Postmodernist Myths about Science* (New York: Oxford University Press, 1998); Alan D. Sokal, Jean Bricmont, eds., *Fashionable Nonsense: Postmodern Intellectuals' Abuse of Science* (New York: Picador, 1998); Paul D. Gross, Norman Levitt, eds., *Higher Superstition: The Academic Left and Its Quarrels with Science* (Baltimore: Johns Hopkins University Press, 1997). Another book in this vein, previously cited, is *The Flight from Science and Reason.*

50. Both the skeptic and postmodern viewpoints are seductive in their own right. One wonders, though, if giving tacit encouragement to website theorists with genuinely noxious claims—as, for example, those that claim the Nazi holocaust never happened—will ultimately benefit "the people." Likewise, even if "elitist" in stance, the skeptics avoid the elitist jargon common to the cultural studies theorists.

51. Alan D. Sokal, in "What the *Social Text* Affair Does and Does Not Prove" (online article dated April 8, 1987), made the more moderate case that while science should be subject to social critique, only "bad" science then will perish. His answer to the problem that historically such "bad" science often was accepted as "good" is that such work tends to fall into categories that can now be detected, such as racist science, eugenics, and "masculinist" medical theory. See http://www.physics.nyu.edu/faculty/sokal/noretta.html#10 (accessed December 20, 2003).

52. Park, *Voodoo,* 193–194.
53. Park, *Voodoo,* 212.
54. Sagan, *Demon-Haunted World,* 29.
55. Park, *Voodoo,* 213.
56. In a closely related argument, skeptics often co-opt the millennialist premise that human sensory powers will unfold alongside new technologies. In an article that began with a critique of popular belief in supernatural and ESP, scientist Neil de Grasse Tyson went on to describe how science has enhanced the human sensory apparatus with the development of sensitive scientific instruments. New scientific instrumentation, then, could be thought of as providing "non-biological" senses. De Grasse Tyson concluded that these "new ways of knowing are new windows on the universe. . . . Whenever this happens, a new level of majesty and complexity in the universe reveals itself to us, as though we were technologically evolving into supersentient beings, always coming to our senses." Neil de Grasse Tyson, "Coming to Our Senses," *Natural History* (March 2001): 84–87.

INDEX

ABOUT THE AUTHOR

Fred Nadis writes about American cultural history, science and technology, and the popular arts. He earned a Ph.D. in American studies from the University of Texas in 2002, and he currently lives with his wife and two children in Kyoto, Japan, where he teaches in the Graduate School of American Studies at Doshisha University.